Midwest Ferns

A FIELD GUIDE TO THE FERNS AND FERN RELATIVES OF THE NORTH CENTRAL UNITED STATES

STEVE W. CHADDE

"The bare outline of the polypody thrills me strangely. It is a strange type which I cannot read. It only piques me. Simple as it is, it is as strange as an Oriental character. It is quite independent of my race, and of the Indian, and all mankind. It is a fabulous, mythological form, such as prevailed when the earth and air and water were inhabited by those extinct fossil creatures that we find."

—Henry David Thoreau, 1857

MIDWEST FERNS
A Field Guide to the Ferns and Fern Relatives of the North Central United States

Steve W. Chadde

ISBN: 978-1-951682-04-0

CREDITS

Grateful acknowledgment is given Alan Cressler and Chris Hoess for use of selected images. The Biota of North America Program (*www.bonap.org*) provided permission to use their data to generate the distribution maps. Permission graciously provided by the New York State Museum for use of illustrations published in *Field Guide to Northeastern Ferns* (Bulletin 444) by Eugene C. Ogden (1981).

Author email: steve@chadde.net
VERSION 1.5 (10/11/2019)

Contents

FERN RELATIVES

Cystopteris bulbifera
BULBLET BLADDER FERN

Fertile fronds (center) surrounded by sterile fronds of **cinnamon fern** (*Osmunda cinnamomea*), a common, large fern of wetlands throughout most of the Midwest region.

Introduction

MIDWEST FERNS is a comprehensive field guide to the ferns and fern relatives found in the states of Illinois, Indiana, Iowa, Minnesota, Michigan, Ohio, and Wisconsin. Included are keys, descriptions, and distribution maps for 18 families and 37 genera of **true ferns**, and 3 families and 8 genera of **fern relatives** (quillworts, clubmosses and spike-mosses). Note that in the following discussion, the term 'fern' encompasses true ferns as well as fern relatives.

Ferns occur across the Midwest in a wide variety of habitats, a region characterized by conifer forests, lakes, and extensive wetlands in the glaciated north; rich deciduous forests and exposed bedrock in central and southern areas, (some of which lie south of Pleistocene glaciation), and open woods and prairies in the west. While most of our ferns prefer somewhat shaded locations in forests, a few species are adapted to growing in water or in saturated soils, while others survive on dry rock cliffs. Habitat preferences can be a useful tool to identify ferns, see page 17 for a listing of Midwest ferns and their typical habitat.

Background

Ferns are vascular plants that do not produce flowers or seeds, reproducing instead via the production of spores. Worldwide, ferns are the most abundant group of seedless vascular plants, with an estimated 9,200 to 12,000 species. The fossil record indicates that ferns originated during the Devonian period which began about 416 million years ago, and became abundant and with a diversity of forms during the Carboniferous Period (360-299 million years ago). During that period, giant ferns, horsetails resembling large bamboos, and clubmosses the size of trees were predominant in the warm, moist tropical forests of the lower latitudes. Sea level changes brought on by glacial cycles at the poles resulted in periodic inundation and subsequent burial of these forests. Over time, and with heat and pressure, the remains transformed into coal.

Toward the end of the Carboniferous Period, climatic changes led to a decline in many of the period's ferns. However, ferns again rose in importance in the Mesozoic Era (251-65 million year ago), and many of our modern species originated at that time. In the Cretaceous Period of the latter part of the Mezozoic Era (about 145 million years ago), flowering plants first appeared, gradually becoming predominant over the Earth's land surface.

Today, ferns occur in most terrestrial habitats and are also present in some aquatic situations. About 75% of fern species occur in the tropics. Ferns are also an important part of the understory vegetation in many forest communities and, with about one-third of the species growing on trunks and branches of trees, they are an important component of many epiphytic plant communities. Some species are very beneficial to humans, while some ferns such as the aquatic Kariba weed (*Salvinia molesta*) are troublesome, invasive weeds.

Ferns range in size from the very small, as in the mosquito ferns (*Azolla*) to tree-like, as in the tree ferns (tropical species not present in our region) which can grow to more than 20 meters tall and with leaves to 5 meters or more long.

Fern Reproduction

The fern life cycle usually involves an alternation of 2 free-living generations – **sporophyte** and **gametophyte** (illustration, page 3). **Sporophytes** are the typical, larger leafy form we most often see, and is the spore-producing stage of the life cycle. The gametophyte stage is the sexual phase of a fern. Both stages are photosynthetic. In contrast to nonvascular plants like mosses and liverworts, in ferns, the sporophyte generation is generally predominant and more developed. In seed plants, the gametophyte is no longer free-living but remains enclosed in the sporophyte.

Sporangia, masses of spore cases, are produced on the leaves (often termed **fronds** in ferns) of the sporophyte, or sometimes in cone-like **strobili** (as in the horsetails, *Equisetum*). In true ferns, the sporangia are most often on the underside of the leaf blade, and are often grouped into clusters called **sori**.

Within each sporangium, specialized cells undergo a series of mitotic (structural) divisions followed by meiosis (sexual division) that results in **spores** with half as many chromosomes as in the original sporophyte. More advanced ferns usually have 64 spores per sporangium, but more primitive ferns and fern relatives may have many more. When mature, the sporangium dries and opens, dispersing the spores into the air.

When a spore lands on a suitable surface, it germinates, the cells dividing and forming first a filament and then a heart-shaped **gametophyte** (or sometimes other shapes in some groups). Gametophytes resemble a moss, are small (usually less than 1 cm wide), and have a life span usually of only several months. They are the sexual phase of the life cycle in that they produce multicellular sex organs at maturity on the side of the gametophyte located away from light. **Antheridia** (male gametangia) are produced among the root-like rhizoids near the base of the plant. At maturity they open to release tailed spermatozoids.

Archegonia (female gametangia) are usually formed near the notch at the opposite end of the gametophyte. They are flask-shaped and contain a single egg cell. A droplet of water is needed for the spermatozoids to swim to an archegonium. When mature, the neck of the archegonium opens, and the spermatozoid swims down the opening to fuse with the egg. This forms a **zygote** with twice the number of chromosomes as the gametophyte. The zygote grows and develops into a new sporophyte, the gametophyte withers away, and the cycle is completed.

VARIATIONS

In some ferns and fern relatives, including members of the clubmosses (Lycopodiaceae) and moonworts and grape ferns (Ophioglossaceae), the gametophytes are not surface-dwelling and green. Instead, they are below the soil surface and nonphotosynthetic, often appearing as pale brown or yellow fuzzy cylinders or pads of tissue. These gametophytes are mycotrophic; that is, they receive their nutrients from soil fungi that establish connections with their rhizoids. Male and female organs are produced as in typical ferns, and the life cycle continues in the normal manner.

Other ferns and fern relatives, including spike mosses (Selaginellaceae), quillworts (Isoetaceae), and the aquatic ferns (Azollaceae, Marsileaceae, Salviniaceae), depart from the typical life cycle in producing 2 different types of sporangia (heterospory). One sporangia produces numerous microscopic microspores that germinate to produce male gametophytes. The other form of

Fern Life Cycle

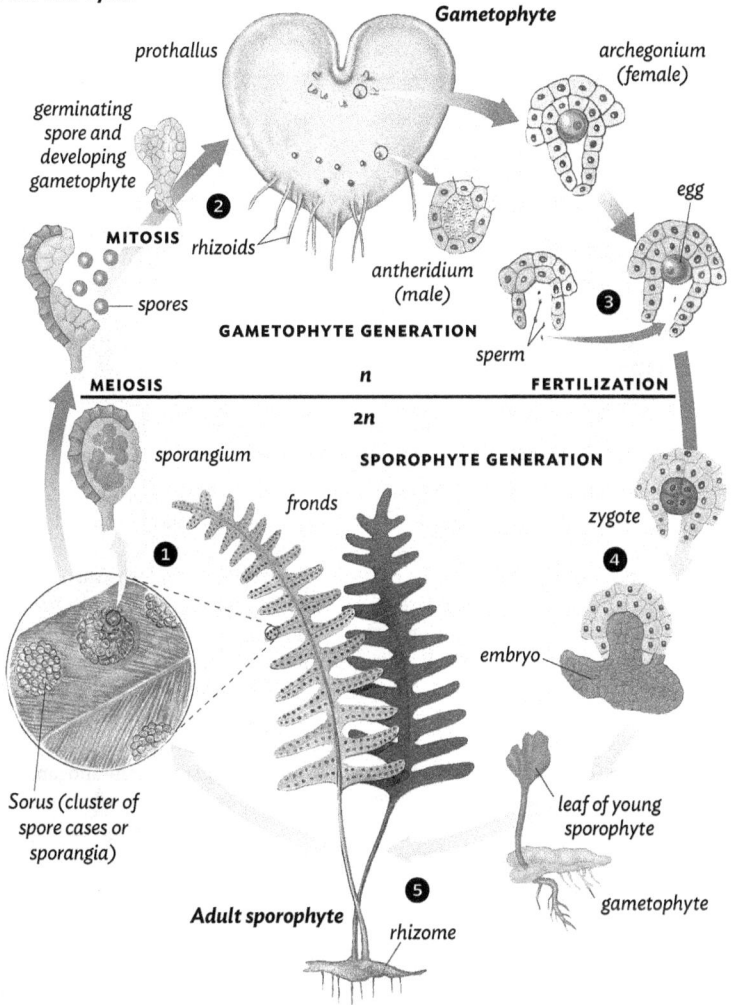

Gametophyte

prothallus

archegonium
(female)

germinating
spore and
developing
gametophyte

egg

② MITOSIS

rhizoids

antheridium
(male)

spores

③

GAMETOPHYTE GENERATION

sperm

MEIOSIS   ***n***   **FERTILIZATION**

2n

sporangium   **SPOROPHYTE GENERATION**

fronds

zygote

①

④

embryo

Sorus (cluster of
spore cases or
sporangia)

leaf of young
sporophyte

gametophyte

Adult sporophyte

⑤

rhizome

FERN LIFE CYCLE
❶ Clusters (sori) of sporangia (spore cases) grow on the undersurface of mature fern fronds. ❷ Released from its spore case, the haploid spore is carried to the ground, where it germinates into a tiny, usually heart-shaped, gametophyte, anchored to the ground by rhizoids (rootlike projections). ❸ Under moist conditions, mature sperm are released from the antheridia and swim to the egg-producing archegonia that have formed on the underside of the gametophyte. ❹ When fertilization occurs, a zygote forms and develops into an embryo within the archegonium. ❺ The embryo eventually grows larger than the gametophyte and becomes a sporophyte.

sporangia produces many fewer and much larger megaspores (usually visible to the naked eye), which develop into female gametophytes.

VEGETATIVE REPRODUCTION

Many ferns display some form of vegetative reproduction. This may be simply the breaking of a rhizome into smaller pieces that become established as separate plants. This is common in horsetails (Equisetaceae), and if growing along a river, pieces of the rhizome may be widely dispersed during high-water periods.

Other ferns develop specialized structures for vegetative propagation. These include **stolons** (long, spreading stems that root at their tips and form new plants), **bulbils** on the leaves that can germinate to form new plantlets, and fronds that can produce roots where they come into contact with soil. A few species have specialized underground structures, such as tubers and similar offsets. Some species produce **gemmae**, which are specialized fragments of plants that break off and to form new plants. In a few species of filmy ferns (Hymenophyllaceae), and shoestring ferns (Vittariaceae), the ability to produce sporophytes has been lost and plants exist only as colonies of gametophytes spreading through the production of tiny air-dispersed gemmae.

APOGAMY

Apogamy is a form of asexual reproduction where the sporophyte grows directly from the gametophyte rather than from the fertilized egg. Apogamy is a widespread and important mechanism of reproduction, more common in ferns than in any other group of plants. Apogamous ferns are most common in environments with seasonal extremes of heat, cold, or drought. In the sporangia of these plants, a mechanism during the series of cell divisions results in spores with the same genetic makeup as the sporophyte plant. These 'diplospores' grow into gametophytes that produce new sporophytes directly from tissue near the notch of the gametophyte. Advantages of apogamy include faster development of the gametophyte and that standing water is no longer needed for fertilization to take place. Many apogamous ferns continue to produce antheridia with functional spermatozoids, which can be released and fertilize eggs on nearby gametophytes of related sexual species. Once formed, such hybrids are always apogamous and able to reproduce themselves.

Fern Structure

The key to identifying ferns is to know a few basic terms that describe the parts of the plant; these are discussed below and are used in the descriptions throughout the book. See the **Glossary** (page 433) for a complete listing of botanical terms.

STEMS

Most ferns have specialized stems called **rhizomes** (synonym **rootstock**). Fern rhizomes are inconspicuous because they generally grow on or below the surface of the substrate (soil, moss or duff) on which the fern grows. Because the stems are in direct contact with the soil people often confuse stems with roots. Fern **roots** are generally thin and wiry in texture and grow along the stem. They absorb water and nutrients and help secure the fern to its substrate.

Rhizomes vary greatly in size, thickness and orientation. These modified stems are sometimes smooth but more often have scales or hairs, at least towards the growing tip. Stems can be short-creeping, with fronds that are somewhat scattered along the stem, such as in fragile fern (*Cystopteris*); stems can be long-creeping resulting in fronds scattered along the stem, as in bracken fern (*Pteridium*). Stems can also be vertical, producing circular clusters of leaves, as in most wood ferns (*Dryopteris*). In cross-section, most fern rhizomes appear as an irregular ring of vascular bundles. In some primitive ferns and fern relatives, the vascular system is a solid uninterrupted cylinder.

In some of the primitive ferns and in most fern relatives, other types of stems occur. Moonworts and grape ferns (Ophioglossaceae) usually have somewhat tuberous stems. Quillworts (Isoetaceae) have very short stout stems with the nodes very close together to form a swollen, cormlike base. Most clubmosses (Lycopodiaceae) have relatively unspecialized stems.

In horsetails (Equisetaceae), the underground stem is short to long-creeping, with wiry roots radiating from the underground stem's joints. The aboveground stems are hollow and vertically grooved or ridged. In most species, silica is incorporated into the stem's outside surface, giving plants a rough texture. Leaves are tiny and arise at the joints of the aboveground stem. Since the leaves are fused to each other around the stem, they form a sheath extending a short way up the stem from the joint. The color, number and tip shape of these sheathing leaves are good characters for distinguishing species of horsetails from each other (these are illustrated on page 183). Some horsetails also have slender branches radiating from the joints. These branches give many *Equisetum* their 'horsetail' look.

Horsetails can have one kind of aboveground stem (monomorphic), or be dimorphic, with fertile and sterile aboveground stems. Spores are produced by cones situated at the tip of the aboveground stem.

FRONDS

The fern frond (synonym **leaf**) includes the entire aboveground leaf, including the **stipe** (synonyms **petiole, stalk**) and **leaf blade** (synonym **lamina**) and have a wide variety of forms. In most true ferns, the fronds are the dominant feature of the sporophyte and can be complex in their pattern of division. In most fern relatives and a few primitive ferns, however, the leaves are reduced and scale-like, needle-like or grass-like, with only a single vein.

In most ferns, the development of the leaf follows a pattern known as **circinate vernation**. This produces a characteristic **crozier** or **fiddlehead** or as the leaf uncurls and is the first structure to appear aboveground in spring (a crozier is the stylized staff of office carried by high-ranking church officials). The fern relatives and the moonworts and grape ferns (Ophioglossaceae) do not exhibit circinate vernation but expand by unfolding or they develop in an indefinite pattern.

The **stipe** of fern leaves may be circular, angled, or U-shaped in cross-section, and is sometimes hairy or scaly. There are one to several vascular strands or bundles of conductive tissue, and the number and shape of these in the stipe are often diagnostic for individual families or genera.

The **leaf blade** varies from **entire** (synonym **simple**) as in hart's-tongue fern (*Asplenium scolopendrium*) to highly divided as in lady fern (*Athyrium filix-femina*). In true ferns, the most common type of division is is one or more times pinnately compound, less common are pedate and palmate patterns of division. In divided blades, the primary segments are called **pinnae** (synonym **leaflets**); a single segment is a **pinna**.

A pinna may be further divided into smaller segments (2-pinnate or bipinnate); these are called **pinnules** (synonym **subleaflets**) if fully divided from the pinna.

Pinnate means that the blade has distinct pinnae joined to the axis of the pinna

Parts of a Fern

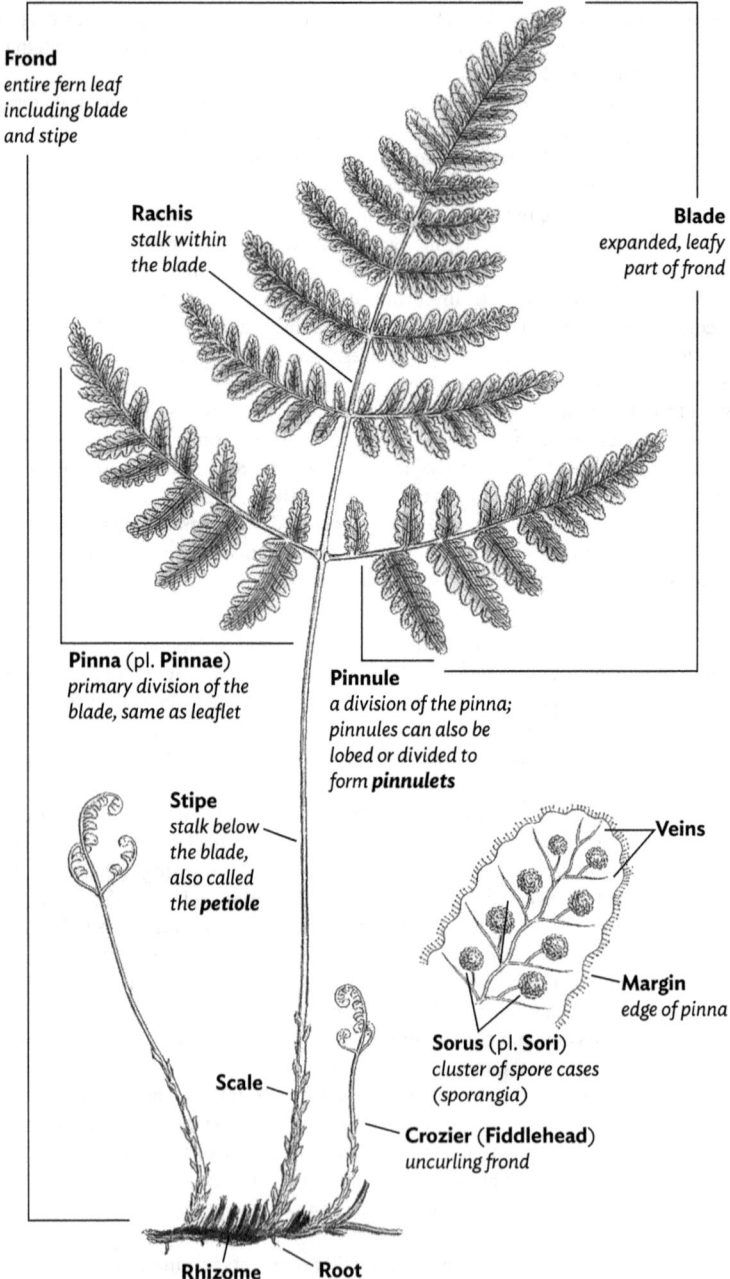

Frond
*entire fern leaf
including blade
and stipe*

Rachis
*stalk within
the blade*

Blade
*expanded, leafy
part of frond*

Pinna (pl. **Pinnae**)
*primary division of the
blade, same as leaflet*

Pinnule
*a division of the pinna;
pinnules can also be
lobed or divided to
form **pinnulets***

Stipe
*stalk below
the blade,
also called
the **petiole***

Veins

Margin
edge of pinna

Sorus (pl. **Sori**)
*cluster of spore cases
(sporangia)*

Scale

Crozier (**Fiddlehead**)
uncurling frond

Rhizome　**Root**

Vein Types

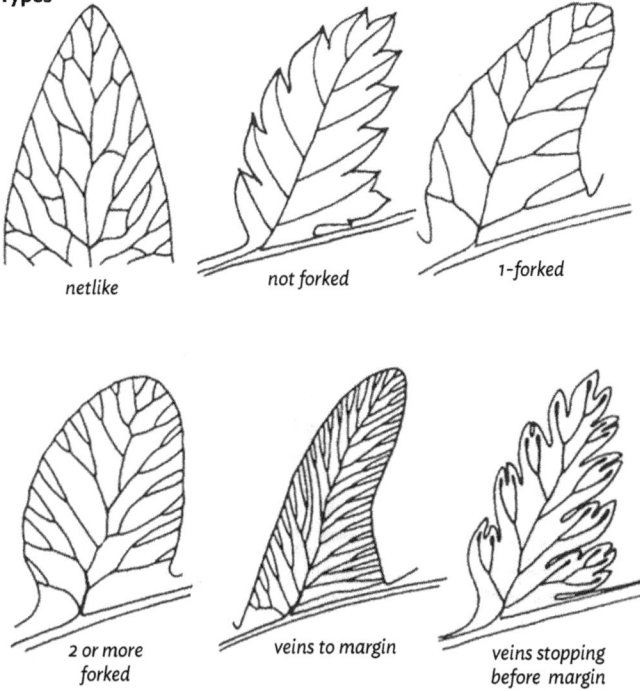

netlike

not forked

1-forked

2 or more
forked

veins to margin

veins stopping
before margin

by a narrow stalk. In many ferns, the pinnules are not cut to their midrib (only deeply lobed), and the term **pinnatifid** is used to describe this type of blade dissection.

The continuation of the stipe as the central axis of the leaf blade is known as the **rachis**. Pinnae are attached to the rachis. The midrib of the pinna and the pinnule is termed the **costa** (plural **costae**). The smaller veins of the pinnule may be unbranched or branched, and free (not rejoining) to variously anastomosed (joined to form a netlike pattern). Most of our ferns have free veins, but several genera have netlike venation as in the chain ferns (*Woodwardia*).

Fern fronds may be smooth or variously covered with hairs, scales, or both. In some species, the leaves are glandular and sticky. Other species secrete a powdery **farina**, usually on the leaf underside, which may be white or bright yellow or orange. Fronds also vary greatly in thickness and texture. The thinnest leaves occur in the filmy ferns (Hymenophyllaceae), with leaves often only 1-2 cell layers thick. In contrast, other species have thick leathery leaves, or leaves with dense coverings of hairs, scales, glands, or farina. These are believed to be adaptations to dry, harsh habitats such as those found on exposed rock cliffs.

SPORANGIA

In most ferns, the **sporangia** (singular **sporangium**) are spore-cases produced on the underside of normal fronds. In fern relatives and a few primitive ferns, relatively large sporangia are borne in either cone-like **strobili** at the tips of stems

Frond Division

Simple
undivided, not compound

Pinnatifid
blade cut partially to rachis

1-Pinnate
blade divided into pinnae

1-Pinnate Pinnatifid
*blade divided into pinnae
and the pinnae lobed*

2-Pinnate
*blade divided into pinnae,
the pinnae divided into
pinnules*

3-Pinnate
*blade divided into pinnae,
the pinnae divided into
pinnules, the pinnules
again divided (pinnulets)*

pinna

pinnule

2-Pinnate Pinnatifid
*blade divided into pinnae, the pinnae divided
into pinnules, and the pinnules lobed*

Pedate
*blade palmately divided,
the segments divided
again into 2*

Sori

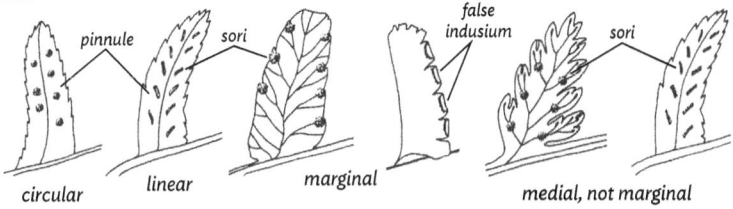

pinnule / sori / false indusium / sori

circular   linear   marginal   medial, not marginal

Indusia

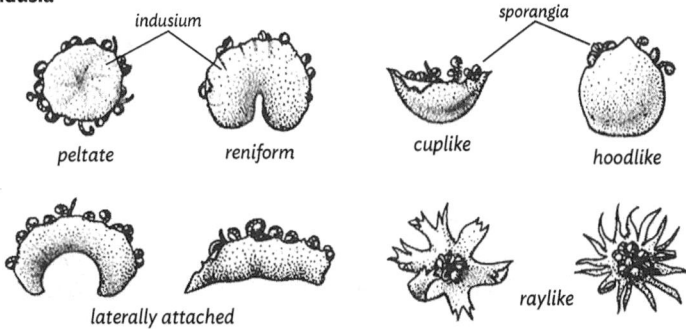

indusium   sporangia

peltate   reniform   cuplike   hoodlike

laterally attached   raylike

ABOVE

Fern Sori and Indusia. *Sori* (singular *sorus*) are groups of *sporangia* (singular *sporangium*), which contain spores. Sori are usually found on the underside of the blade. Young sori are often covered by a flap of protective tissue called the *indusium* (plural *indusia*). The shape and arrangement of the sori and indusia can be helpful characters for identifying unknown ferns. However, depending on the time of year, sori and indusia may be too immature or may have already been shed, and are of little use for identification.

LEFT

Fern Division

• **1-pinnate** leaf blades divided into pinnae, the pinnae more or less entire and not deeply lobed.

• **1-pinnate-pinnatifid** leaf blades divided into pinnae, and the pinnae deeply lobed but not completely divided to midrib.

• **2-pinnate** leaf blades divided into pinnae, and the pinnae divided into pinnules.

• **2-pinnate-pinnatifid** leaf blades divided into pinnae, and the pinnae divided into pinnules, but the pinnules deeply lobed and not completely divided to the midrib.

• **3-pinnate** leaf blades thrice pinnately divided.

• **3-pinnate-pinnatifid** leaf blades thrice pinnately divided, and the pinnulets (smallest segments) deeply lobed but not completely divided to midrib.

Sorus of Male Fern

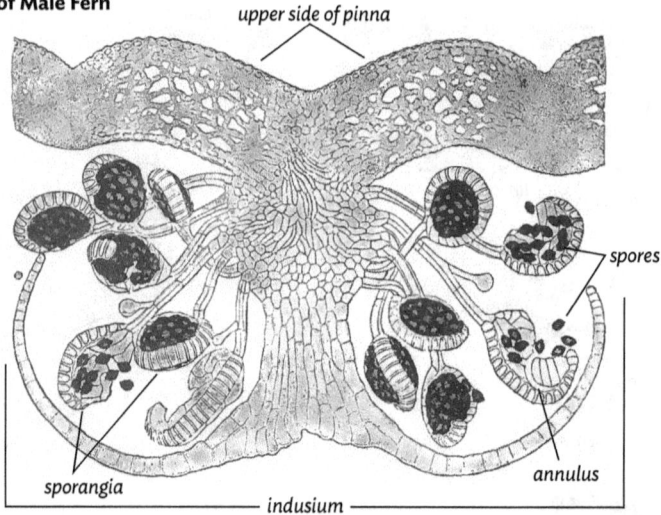

upper side of pinna

spores

sporangia

annulus

indusium

Cross-section through single **sorus** of male fern (*Dryopteris filix-mas*). The **indusium** is shaped like an inverted open umbrella and covers the stalked **sporangia** which are opening to release their spores. Circled area at right outlines a single sorus on the pinna underside.

or branches, or in the axil at the leaf base. Some ferns have **dimorphic fronds**, with **sterile or vegetative fronds** (the trophophyll), and **fertile fronds** (sporophyll) of different size and shape. In other ferns, the frond is divided into specialized fertile and sterile portions.

The position, shape, and other details of the sporangia on the leaf underside are often characteristic of particular families or genera and are an important factor in fern classification. In most ferns, the sporangia are grouped into discrete clusters known as **sori** (singular **sorus**). Sori may be round to linear, positioned along the margin or towards the midvein (costa), on the blade surface or in a groove or channel. In some species, sporangia may entirely cover the blade underside (a condition termed **acrostichoid**). In some primitive ferns, sporangia are only sparsely scattered along a few veins.

In some species, such as maidenhair fern (*Adiantum*), the developing sori are protected by a recurved leaf margin termed a **false indusium**. In other ferns, a covering of deciduous scales, or a more permanent small flap of tissue, the **indusium** is present. Indusia vary in shape, size, texture and persistence, ranging from umbrella-shaped to round to linear. In the water ferns (Azollaceae,

Representative Fern Spores

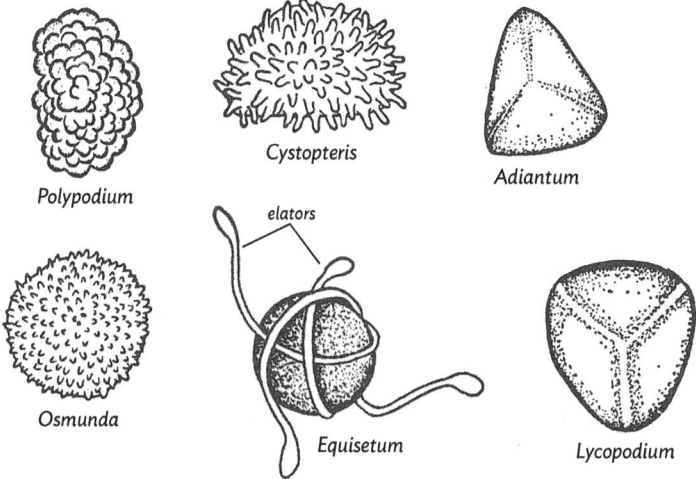

Polypodium

Cystopteris

Adiantum

Osmunda

elators

Equisetum

Lycopodium

Marsileaceae, Salviniaceae), sporangia are contained within hard capsules called **sporocarps**.

The sporangia themselves are usually positioned on a thickened vein tip or along a portion of a vein. In most ferns, the sporangium consists of a stalk and a capsule. The capsule may have an **annulus**, a ring or region of cells with only some of the walls thickened; the annulus acts somewhat like a spring to project the spores out of the sporangia.

SPORES

Spores are the main means by which ferns are dispersed to form new populations. In most ferns, spores are relatively long-lived and impervious, and fern spores have been successfully germinated after more than a hundred years of storage. After the sporangia opens to shed the mature spores, most fall within a few meters of the parent. However, spores have been recovered from the stratosphere during high-altitude atmospheric sampling, and ferns are among the most successful colonists of isolated ocean islands, testimony to their ability to successfully travel very long distances. In a few ferns, the spores are shorter-lived, relatively thin-walled, green and photosynthetically active, allowing for rapid establishment of new plants following dispersal.

Developmentally, spores are the direct products of meiosis, which begins with a single spore mother cell undergoing 2 separate rounds of division, and yields a **tetrad** that breaks into 4 individual spores. Two main spore types are recognized: **trilete (tetrahedral) spores** vary from nearly spherical to somewhat three-angled and have a three-branched scar where each spore was attached to the others in the tetrad. **Monolete (bilateral) spores** are ellipsoid to bean-shaped and have a linear attachment scar along one side. The attachment scar is usually where the spore ruptures during germination.

Mature spores vary greatly in size and surface sculpturing (see figure above). Small spores of some species are only about 15 microns in diameter, whereas megaspores of some *Selaginella* species may approach 1 mm. The surface sculpturing is often diagnostic for various species, genera or families (as in the quillworts, *Isoetes*), ranging from smooth to wrinkled, spiny, or with wing-like ridges.

GAMETOPHYTES

Upon germination of the spore, cell division produces first a threadlike structure. In most ferns, this continues to divide and eventually develops into the mature **gametophyte**. The typical fern gametophyte is flat and heart-shaped, with a notch and 2 lobes at one end and with the other end narrowed or rounded. Gametophytes vary in size from a few millimeters to about 1 cm wide. Other shapes occur but are not common; these include gametophytes that are filamentous, strap-shaped, or irregularly lobed. In ferns with underground mycorrhizal gametophytes, these mature mostly to either tubular or cushion-shaped structures.

Similar to mosses, gametophytes lack vascular tissue and roots. However, hair-like structures called **rhizoids** absorb water and nutrients and help anchor the gametophyte to the substrate. The **gametangia** (male and female sex organs) generally are formed on the side of the gametophyte away from the light (except in subterranean gametophytes). The male organs (**antheridia**) are positioned among the rhizoids, are spherical, and consist of a covering of cells enclosing the spermatozoids. The female organs (**archegonia**) are usually located near the notch on a slightly thickened pad of tissue. They are flask-shaped and somewhat sunken into the tissue. The neck of the archegonium consists of 4 columns of cells that separate at maturity, opening a canal and exposing the egg cell at the base for fertilization by the spermatozoid.

Fern Classification

At the genus and species level, fern names have remained fairly stable in recent years. Higher level classifications of ferns, however, remain somewhat controversial, as relationships between large groups continue to be studied and clarified. For our purposes, an emphasis is placed on family divisions and the fern families are arranged alphabetically in the book, first by the true ferns followed by the fern relatives. See page 13 for the higher level (subclass, order) arrangement of Midwest ferns and fern relatives.

In the past, ferns had been loosely grouped with other spore-bearing vascular plants, these often called "fern allies" or "lycophytes." However, recent studies suggest an important dichotomy within vascular plants, separating the fern relatives or lycophytes (less than 1% of all vascular plant species) from a group termed the euphyllophytes. Euphyllophytes comprise two major groups: the **spermatophytes** (seed plants), which number more than 260,000 species, and the **monilophytes** (ferns), with over 9,200 species, including horsetails, whisk ferns, and all eusporangiate and leptosporangiate ferns.

Genetic studies also reveal surprises about the relationships among true ferns and fern allies. True ferns appear to be closely related to horsetails, and in fact these plants are now grouped within the true ferns. Also, plants commonly called fern relatives (clubmosses, spike-mosses and quillworts) are not closely related to the true ferns. However, in the tradition of past fern guides, both true ferns and fern relatives are included so that all spore-bearing vascular plants in the region can be identified.

FAMILY RELATIONSHIPS

Chase and Reveal (2009) include all ferns (and all land plants) within Class Equisetopsida. In this Flora, 3 subclasses, 7 orders, and 18 families of **true ferns** are treated; for **fern relatives**, 1 subclass, 3 orders and 3 families are described. Subclass divisions follow Stevens (2001); order and family groupings follow Smith (2006) and Rothfels et al. (2012).

True Ferns

SUBCLASS OPHIOGLOSSIDAE
- ORDER Ophioglossales
   OPHIOGLOSSACEAE | Adder's-Tongue Family

SUBCLASS EQUISETIDAE
- ORDER Equisetales
   EQUISETACEAE | Horsetail Family

SUBCLASS POLYPODIIDAE
- ORDER Hymenophyllales
   HYMENOPHYLLACEAE | Filmy Fern Family

- ORDER Osmundales
   OSMUNDACEAE | Royal Fern Family

- ORDER Polypodiales
   ASPLENIACEAE | Spleenwort Family
   ATHYRIACEAE | Lady Fern Family
   BLECHNACEAE | Chain Fern Family
   CYSTOPTERIDACEAE | Bladder Fern Family
   DENNSTAEDTIACEAE | Bracken Fern Family
   DRYOPTERIDACEAE | Wood Fern Family
   ONOCLEACEAE | Sensitive Fern Family
   POLYPODIACEAE | Polypody Fern Family
   PTERIDACEAE | Maidenhair Fern Family
   THELYPTERIDACEAE | Maiden Fern Family
   WOODSIACEAE | Cliff Fern Family

- ORDER Salviniales
   MARSILEACEAE | Water-Clover Family
   SALVINIACEAE | Water Fern Family

- ORDER Schizaeales
   LYGODIACEAE | Climbing Fern Family

Fern Relatives

SUBCLASS LYCOPODIIDAE
- ORDER Lycopodiales
   LYCOPODIACEAE | Clubmoss Family

- ORDER Isoetales
   ISOETACEAE | Quillwort Family

- ORDER Selaginellales
   SELAGINELLACEAE | Spike-Moss Family

Economic Importance

Only a handful of fern species are economically important. Perhaps the best known use is horticultural, and many species are used in landscaping, as houseplants, and by florists. Pieces of the root mantles covering stems of tree ferns ('orchid bark') are used as a substrate for growing orchids and epiphytic plants.

An interesting phenomenon called **pteridomania** ("fern craze") occurred in Great Britain in the Victorian period of the mid- to late-1800s. Anything fern-related was highly popular, including illustrated fern books, growing ferns in gardens and greenhouses, and objects and art with fern motifs.

Ferns have been used in handicrafts by Native Americans. For example, the twining stipe of some members of the Climbing Fern Family (Lygodiaceae) and the lustrous stipes of maidenhair (*Adiantum*), were used in basket-making and in bracelets. Leaves of bracken fern (*Pteridium*) have been used to make a green dye.

Clubmosses (Lycopodiaceae) have an interesting history of uses. The spores contain nonvolatile oils that made them useful as a dry lubricant. They have also been used to keep latex products like surgical gloves from sticking together (now mostly discontinued since the spores can be a skin irritant to some people). Other uses of the spores have been in flash powder for photography and in fingerprint powder in forensic investigations.

Various ferns are eaten as food, with the young fronds prepared like a vegetable. The most important edible species is **ostrich fern** (*Matteuccia struthiopteris*), whose fiddleheads are available in markets of the northeastern United States in late spring. In parts of Asia, *Diplazium esculentum* is grown for food. Formerly, fiddleheads of **bracken fern** (*Pteridium aquilinum*) were eaten, but studies have linked this species to stomach cancer, and it should be avoided.

The most economically valuable species of fern is **mosquito fern** (*Azolla*), a tiny floating aquatic fern. For many centuries, farmers in parts of eastern Asia have inoculated rice paddies with this plant to increase the yield of their harvest. The hollow chambers in *Azolla* leaves contain symbiotic cyanobacteria (*Anabaena azollae*) that convert atmospheric nitrogen into the nitrate form, an important plant nutrient. Research is ongoing to identify superior strains of this fern.

Several ferns have negative economic impacts because of their weediness. The most notable are species of *Salvinia* and *Pteridium*. **Kariba weed** (or giant salvinia, *Salvinia molesta*) is a floating aquatic fern that is aggressively weedy throughout warmer parts of the world. Native to Brazil, it is now present in California, Texas, and the southeastern states. Kariba weed can form a surface mat several inches thick, preventing light and oxygen from penetrating into the water, choking out other aquatic plants as

Pteridium aquilinum
Bracken fern

well as devastating fisheries. **Bracken fern** (*Pteridium aquilinum*) is widely distributed, and has an extensive, deep, creeping rhizome sometimes several hundred meters in length. Bracken fern quickly invades open habitats, outcompeting other plants. Plants are toxic to livestock, rendering large areas unusable, especially in parts of Europe.

Fern Conservation

A number of our ferns are rare due to uncommon habitats or to small population sizes. Some of our species are afforded legal protection under state law (endangered or threatened); in the species descriptions, the status is noted under the MIDWEST RANGE heading. In general, when 'botanizing', disturb habitats as little as possible, and refrain from collecting plants; in most cases, photographs will serve as adequate documentation.

The following terms are used to describe rarity within a state. **Endangered** and **threatened** status provides legal protection to a species; **special concern** status does not, but includes those species which are apparently rare within a state and possibly at risk of becoming threatened or endangered.

• **Endangered**: a species threatened with "extirpation" (extinction within a single state or area and not across its entire range).

• **Threatened**: a species whose survival within a state is not in immediate danger, but for which a threat exists; continued stress on the species may result in its becoming endangered.

• **Special concern**: a species which may become threatened within a state if it is under continued or increased stress; includes species for which there is some concern, but not enough information is available to assess its status.

Every state in the region uses the endangered and threatened categories. **Special concern** status is used in Iowa, Minnesota, Michigan, and Wisconsin. Comparable designations in Indiana are **rare** and **watch list**, and in Ohio, **potentially threatened**.

Each state maintains lists of species of conservation concern or has a natural heritage program to monitor populations of rare species; more information is available at the websites below.

NATURAL HERITAGE INFORMATION

• INDIANA www.in.gov/dnr/naturepreserve/
• ILLINOIS www.dnr.illinois.gov/espb
• IOWA www.iowadnr.gov/environment/
• MINNESOTA www.dnr.state.mn.us
• MICHIGAN mnfi.anr.msu.edu/
• OHIO www2.ohiodnr.gov/
• WISCONSIN dnr.wi.gov/topic/endangeredresources/

The Fern Gatherer. By Charles Sillem Lidderdale, 1877; painted during the fern craze that swept England in the 2nd half of the 19th century.

Key to Habitat Groups

A diversity of habitats occur in the Midwest, ranging from aquatic to to terrestrial to rock. Large differences exist between the flora of northern vs. southern areas, as well as eastern vs. western areas. Within our region, most ferns and fern relatives occupy specific habitats, and while many species are geographically wide-ranging, others are limited in their Midwest occurrence. The following groupings can be helpful to quickly identify unknown species; check the regional map to verify presence in a particular state:

(A) **Aquatic Habitats** (lakes, ponds, rivers), page 17

(B) **Wetlands** (swamps, marshes, bogs, low-lying places), page 18

(C) **Rock Cliffs and Boulders** (both calcareous and non-calcareous rock), p. 20

(D) **Calcareous Rock** (species usually found on limestone or dolomite), page 21

(E) **Non-Calcareous Rock** (e.g. sandstone, gneiss), page 22

(F) **Rocky Ravines and Slopes** (where shaded), page 23

(G) **Moist Forests** (loamy soils), page 24

(H) **Drier Forests, Fields, Prairies and Dunes** (sandy soils), page 27

(I) **Wide-ranging species** (found in many types of moist to dry habitats), p. 30

---------------------------------**(A)**---------------------------------

FERNS OF AQUATIC HABITATS (LAKES, PONDS, RIVERS)

Azolla cristata (CRESTED MOSQUITO FERN) . . . . page 332
 • floating on quiet water of river backwaters and ponds
 • Ill, Ind, Iowa, s Mich, se Minn, Ohio, Wisc

....

Marsilea quadrifolia (EUROPEAN WATER-CLOVER) . . . . page 214
 • in mud or shallow water of quiet ponds and rivers
 • Ill, s Ind, s Iowa, Mich Lower Peninsula, Ohio

Marsilea vestita (HAIRY WATER-CLOVER) . . . . page 216
 • margins of shallow pools on prairies and on rock outcrops
 • nw Iowa, w Minn

B

FERNS OF WETLANDS
(SWAMPS, MARSHES, BOGS, LOW-LYING PLACES)

Isoetes melanopoda (BLACK-FOOT QUILLWORT) . . . . . . page 375
- vernal pools on rock outcrops (Minn); shores, ditches
- Ill, s Ind, e Iowa, sw Minn

Lycopodiella inundata (BOG CLUB-MOSS) . . . . page 408
- sandy depressions, bog mats, sandy lake and pond shores
- n Ill, Ind, e Iowa, Mich, Minn, Ohio, Wisc

Matteuccia struthiopteris (OSTRICH FERN) . . . . page 220
- streambanks, ravines, seeps , wet woods and swamps
- regionwide

Onoclea sensibilis (SENSITIVE FERN) . . . . page 222
- swamps, streambanks, thickets, marshes, moist woods
- regionwide

Osmunda cinnamomea (CINNAMON FERN) . . . . page 286
- bogs, acid swamp forests, wet shrub thickets
- regionwide

Osmunda regalis (ROYAL FERN) . . . . page 290
- swamps, low woods, marshy meadows, cedar bogs
- Ill, Ind, e Iowa, Mich, Minn, Ohio, Wisc

Parathelypteris simulata (BOG FERN) . . . . page 340
- swamp and bog margins, wet woods
- west-central and sw Wisc

Selaginella eclipes (HIDDEN SPIKE-MOSS) . . . . page 423
- fens, wet sandy or marly lakeshores
- Ill, Ind, Iowa, Mich, Ohio, Wisc

Selaginella selaginoides (NORTHERN SPIKE-MOSS) . . . . page 426
- fens and fen-like shores, especially along the Great Lakes
- n Mich, ne Minn, Wisc (Door County)

Thelypteris palustris (EASTERN MARSH FERN) . . . . page 346
- wet meadows, swamps, ditches, fens, bogs, shores, seeps
- regionwide

Woodwardia areolata (NETTED CHAIN FERn) . . . . page 96
- wet woods, acidic bogs, seeps
- s Ill, Ind, Ohio

Woodwardia virginica (VIRGINIA CHAIN FERN) . . . . page 98
- acidic swamps, bogs, wet places in woods, lake shores
- Ind, Mich, Ohio

C

ROCK CLIFFS AND BOULDERS (ALL ROCK TYPES)

Asplenium platyneuron (EBONY SPLEENWORT) . . . . page 64
• limestone or sandstone cliffs and ledges
• Ill, Ind, Iowa, Mich, se Minn, Ohio, Wisc

Asplenium trichomanes (MAIDENHAIR SPLEENWORT) . . page 72
• damp, shaded, mossy cliffs and boulders
• Ill, Ind, Mich, Minn, Ohio, Wisc

Cheilanthes feei (SLENDER LIP FERN) . . . . page 310
• cliffs and ledges
• Ill, Iowa, Minn, Wisc

Cryptogramma stelleri (FRAGILE ROCKBRAKE) . . . . page 318
• moist, shaded, rocks and cliffs; seepy crevices and ledges
• n Ill, Iowa, n Mich, Minn, Wisc

Cystopteris bulbifera (BULBLET BLADDER FERN) . . . . . . page 106
• cliffs, shaded talus slopes, rarely on soil
• regionwide

Cystopteris fragilis (BRITTLE BLADDER FERN) . . . . page 108
• cliff faces, moist talus slopes; sometimes in moist woods
• ne Ill, Iowa, Mich, Minn, Wisc

Cystopteris tenuis (UPLAND BRITTLE BLADDER FERN) . . page 116
• shaded, mossy rock outcrops; soil of moist forests
• regionwide

Huperzia appressa (MOUNTAIN FIR-MOSS) . . . . page 398
• open and exposed cliffs, talus slopes
• Mich, ne Minn, n Wisc

Pellaea atropurpurea (PURPLE-STEM CLIFFBRAKE) . . . . page 322
• dry outcrops, ledges and talus slopes
• Ill, Ind, e Iowa, n Mich, s Minn, Ohio, Wisc

Pellaea glabella (SMOOTH CLIFFBRAKE) . . . . page 324
• ledges and crevices of cliffs
• Ill, Ind, Iowa, n Mich, Minn, Ohio, Wisc

Polypodium virginianum (ROCK POLYPODY) . . . . page 300
• bluffs, rock outcrops, ledges, boulders and mossy talus
• regionwide

Woodsia oregana (OREGON CLIFF FERN)............. page 362
  • rock outcrops, ledges, crevices and talus slopes
  • Iowa, Mich Upper Peninsula, Minn, Wisc

---------------------------------- **D** ----------------------------------

CALCAREOUS ROCK (LIMESTONE, DOLOMITE)

Asplenium rhizophyllum (WALKING FERN)........... page 66
  • mossy limestone or dolomite rocks, ledges and crevices
  • regionwide

Asplenium ruta-muraria (WALL-RUE)................ page 68
  • crevices in limestone cliffs and mossy, calcareous talus slopes
  • Ind, Iowa, Mich, Ohio

Asplenium scolopendrium (HART'S-TONGUE FERN) .... page 70
  • shaded dolomite woodlands, boulders and crevices
  • Mich e Upper Peninsula only

Asplenium viride (BRIGHT-GREEN SPLEENWORT)....... page 74
  • shaded limestone crevices and ledges, talus slopes and boulders
  • n Mich, n Wisc

....

Cystopteris laurentiana (ST. LAWRENCE BLADDER FERN) page 110
  • cracks and ledges on calcareous rock outcrops
  • n Ill, e Iowa, n Mich, Minn, Wisc

Cystopteris tennesseensis (TENNESSEE BLADDER FERN) page 114
  • moist calcareous cliffs
  • Ill, Ind, Iowa, se Minn, Ohio, Wisc

....

Dryopteris filix-mas (MALE FERN) .... page 152
  • shaded rocky woods and outcrops
  • n Mich, Minn, n Wisc

....

Polystichum lonchitis (NORTHERN HOLLY FERN)...... page 174
  • calcareous rock outcrops and boulders
  • n Mich, n Wisc

....

Woodsia obtusa (BLUNT-LOBE CLIFF FERN) .... page 360
  • open to partly shaded outcrops and cliff ledges
  • regionwide

E

NON-CALCAREOUS ROCK (E.G. SANDSTONE)

Asplenium pinnatifidum (LOBED SPLEENWORT) . . . . page 62
• dry to moist, shaded crevices, especially sandstone and gneiss
• Ill, Ind, Ohio, s Wisc

Cheilanthes lanosa (HAIRY LIP FERN) . . . . page 312
• dry, exposed, sandstone bluffs and ledges
• Ill, s Ind, Wisc

Cryptogramma acrostichoides (AMERICAN ROCKBRAKE) p. 316
• crevices of acidic rock usually in dry, open, sunny settings
• Mich Upper Peninsula (Isle Royale)

Dryopteris fragrans (FRAGRANT WOOD FERN) . . . . page 154
• dry, sunny or partially shaded cliffs, boulders, and talus slopes
• Mich Upper Peninsula, n Minn, Wisc

Huperzia porophila (ROCK FIR-MOSS). . . . . page 402
• sandstone outcrops and rocky woods, where moist and shaded
• Ill, s Ind, ne Iowa, n Mich, Minn, Ohio, sw Wisc

Selaginella rupestris (SAND CLUBMOSS). . . . . page 424
• dry rocky balds, sand dunes and open sandy areas
• Ill, Ind, Iowa, Mich, Minn, Wisc

Woodsia alpina (NORTHERN CLIFF FERN). . . . . page 354
• crevices and ledges of moist, partially shaded cliffs
• Mich Upper Peninsula, Minn

Woodsia glabella (SMOOTH CLIFF FERN) . . . . page 356
• crevices in moist, north-facing cliffs
• n Minn

Woodsia ilvensis (RUSTY CLIFF FERN) . . . . page 358
• dry, exposed or partly shaded cliffs and ledges
• Ill, Iowa, Mich Upper Peninsula, Minn, Wisc

Woodsia scopulina (ROCKY MOUNTAIN CLIFF FERN). . . page 364
• north-facing cliffs of slate rock
• ne Minn

ROCKY RAVINES AND SLOPES (WHERE SHADED)

Adiantum pedatum (NORTHERN MAIDENHAIR) . . . . page 306
- rich deciduous forests
- regionwide

Asplenium scolopendrium (HART'S-TONGUE FERN) . . . . page 70
- rocky woodlands underlain by dolomite
- Mich e Upper Peninsula only

Asplenium trichomanes (MAIDENHAIR SPLEENWORT) . . page 72
- damp, shaded, mossy limestone cliffs and boulders
- Ill, Ind, Mich, Minn, Ohio, Wisc

Botrypus virginianus (RATTLESNAKE FERN) . . . . page 258
- moist deciduous forests, also in drier woods
- regionwide

Cystopteris bulbifera (BULBLET BLADDER FERN) . . . . . . page 106
- moist cliffs, talus slopes, cedar swamps, near seeps
- regionwide

Dryopteris expansa (SPREADING WOOD FERN) . . . . page 150
- cool moist hardwood or mixed hardwood-conifer forests
- Mich Upper Peninsula, Minn, n Wisc

Dryopteris filix-mas (MALE FERN) . . . . page 152
- moist, rocky woods and outcrops; typically over limestone
- n Mich, Minn, n Wisc

Dryopteris goldiana (GOLDIE'S WOOD FERN) . . . . page 156
- moist shaded woods, especially in ravines and near seeps
- regionwide

Dryopteris intermedia (EVERGREEN WOOD FERN) . . . . . page 158
- moist to dry forests
- regionwide

Dryopteris marginalis (MARGINAL WOOD FERN) . . . . . . page 160
- rocky wooded slopes and ravines, streambanks
- Ill, Ind, Iowa, Mich, se Minn, Ohio, Wisc

Gymnocarpium dryopteris (NORTHERN OAK FERN) . . . page 122
- moist woods of all types
- Ill, Iowa, Mich, Minn, Ohio, Wisc

Gymnocarpium robertianum (LIMESTONE OAK FERN). page 124
• shade of calcareous woods and on swamp hummocks
• nw Ill, e Iowa, Mich, Minn, Wisc

Huperzia porophila (ROCK FIR-MOSS). . . . . page 402
• sandstone outcrops, rocky woods; where moist and shaded
• Ill, s Ind, ne Iowa, n Mich, Minn, sw Wisc

Phegopteris connectilis (NARROW BEECH FERN) . . . . . . page 342
• moist, rich substrates of stream banks and woodlands
• Ill, e Iowa, Mich, Minn, Ohio, Wisc

Phegopteris hexagonoptera (BROAD BEECH FERN) . . . . page 344
• rich, moist deciduous forests
• Ill, Ind, e Iowa, Mich, se Minn, Ohio, Wisc

Polystichum acrostichoides (CHRISTMAS FERN). . . . . . page 170
• shaded, moist to dry forests, rocky slopes and ravines
• Ill, Ind, Iowa, Mich, se Minn, Ohio, Wisc

Polystichum braunii (BRAUN'S HOLLY FERN). . . . . page 172
• cool, moist, shaded woods; especially where rocky
• Mich, ne Minn, n Wisc

G

MOIST FORESTS (LOAMY SOILS)

Adiantum pedatum (NORTHERN MAIDENHAIR) . . . . page 306
• moist deciduous forests
• regionwide

Asplenium rhizophyllum (WALKING FERN). . . . . page 66
• moist deciduous woods; calcareous rocks
• regionwide

Athyrium filix-femina (LADY FERN). . . . . page 82
• moist to dry woods, swamps, shores, meadows, thickets, ravines
• regionwide

Botrychium lanceolatum (LANCE-LEAF MOONWORT). . page 240
• moist deciduous and mixed conifer-hardwood forests
• Mich, Minn, Ohio, Wisc

Botrychium minganense (MINGAN MOONWORT) . . . . . page 246
- moist woods, old clearings
- Mich, n Minn, n Wisc

Botrychium mormo (LITTLE GOBLIN MOONWORT) . . . . page 248
- mature deciduous forests, eastern hemlock often present
- n Mich, Minn, Wisc

Botrypus virginianus (RATTLESNAKE FERN) . . . . page 258
- moist to dry woods, hummocks in cedar swamps
- regionwide

Cystopteris protrusa (LOWLAND BLADDER FERN) . . . . . page 112
- rich, moist forests and alluvial woods
- regionwide

Cystopteris tenuis (UPLAND BRITTLE BLADDER FERN) . . page 116
- shaded, mossy rock outcrops and soil of moist forests
- regionwide

Deparia acrostichoides (SILVERY-SPLEENWORT) . . . . page 88
- moist, shaded forests; ravines, shaded streambanks and seeps
- regionwide

Diplazium pycnocarpon (GLADE FERN) . . . . page 92
- moist, open to partially shaded woods, meadows and slopes
- Ill, Ind, e Iowa, Mich, s Minn, Ohio, Wisc

Dryopteris carthusiana (SPINULOSE WOOD FERN) . . . . . page 142
- moist to wet woods, streambanks, swamp hummocks
- regionwide

Dryopteris expansa (SPREADING WOOD FERN) . . . . page 150
- cool moist hardwood or mixed hardwood-conifer forests
- Mich Upper Peninsula, Minn, n Wisc

Dryopteris goldiana (GOLDIE'S WOOD FERN) . . . . page 156
- shaded woods, especially in ravines and near seeps and springs
- regionwide

Dryopteris intermedia (EVERGREEN WOOD FERN) . . . . . page 158
- moist to dry forests
- regionwide

Dryopteris marginalis (MARGINAL WOOD FERN) . . . . . . page 160
- rocky wooded slopes and ravines, streambanks
- Ill, Ind, Iowa, Mich, se Minn, Ohio, Wisc

Equisetum laevigatum (SMOOTH SCOURING-RUSH) . . . page 192
- moist to dry sandy shores, meadows, dunes, prairies, ditches
- regionwide

Equisetum scirpoides (DWARF SCOURING-RUSH). . . . . . page 198
- cedar swamps, moist mixed forests, usually in shade
- n Ill, ne Iowa, Mich, Minn, Wisc

Huperzia lucidula (SHINING FIR-MOSS) . . . . page 400
- moist deciduous, mixed conifer-hardwood, or conifer forests
- regionwide

Gymnocarpium dryopteris (NORTHERN OAK FERN) . . . page 122
- moist woods
- Ill, Iowa, Mich, Minn, Ohio, Wisc

Gymnocarpium robertianum (Limestone Oak Fern). . . page 124
- shade of calcareous woods and on swamp hummocks
- nw Ill, e Iowa, Mich, Minn, Wisc

Lygodium palmatum (AMERICAN CLIMBING FERN) . . . . page 210
- woodland margins, wet thickets; soils moist and acidic
- s Ind, sw Mich, Ohio

Ophioglossum vulgatum (SOUTHERN ADDER'S-TONGUE) p. 266
- moist deciduous forests, moist meadows, thickets and clearings
- Ill, Ind, se Iowa, s Mich, Ohio

Osmunda claytoniana (INTERRUPTED FERN). . . . . page 288
- low places in mesic to wet forests, on hummocks in swamps
- regionwide

Parathelypteris noveboracensis (NEW YORK FERN) . . . . page 338
- moist shady woods, especially near swamps, streams, and seeps
- Ill, Ind, Mich, Ohio

Phegopteris connectilis (NARROW BEECH FERN) . . . . . . page 342
- moist, rich substrates of stream banks and woodlands
- Ill, e Iowa, Mich, Minn, Ohio, Wisc

Phegopteris hexagonoptera (BROAD BEECH FERN) . . . . page 344
- rich, moist deciduous forests
- Ill, Ind, e Iowa, Mich, se Minn, Ohio, Wisc

Polystichum acrostichoides (CHRISTMAS FERN). . . . . . page 170
 • shaded, moist to dry forests, rocky slopes and ravines
 • Ill, Ind, Iowa, Mich, se Minn, Ohio, Wisc

Polystichum lonchitis (NORTHERN HOLLY FERN). . . . . . page 174
 • northern hardwood and mixed conifer-hardwood forests
 • n Mich, n Wisc

Sceptridium biternatum (SPARSE-LOBE GRAPE FERN). . page 272
 • moist, shaded situations; low woods, ravines, thickets, forested
 floodplains
 • s Ill, s Ind and Ohio

Sceptridium oneidense (BLUNT-LOBE GRAPE FERN) . . . page 278
 • shaded, wet to dry, hardwood or mixed forests, occasionally
 in conifer swamps
 • n Ill, Ind, Mich, Minn, Ohio, Wisc

H

DRIER FORESTS, FIELDS, PRAIRIES AND DUNES (SANDY SOILS)

Asplenium platyneuron (EBONY SPLEENWORT) . . . . page 64
 • open woods, old fields, clearings
 • Ill, Ind, Iowa, Mich, se Minn, Ohio, Wisc

Athyrium filix-femina (LADY FERN). . . . . page 82
 • moist to dry woods, swamps, shores, meadows, thickets, ravines
 • regionwide

Botrychium acuminatum (TAILED MOONWORT). . . . . . page 244
 • sand dunes, old fields, grassy railroad sidings, roadside ditches
 • Mich, ne Minn

Botrychium campestre (IOWA MOONWORT). . . . . page 236
 • prairies and dunes and other open, calcareous sites
 • n Ill, Iowa, Mich, Minn, Wisc

Botrychium hesperium (WESTERN MOONWORT). . . . . . page 238
 • sand dunes, sandy open fields and open woodlands
 • n Mich

Botrychium lunaria (COMMON MOONWORT) . . . . page 242
 • open grassy fields and forests, usually on sandy or gravelly soil
 • Mich, n Minn, n Wisc

Botrychium matricariifolium (DAISY-LEAF MOONWORT) p. 244
• sandy fields and clearings, and dry rocky woods
• n Ill, n Ind, e Iowa, Mich, Minn, Ohio, Wisc

Botrychium pallidum (PALE MOONWORT)........... page 250
• open woods and shrubby habitats
• Mich, n Minn, nw Wisc

Botrychium simplex (LEAST MOONWORT)........... page 252
• variety of open to shaded habitats: prairies, clearings,
bracken grasslands, sandy woods, swamps and lakeshores
• n Ill, n Ind, Iowa, Mich, Minn, Ohio, Wisc

Botrychium spathulatum (SPOON-LEAF MOONWORT) . page 254
• sandy, sunny, grassy fields and open woodlands
• n Mich, Minn, Wisc (Door County)

Dendrolycopodium dendroideum (PRICKLY TREE-CLUBMOSS) p. 380
• moist to dry deciduous and mixed forests
• Ill, s Ind, ne Iowa, Mich, Minn, Ohio, Wisc

Dendrolycopodium hickeyi (PENNSYLVANIA TREE-CLUBMOSS) p. 382
• drier deciduous, mixed, and conifer forests
• Ill, n Ind, Mich, Minn, Ohio, Wisc

Dendrolycopodium obscurum (PRINCESS-PINE)...... page 384
• moist to dry deciduous and mixed conifer-deciduous forests
• ne Ill, Ind, Mich, Minn, Ohio, Wisc

Dennstaedtia punctilobula (HAY-SCENTED FERN)..... page 130
• partially shaded to sunny woods, roadsides, fields, clearings
and rocky slopes
• Ill, Ind, Mich, Ohio, s Wisc

Diphasiastrum complanatum (TRAILING CREEPING-CEDAR) p. 388
• woods, thickets, clearings; soils typically sandy and acidic
• e Iowa, Mich, Minn, Wisc

Diphasiastrum digitatum (FAN CREEPING-CEDAR) .... page 390
• sandy woods, clearings, fields, thickets, where sometimes
forming extensive patches from the shallow rhizomes
• regionwide

Diphasiastrum sabinifolium (SAVIN-LEAF CREEPING-CEDAR) p. 392
• Sandy disturbed places
• n Mich

Diphasiastrum tristachyum (DEEP-ROOT CREEPING-CEDAR) p. 394
- Dry, sandy fields and open woods; sandy disturbed places
- nw Ind, Mich, Minn, Ohio, Wisc

Equisetum hyemale (TALL SCOURING-RUSH) . . . . page 190
- Sandy meadows and old fields, shores, dunes, roadsides, ditches, railroad embankments; sites wet to dry
- regionwide

Equisetum laevigatum (SMOOTH SCOURING-RUSH) . . . page 192
- Moist to dry sandy river terraces, shores and meadows, open swamps, moist dunes, prairies, ditches, roadsides, along railways
- regionwide

Equisetum variegatum (VARIEGATED SCOURING-RUSH)page 202
- Moist sand, shores, beaches, interdunal flats, roadside ditches, borrow pits
- Ill, n Ind, Mich, Minn, n Ohio, Wisc

Lycopodium clavatum (RUNNING GROUND-PINE) . . . . . page 412
- Dry to moist woods and clearings; soils usually sandy
- n Ill, n Ind, e Iowa, Mich, Minn, Ohio, Wisc

Lycopodium lagopus (ONE-CONE GROUND-PINE). . . . . . page 414
- Sterile, sandy acid soils in open mixed conifer-hardwood or coniferous forests, sandy borrow pits, fields, roadsides
- ne Ill, Iowa, Mich, Minn, Ohio, n Wisc

Ophioglossum engelmannii (LIMESTONE ADDER'S-TONGUE) p. 262
- Dry barrens and glades over limestone; dry limestone and dolomite open woods and rocky openings
- s Ill, s Ind, s Ohio

Ophioglossum pusillum (NORTHERN ADDER'S-TONGUE) page 264
- Moist sandy fields, wet meadows, ditches, also in drier situations; soils typically sandy
- n Ill, n Ind, Iowa, Mich, Minn, n Ohio, Wisc

Polystichum acrostichoides (CHRISTMAS FERN). . . . . . page 170
- Shaded, moist to dry forests, rocky slopes and ravines
- Ill, Ind, Iowa, Mich, se Minn, Ohio, Wisc

Pteridium aquilinum (NORTHERN BRACKEN FERN). . . . page 132
- Dry, acidic, sandy woods, clearings, waste places, burned areas
- regionwide

Sceptridium dissectum (CUT-LEAF GRAPE FERN). . . . . . page 274
- Dry hilltops, dry to moist sandy fields, pastures and open forests; sometimes along old trails and roads
- Ill, Ind, Iowa, Mich, Minn, Ohio, Wisc

Sceptridium multifidum (LEATHERY GRAPE FERN) . . . . page 276
• Dry to moist sandy barrens, openings in oak and pine woods, old fields and clearings, along trails and old roads, grassy meadows and lawns
• Ill, e Iowa, Mich, Minn, Ohio, Wisc

....

Selaginella rupestris (SAND CLUBMOSS) . . . . page 424
• Dry rocky balds, sand dunes and open sandy areas; soils acidic
• Ill, Ind, Iowa, Mich, Minn, Wisc

....

Spinulum annotinum (INTERRUPTED CLUB-MOSS) . . . . page 418
• Wet to dry forests and clearings; soils often sandy or rocky
• Mich, Minn, Wisc

— — — — — — — — — ❶ — — — — — — — — —

FOUND IN MANY TYPES OF MOIST TO DRY HABITATS

Athyrium filix-femina (LADY FERN) . . . . page 82
• Moist to dry woods, swamps, thickets, meadows
• regionwide

....

Equisetum arvense (FIELD HORSETAIL) . . . . page 186
• Variety of sunny or shaded habitats (apart from where extremely wet or dry), including fields, woods, ditches, shores, roadsides, railroad banks
• regionwide

....

Huperzia selago (FIR-MOSS) . . . . page 404
• Openings in cedar swamps and moist to dry conifer woods; mossy, shaded streambanks, boulders, trails and old roads, logs, moist sandy borrow pits
• n Mich, n Minn, n Wisc

....

Pteridium aquilinum (NORTHERN BRACKEN FERN) . . . . page 132
• Sandy woods, fields, clearings, disturbed areas
• regionwide

....

Sceptridium rugulosum (TERNATE GRAPE FERN) . . . . . . page 280
• Dry to moist brushy or grassy meadows, sandy open woods, mossy places in forests of jack pine and red pine
• Mich, Minn, Wisc

Illustrated Key to Families & Genera

1 Plant a free-living gametophyte, resembling a thallose liverwort; rare in dark, moist recesses in rock overhangs in s Ind and Ohio **PTERIDACEAE**, page 302
.... *Vittaria appalachiana*, page 328

1 Plant a sporophyte, consisting of a stem and well-developed leaves more than 1 cell thick (except in *Trichomanes*), generally reproducing by spores **2**

2 Plants aquatic, either floating and unattached, or rooting and often completely submersed .... **Key Ⓐ**, page 32

> *Includes members of*
> * **ISOETACEAE**   *Isoetes*
> * **MARSILEACEAE**   *Marsilea*
> * **SALVINIACEAE**   *Azolla*

2 Plants of various wetland, upland, and rock habitats **3**

3 Leaves not "fern-like;" unlobed, variously awl-shaped, scalelike, or grasslike (the **fern relatives** will key here plus two species of *Asplenium*) .... **Key Ⓑ**, page 33

> *Includes members of*
> * **ASPLENIACEAE**   *Asplenium rhizophyllum, A. scolopendrium*
> * **EQUISETACEAE**   *Equisetum*
> * **ISOETACEAE**   *Isoetes*
> * **LYCOPODIACEAE**   *Dendrolycopodium, Diphasiastrum,*
>   *Huperzia, Lycopodiella, Lycopodium*
> * **SELAGINELLACEAE**   *Selaginella*

3 Leaves "fern-like," variously lobed or divided (**true ferns** will key here; the following sub-keys are based on size of frond and habitat, either soil or rock).. **4**

4 Plants small, leaf blades (not including the stipe) small, less than 30 cm long or wide (some species will key both here and in the next lead)................. **5**

4 Plants larger, leaf blades medium to large, more than 30 cm long or wide **6**

5 Plants growing on rock, rock walls, or over rock in thin soil, or rarely on the bark of trees... **Key Ⓒ**, page 36

> *Includes members of*
> * **ASPLENIACEAE**   *Asplenium*
> * **CYSTOPTERIDACEAE**   *Cystopteris, Gymnocarpium*
> * **DRYOPTERIDACEAE**   *Cyrtomium*
> * **HYMENOPHYLLACEAE**   *Trichomanes*
> * **POLYPODIACEAE**   *Pleopeltis, Polypodium*
> * **PTERIDACEAE**   *Adiantum, Cheilanthes, Cryptogramma,*
>   *Pellaea, Pteris*
> * **THELYPTERIDACEAE**   *Phegopteris*
> * **WOODSIACEAE**   *Woodsia*

5 Plants terrestrial, growing in soil, not associated with rock outcrops .... **Key Ⓓ**
.... page 40

Includes members of

• **ASPLENIACEAE**	*Asplenium*
• **BLECHNACEAE**	*Woodwardia*
• **CYSTOPTERIDACEAE**	*Gymnocarpium*
• **DRYOPTERIDACEAE**	*Polystichum*
• **ONOCLEACEAE**	*Onoclea*
• **OPHIOGLOSSACEAE**	*Botrychium, Botrypus, Sceptridium*
• **PTERIDACEAE**	*Adiantum pedatum*
• **THELYPTERIDACEAE**	*Phegopteris, Thelypteris*

6 Plants growing on rock, or over rock in thin soil mats or pockets of soil . **Key Ⓔ**
.... page 43

Includes members of

• **ASPLENIACEAE**	*Asplenium platyneuron*
• **ATHYRIACEAE**	*Athyrium*
• **CYSTOPTERIDACEAE**	*Cystopteris*
• **DENNSTAEDIACEAE**	*Dennstaedtia, Pteridium*
• **DRYOPTERIDACEAE**	*Cyrtomium, Dryopteris, Polystichum*
• **LYGODIACEAE**	*Lygodium palmatum*
• **PTERIDACEAE**	*Adiantum*
• **THELYPTERIDACEAE**	*Parathelypteris, Thelypteris*
• **WOODSIACEAE**	*Woodsia*

6 Plants growing in soil, not associated with rock outcrops ...... **Key Ⓕ**, page 47

Includes members of

• **ASPLENIACEAE**	*Asplenium platyneuron*
• **ATHYRIACEAE**	*Athyrium, Deparia, Diplazium*
• **BLECHNACEAE**	*Woodwardia*
• **CYSTOPTERIDACEAE**	*Cystopteris bulbifera*
• **DENNSTAEDIACEAE**	*Dennstaedtia, Pteridium*
• **DRYOPTERIDACEAE**	*Dryopteris, Polystichum acrostichoides*
• **LYGODIACEAE**	*Lygodium palmatum*
• **ONOCLEACEAE**	*Matteucia, Onoclea*
• **OPHIOGLOSSACEAE**	*Botrypus virginianus*
• **OSMUNDACEAE**	*Osmunda*
• **PTERIDACEAE**	*Adiantum pedatum*
• **THELYPTERIDACEAE**	*Thelypteris*
• **WOODSIACEAE**	*Woodsia obtusa*

─────────────── **KEY Ⓐ** ───────────────

FERNS GROWING IN WATER, FLOATING
OR ROOTED AND SUBMERSED

1 Plants aquatic, floating .... 1. **SALVINIACEAE** (*Azolla*), page 330
1 Plants aquatic, rooted, sometimes exposed on shores as water levels drop..... **2**

2 Plants resemble a clover, with 4 terminal leaf segments .. 2. **MARSILEACEAE**
··· *(Marsilea)*, page 212

2 Leaves linear, from a swollen, corm-like base. 3. **ISOETACEAE** *(Isoetes)*, p. 370

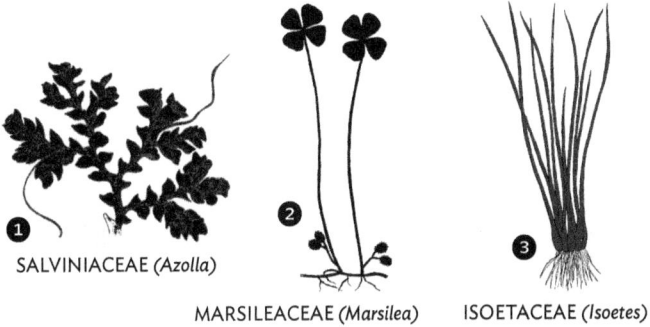

SALVINIACEAE *(Azolla)*

MARSILEACEAE *(Marsilea)*   ISOETACEAE *(Isoetes)*

──────────────── **KEY B** ────────────────

FERNS WITH LEAVES UNLOBED, AWL-SHAPED, SCALELIKE, OR GRASSLIKE, AND NOT "FERN-LIKE"

1 Stems obviously jointed; leaves small and scalelike, in a whorl from the joints or nodes, or sometimes absent; spores borne in a terminal conelike strobilus covered with peltate scales (i.e., scales more or less round and attached at middle like an umbrella)
EQUISETACEAE, page 178
Equisetum, page 178

Equisetum

1 Stems not jointed; leaves scalelike or larger, but if scalelike not in whorls from the nodes; spores borne variously, but if in a terminal strobilus the scales not peltate **2**

2 Leaves linear, grasslike, 1-50 cm long
ISOETACEAE, page 370
Isoetes, page 370

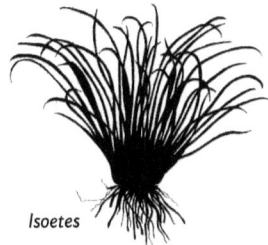

Isoetes

2 Leaves various (scalelike, awl-like, moss-like, or flat), but not linear and grasslike **3**

3 Leaves very numerous and overlapping along creeping, ascending, or erect stems; the leaves usually scalelike or awl-like, 0.5-2 (-3) mm wide, typically sharp- or hair-tipped; sporangia borne in strobili **4**

> *Includes members of*
> • **LYCOPODIACEAE**   *Dendrolycopodium, Diphasiastrum, Huperzia, Lycopodiella, Lycopodium, Spinulum*
> • **SELAGINELLACEAE**   *Selaginella*

3 Leaves not as above. **10**

> *Includes members of*
> • **ASPLENIACEAE**   *Asplenium rhizophyllum, A. scolopendrium*
> • **OPHIOGLOSSACEAE**   *Ophioglossum*

4 Sporangia borne in flattened or 4-sided strobili sessile at tips of leafy branches; spores and sporangia of two sizes, the megasporangia larger and borne at base strobili
> **SELAGINELLACEAE**, page 420
> *Selaginella*, page 420

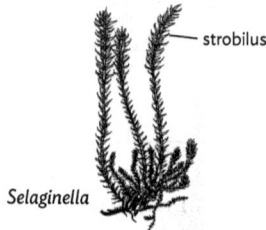

strobilus

Selaginella

4 Sporangia borne either in axils of normal foliage leaves, or in strobili sessile at tips of leafy branches, or stalked on specialized branches with fewer and smaller leaves; spores and sporangia of one size. **5**

5 Leafy stems erect, simple or dichotomously branched, the ultimate branches upright; sporophylls like the sterile leaves or only slightly smaller, in annual bands along the stem; vegetative reproduction by leafy gemmae near tip of stem
> **LYCOPODIACEAE**, page 376
> *Huperzia*, page 396

Huperzia

5 Leafy stems prostrate or erect, if erect then generally branched, the ultimate branches spreading (horizontal) or ascending; sporophylls differing from sterile leaves, either broader and shorter, or more spreading, grouped into terminal cones; lacking vegetative reproduction by gemmae. **6**

6 Plants of wetlands, mostly on moist or wet sand or peat; leaves herbaceous, pale or yellow-green, dull, deciduous; leafy stems creeping; rhizome dying back annually to an underground vegetative tuber at tip
> **LYCOPODIACEAE**, page 376
> *Lycopodiella*, page 406

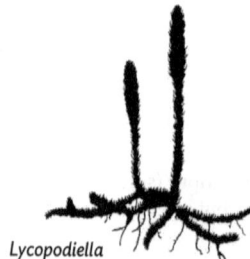

Lycopodiella

6 Plants of uplands, mostly in moist to dry soils; leaves stiff, bright to dark green, shiny, evergreen; leafy stems mainly erect, treelike, fanlike, or creeping (if creeping, then the leaves with a hairlike tip); rhizome trailing, perennial **7**

7 Shoots flat-branched, 1-5 mm wide (including the leaves); leaves scalelike, dimorphic, overlapping and appressed to stem, in 4 ranks; strobili on long, branched stalks
      LYCOPODIACEAE, page 376
        Diphasiastrum, page 386   *Diphasiastrum*

7 Shoots round-branched, usually 5-8 mm wide (including the leaves), leaves awl-shaped, monomorphic (though sometimes differing in size), separate, spreading or ascending, in 6 ranks; strobili sessile at stem tips **8**

8 Strobili on elongate, sparsely leafy stalks at tips of leafy upright branches; leaves with hairlike tip
      LYCOPODIACEAE, page 376
        Lycopodium, page 410   *Lycopodium*

8 Strobili not stalked, borne immediately above densely leafy portion of upright branches; leaves tapered to a narrow tip **9**

9 Upright leafy stems 10 mm or more wide (including the leaves), branched 1-4 times; leaves tipped with a small sharp spine; horizontal shoots creeping on ground surface
      LYCOPODIACEAE, page 376
        Spinulum, page 416   *Spinulum*

9 Upright leafy stems 3-8 mm wide (including the leaves), treelike or fanlike; leaves tapered to tip, spine absent; horizontal shoots underground
      LYCOPODIACEAE, page 376
        Dendrolycopodium, page 378   *Dendrolycopodium*

10 Plants with 1 (-several) leaves, the sterile leaf blade ovate, entire-margined, obtuse, the longer fertile portion with 2 rows of sporangia somewhat embedded in it
      OPHIOGLOSSACEAE, page 226
        Ophioglossum, page 260   *Ophioglossum*

10 Plant with many leaves, generally 5 or more, not divided into separate sterile and fertile segments, the leaves strap-shaped, or lance-shaped and with a long-tapering tip (this often with a plantlet which can root to form new plants) **ASPLENIACEAE**, page 54
Asplenium rhizophyllum, page 66
Asplenium scolopendrium, page 70

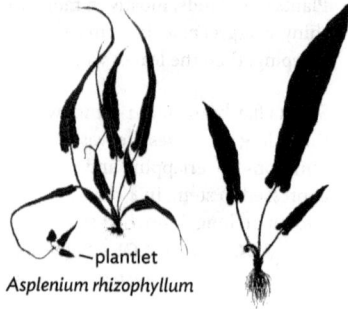

plantlet
Asplenium rhizophyllum

Asplenium scolopendrium

KEY **C**

SMALL FERNS, GROWING ON ROCK, ON ROCK IN THIN SOIL, OR (RARELY) ON TREES

1 Fronds pinnatifid or 2-pinnatifid, most of the pinnae not fully divided from one another (the rachis winged by leaf tissue for most or all its length) **2**

Includes members of
- **ASPLENIACEAE** *Asplenium*
- **POLYPODIACEAE** *Pleopeltis, Polypodium*
- **PTERIDACEAE** *Pteris multifida*
- **THELYPTERIDACEAE** *Phegopteris*

1 Fronds pinnate, pinnate-pinnatifid, 2-pinnate, or even more divided (rachis naked for most of its length, but often winged in upper portion) **6**

Includes members of
- **ASPLENIACEAE** *Asplenium*
- **CYSTOPTERIDACEAE** *Cystopteris, Gymnocarpium*
- **DRYOPTERIDACEAE** *Cyrtomium*
- **HYMENOPHYLLACEAE** *Trichomanes*
- **PTERIDACEAE** *Adiantum, Cheilanthes, Cryptogramma, Pellaea*
- **WOODSIACEAE** *Woodsia*

2 Fronds 2-pinnatifid, at least the lowermost pinnae deeply lobed **3**
2 Fronds 1-pinnatifid. **4**

3 Lowermost (and other) pinnae with numerous, nearly even lobes; native **THELYPTERIDACEAE**, page 334
Phegopteris, page 342

3 Lowermost pinnae with a few, irregular lobes (the upper pinnae unlobed); introduced, uncommon in s Ill and possibly s Ind **PTERIDACEAE**, page 302
Pteris multifida, page 303

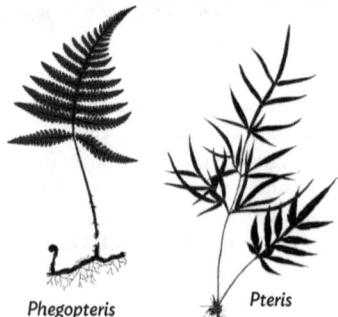

Phegopteris

Pteris

4 Blades with long, narrow tapering tip, upper portion of blade unlobed or only slightly lobed; sori elongate
 ASPLENIACEAE, page 54
 Asplenium (in part), page 54

Asplenium

4 Fronds without a long, narrow tapering tip; blade lobed for most of its length; sori round **5**

5 Blade densely scaly on lower surface; margins entire; rhizome 1-2 mm in diameter
 POLYPODIACEAE, page 294
 Pleopeltis, page 295

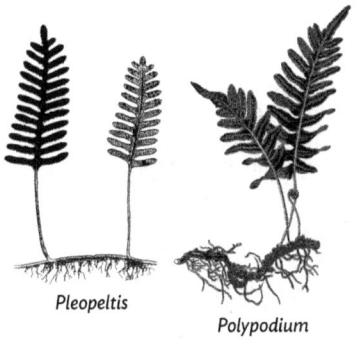

5 Blade without scales on lower surface; margins finely toothed; rhizome 3-6 mm in diameter
 POLYPODIACEAE, page 294
 Polypodium, page 298

Pleopeltis

Polypodium

6 Fronds 1-pinnate or 1-pinnate-pinnatifid **7**
6 Fronds 2-pinnate or more divided **11**

7 Fronds very delicate, only 1 cell thick; sori in cups on leaf margins; plants of rock outcrops with high air humidity
 HYMENOPHYLLACEAE, page 204
 Trichomanes, page 204

Trichomanes

7 Fronds thicker, herbaceous, to leathery in texture, more than 1 cell thick; sori otherwise; plants of various habitats. **8**

8 Pinnae more than 1 cm wide; fronds leathery; veins rejoining to form a netlike pattern; introduced, uncommon in s Ind
 DRYOPTERIDACEAE, page 136
 Cyrtomium fortunei, page 136

Cyrtomium fortunei

8 Pinnae less than 1 cm wide; fronds herbaceous to somewhat leathery; veins free, not rejoining and netlike **9**

9 Sori on the undersurface of the leaf, away from the margins
  ASPLENIACEAE, page 54
  Asplenium, page 54

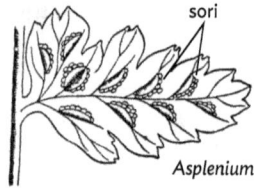

sori

Asplenium

9 Sori on the undersurface of the leaf, along margins and more-or-less hidden beneath either the unmodified inrolled leaf margin or under a modified, reflexed false indusium. **10**

10 Stipes green to straw-colored for at least the upper 1/3, rachis green; fronds dimorphic, the fertile longer than the sterile and with narrower segments **PTERIDACEAE**, page 302
  Cryptogramma, page 316

10 Stipes and rachis dark brown to almost black throughout; fronds similar or somewhat different
  PTERIDACEAE, page 302
  Pellaea, page 320

Pellaea

Cryptogramma

11 Blade broadly triangular in outline
  CYSTOPTERIDACEAE, page 102
  Gymnocarpium, page 120

Gymnocarpium

11 Blade elongate, mostly lance-shaped, generally 4x or more as long as wide, not notably triangular in outline **12**

12 Sori on margins, usually more-or-less hidden under the inrolled margin of the pinnule. **13**
12 Sori not on margins, either naked, or slightly to strongly hidden by indusia. . . **15**

13 Sori round or oblong, distinct and separate along the pinnule margins; fronds bright-green, smooth, herbaceous, delicate, and flexible
  PTERIDACEAE, page 302
  Adiantum pedatum, page 305

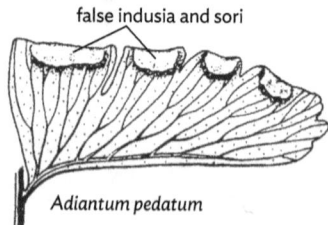

false indusia and sori

Adiantum pedatum

13 Sori continuous along the pinnule margins; fronds mostly dark-green, often hairy, leathery, tough, and stiff **14**

14 Fronds 2-3-pinnate, more or less densely hairy
  PTERIDACEAE, page 302
  Cheilanthes, page 308

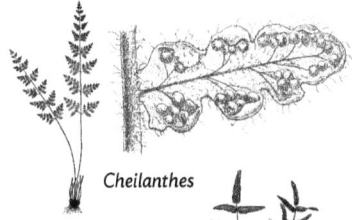

Cheilanthes

14 Fronds 1-2-pinnate, smooth or sparsely and inconspicuously hairy
  PTERIDACEAE, page 302
  Pellaea, page 320

Pellaea

15 Blades 3-12 cm long; sori elongate, covered by a flap-like, entire indusium
  ASPLENIACEAE, page 54
  Asplenium, page 54

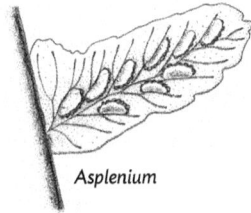

Asplenium

15 Blades 4-30 (-50) cm long; sori round, surrounded or covered by an entire, fringed, or divided indusium. **16**

16 Veins reaching margin; indusium attached under one side of sorus, hoodlike or pocketlike, arching over sorus; stipes smooth or sparsely covered with scales, stipe bases not persistent
  CYSTOPTERIDACEAE, page 102
  Cystopteris, page 103

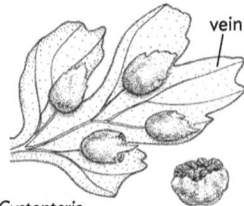

vein

Cystopteris

16 Veins ending short of margin; indusium attached under sorus, cuplike (divided into 3-6 lobes which surround the sorus from below) or of numerous tiny hairs, which extend out from under sorus on all sides; stipes often densely covered with scales, stipe bases persistent
  WOODSIACEAE, page 352
  Woodsia, page 352

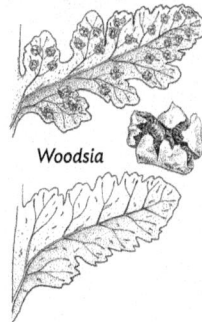

Woodsia

――――――――――――――――――― **KEY D** ―――――――――――――――――――

SMALL FERNS, TERRESTRIAL, GROWING IN SOIL, NOT ASSOCIATED WITH ROCK OUTCROPS

1 Stipe branched once dichotomously, each branch with 3-7 pinnae in one direction only, the outline of the blade fan-shaped, often wider than long
       PTERIDACEAE, page 302
       Adiantum pedatum, page 305

Adiantum

1 Stipe not branched dichotomously, the outline of the blade either longer than wide or triangular and about as wide as long. **2**

2 Fronds pinnatifid or 2-pinnatifid, most of the pinnae not fully divided from one another (the rachis winged by leaf tissue for most or all of its length) **3**

    Includes members of
    • **BLECHNACEAE**   *Woodwardia areolata*
    • **ONOCLEACEAE**   *Onoclea sensibilis*
    • **OPHIOGLOSSACEAE**   *Botrychium, Botrypus*
    • **THELYPTERIDACEAE**   *Phegopteris*

2 Fronds 1-pinnate, 1-pinnate-pinnatifid, 2-pinnate, or even more divided (the rachis naked for most of its length, or often winged in the upper portion). **6**

    Includes members of
    • **ASPLENIACEAE**   *Asplenium platyneuron*
    • **BLECHNACEAE**   *Woodwardia areolata*
    • **CYSTOPTERIDACEAE** *Gymnocarpium*
    • **DRYOPTERIDACEAE**   *Polystichum*
    • **OPHIOGLOSSACEAE**   *Botrychium, Sceptridium*
    • **THELYPTERIDACEAE**   *Thelypteris*

3 Sporangia borne on an erect stalk that arises at or above ground level from stipe of sterile leaf blade (joining stipe of sterile leaf above the rhizome)
    OPHIOGLOSSACEAE, page 226
       Botrychium, page 228
       Botrypus, page 256

Botrychium

Botrypus

3 Sporangia either borne on normal leaf blades or on specialized (fertile) fronds **4**

4 Fronds all alike, sori on normal leaf
blades
   THELYPTERIDACEAE, page 334
               Phegopteris, page 342

Phegopteris

4 Fronds of two types; sori on fronds significantly different than normal fronds . **5**

5 Fertile frond woody, brown, with
bead-like segments; pinnae margins
entire, often wavy or the lowermost
somewhat lobed; pinnae nearly
opposite **ONOCLEACEAE**, page 218
      Onoclea sensibilis, page 219

Onoclea

5 Fertile leaf stiff but herbaceous, green,
the pinnae linear, not at all bead-like;
pinnae margins finely toothed,
otherwise slightly wavy or straight;
pinnae nearly alternate
       BLECHNACEAE, page 94
   Woodwardia areolata, page 96

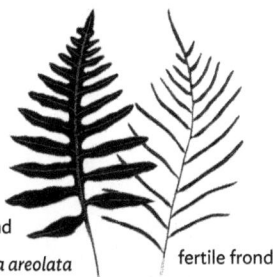

sterile frond
Woodwardia areolata   fertile frond

6 Fronds broadly triangular in outline, about as broad as long; sporangia borne on
an erect stalk that arises at or above ground level from the stipe of the sterile
leaf blade (joining the stipe of the sterile leaf above the rhizome) **7**

6 Fronds lance-shaped in outline, much longer than broad; sporangia either
borne on normal leaf blades, on slightly dimorphic blades, or on an erect stalk
that arises at or above ground level from the stipe of the sterile leaf blade
(joining the stipe of the sterile leaf above the rhizome) **8**

7 Sporangia borne on normal leaf blades
   CYSTOPTERIDACEAE, page 102
       Gymnocarpium, page 120

Gymnocarpium

7 Sporangia borne on an erect stalk that arises at or above ground level from the stipe of the sterile leaf blade (joining the stipe of the sterile leaf above the rhizome)
    OPHIOGLOSSACEAE, page 226
    Sceptridium, page 269

8 Blades 1-8 cm long; sporangia borne on an erect stalk that arises at or above ground level from the stipe of the sterile leaf blade (joining the stipe of the sterile leaf above the rhizome)
    OPHIOGLOSSACEAE, page 226
    Botrychium, page 228

Sceptridium

Botrychium

8 Blades 10-30 (-100) cm long; sporangia either on normal leaf blades or on slightly modified blades **9**

9 Fronds evergreen, dark green, somewhat leathery
    DRYOPTERIDACEAE, page 136
    Polystichum, page 168

Polystichum

9 Fronds light to medium green, herbaceous, deciduous to semi-evergreen **10**

10 Sori elongate; leaf blades somewhat dimorphic, the fertile larger and erect, the sterile smaller and prostrate, the larger leaf blades 2-4 (-6.5) cm wide
    ASPLENIACEAE, page 54
    Asplenium platyneuron, page 64

10 Sori round; leaf blades monomorphic; the larger leaf blades 5-15 cm wide
    THELYPTERIDACEAE, page 334
    Thelypteris, page 346

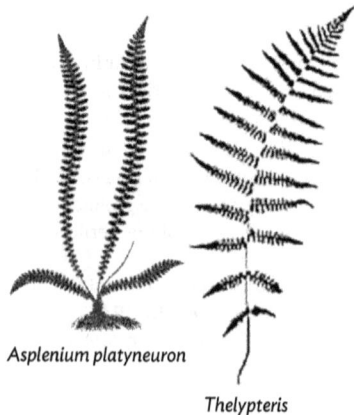

Asplenium platyneuron

Thelypteris

———————————————— **KEY** **E** ————————————————

MEDIUM TO LARGE FERNS, GROWING ON ROCK, OR OVER ROCK IN THIN SOIL

1 Fronds vine-like, 0.3-10 m long, the branching dichotomous, 1 branch of each dichotomy terminating in a pair of pinnae, the pinnae often widely spaced (more than 10 cm apart)
          LYGODIACEAE, page 208
        Lygodium palmatum, page 210

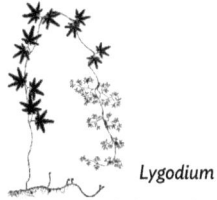

Lygodium

1 Fronds not vine-like, 0.3-1 m long, the branching not as described above, the pinnae regularly and more-or-less closely spaced (mostly less than 10 cm apart) **2**

2 Fronds 1-pinnate-pinnatifid or less divided, the pinnae entire, toothed, lobed or pinnatifid **3**

    Includes members of
• **ASPLENIACEAE**	*Asplenium platyneuron*
• **DRYOPTERIDACEAE**	*Cyrtomium, Dryopteris, Polystichum*
• **THELYPTERIDACEAE**	*Thelypteris*
• **WOODSIACEAE**	*Woodsia*

2 Fronds 2-pinnate or more divided, the pinnae divided to their midribs **8**

    Includes members of
• **ATHYRIACEAE**	*Athyrium*
• **CYSTOPTERIDACEAE**	*Cystopteris bulbifera*
• **DENNSTAEDIACEAE**	*Dennstaedtia punctilobula, Pteridium aquilinum*
• **DRYOPTERIDACEAE**	*Dryopteris*
• **THELYPTERIDACEAE**	*Parathelypteris*
• **WOODSIACEAE**	*Woodsia*

3 Sori elongate, the indusium flap-like, attached along the side; leaf blades (if more than 30 cm long) less than 7 cm wide   **ASPLENIACEAE**, page 54
    Asplenium platyneuron, page 64

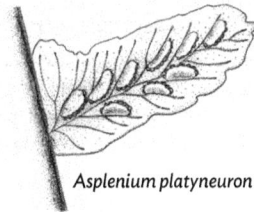

Asplenium platyneuron

3 Sori circular or globular, the indusium peltate, kidney-shaped, or cuplike; leaf blades (if more than 30 cm long) more than 5 cm wide **4**

4 Fronds 1-pinnate, the pinnae toothed and each with a slight to prominent lobe near the base on the side towards the leaf tip, dark green, somewhat leathery; indusia peltate **5**

4 Fronds 1-pinnate-pinnatifid, the pinnae pinnatifid, generally lacking a prominent basal lobe, light green to dark green, herbaceous to slightly leathery; indusium either kidney-shaped or cuplike. **6**

5 Veins anastamosing, rejoining to form a netlike pattern; pinnae 4-25 pairs per leaf; introduced, uncommon in s Ind
DRYOPTERIDACEAE, page 136
Cyrtomium fortunei, page 136

5 Veins branching dichotomously, free, not rejoining to form a netlike pattern; pinnae 25-50 pairs on larger fronds; native species
DRYOPTERIDACEAE, page 136
Polystichum, page 168

Cyrtomium fortunei

6 Vascular bundles in the stipe 3-7
DRYOPTERIDACEAE, page 136
Dryopteris, page 138

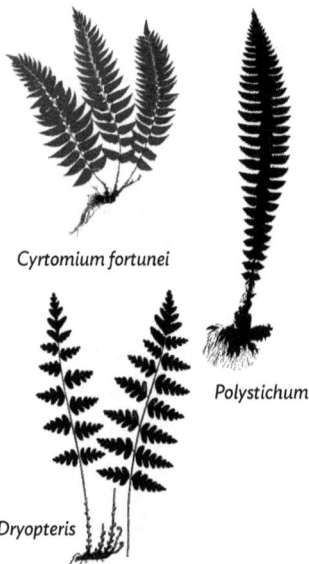

Polystichum

Dryopteris

6 Vascular bundles in the stipe 2, uniting above. **7**

7 Indusium kidney-shaped, arching over the sorus
THELYPTERIDACEAE, page 334
Thelypteris, page 346

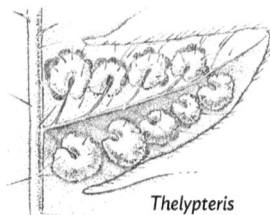

Thelypteris

7 Indusium cuplike, attached beneath sorus and consisting of 3-6 lance-shaped to ovate segments
WOODSIACEAE, page 352
Woodsia, page 352

Woodsia obtusa

8 Sori marginal and borne on underside of the false indusium; stipes and rachis shiny black or reddish-black, glabrous except at the very base of the stipe; pinnules fan-shaped or obliquely elongate
PTERIDACEAE, page 302
Adiantum, page 305

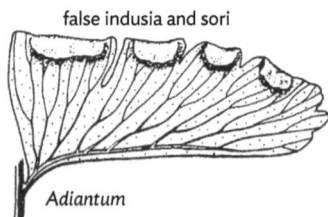

false indusia and sori

Adiantum

8 Sori not marginal, borne on undersurface of leaf blade (or if marginal, as in *Pteridium* and *Dennstaedtia,* borne on undersurface of the leaf); stipes darkened only near base (if at all), rachis green, tan, or reddish; pinnules not notably fan-shaped or obliquely elongate. **9**

9 Blades broadly triangular in outline, about as long as wide
    DENNSTAEDIACEAE, page 128
    Pteridium aquilinum, page 132

Pteridium aquilinum

9 Blades elongate, mostly lanceolate, generally 4x or more as long as wide **10**

10 Outline of leaf blade narrowed to base, the widest point more than 7 pinna pairs above the base, the lowermost pinnae 1/4 or less as long as the longest pinnae; rhizomes long-creeping, the fronds scattered, forming clonal patches
    THELYPTERIDACEAE, page 334
    Parathelypteris, page 338

Parathelypteris

10 Outline of the leaf blade slightly if at all narrowed to the base, the widest point less than 5 pinna pairs from the base, the lowermost pinnae more than 1/2 as long as the longest pinnae; rhizomes short-creeping, the fronds clustered, not forming clonal patches (or with rhizomes long-creeping, fronds scattered, forming clonal patches in *Dennstaedtia punctilobula*). **11**

11 Rhizomes long-creeping, fronds scattered, forming clonal patches; vascular bundles in the stipe 1, U-shaped (even in the lower stipe); sori very small, marginal in sinuses, the indusium cuplike, 2-parted, the outer part a modified tooth of the leaf blade; leaf blades conspicuously finely hairy
    DENNSTAEDIACEAE, page 128
    Dennstaedtia punctilobula, page 130

Dennstaedtia punctilobula

11 Rhizomes short-creeping, the fronds clustered, not forming clonal patches; vascular bundles in the stipe 2-7 (sometimes uniting to 1 in upper stipe); sori mostly larger, mostly not marginal, the indusium not as above (though cuplike in *Woodsia obtusa*); leaf blades either smooth, with flattened scales, or finely hairy with gland-tipped hairs **12**

12 Vascular bundles (3-) 5 (-7) in the
stipe; mostly larger ferns of forests
DRYOPTERIDACEAE, page 136
Dryopteris, page 138

Dryopteris

12 Vascular bundles 2 in the stipe (or joining near the leaf blade into 1); ferns of
forests and rocky habitats **13**

13 Fronds 1-pinnate-pinnatifid;
indusium cuplike, attached beneath
the sorus and consisting of 3-6
lanceolate to ovate segments; mostly
smaller ferns on rock
WOODSIACEAE, page 352
Woodsia, page 352

Woodsia obtusa

13 Fronds 2-pinnate-pinnatifid; indusium flaplike or pocketlike, attached at one
side of the sorus and arching over it **14**

14 Fronds 10-30 cm wide, the tip acute
to acuminate; indusium flaplike
ATHYRIACEAE, page 80
Athyrium filix-femina, page 82

14 Fronds 4-9 cm wide, the tip long-
attenuate; indusium pocketlike or
hoodlike
CYSTOPTERIDACEAE, page 102
Cystopteris bulbifera, page 106

Athyrium filix-femina

Cystopteris bulbifera

─────────── **KEY F** ───────────

MEDIUM TO LARGE FERNS, GROWING IN SOIL
(NOT ON ROCK OUTCROPS)

1 Fronds vine-like, 0.3-10 m long, the branching dichotomous, 1 branch of each dichotomy terminating in a pair of pinnae, the pinnae often widely spaced (more than 10 cm apart)
> **LYGODIACEAE**, page 208
> *Lygodium palmatum*, page 210

Lygodium

1 Fronds not vine-like, 0.3-3 m long, the branching not as described above, the pinnae regularly and more-or-less closely spaced (mostly less than 10 cm apart)
. . . . **2**

2 Blades broadly (about equilaterally) triangular, pentagonal, or flabellate in outline, 0.7-1.3x as long as wide **3**

2 Fronds elongate in outline, mostly ovate, lanceolate, oblanceolate, or narrowly triangular, 1.5-10x or more as long as wide **5**

3 Blades fan-shaped in outline, the stipe branched once dichotomously, each branch bearing 3-7 pinnae
> **PTERIDACEAE**, page 302
> *Adiantum pedatum*, page 305

Adiantum

3 Blades broadly triangular in outline, the stipe not branched dichotomously . . . **4**

4 Sporangia in a stalked, specialized, fertile portion of the blade; texture of mature blades somewhat fleshy; plants solitary from a short underground rhizome with thick, mycorrhizal roots; plants of moist forests
> **OPHIOGLOSSACEAE**, page 226
> *Botrypus virginianus*, page 256

Botrypus virginianus

4 Sporangia in marginal, linear sori, indusium absent, protected by the revolute leaf margin and a minute false indusium; texture of mature leaf blades hard and stiff; plants colonial from deep rhizomes; plants of moist to dry woodlands and openings
> **DENNSTAEDIACEAE**, page 128
> *Pteridium aquilinum*, page 132

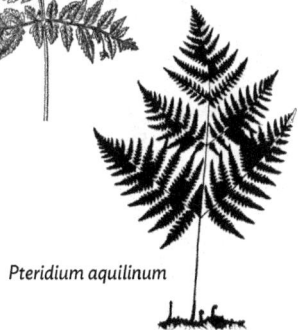

Pteridium aquilinum

5   Fronds 2-pinnate or more divided, the pinnae divided to their midribs **6**

> *Includes members of*
> - **ATHYRIACEAE**   *Athyrium*
> - **CYSTOPTERIDACEAE** *Cystopteris bulbifera*
> - **DENNSTAEDIACEAE**   *Dennstaedtia punctilobula*
> - **DRYOPTERIDACEAE**   *Dryopteris*
> - **OSMUNDACEAE**   *Osmunda regalis*

5   Fronds 1-pinnate-pinnatifid or less divided; the pinnae entire, toothed, lobed or pinnatifid. **10**

> *Includes members of*
> - **ASPLENIACEAE**   *Asplenium platyneuron*
> - **ATHYRIACEAE**   *Deparia, Diplazium*
> - **BLECHNACEAE**   *Woodwardia*
> - **DRYOPTERIDACEAE**   *Dryopteris, Polystichum acrostichoides*
> - **ONOCLEACEAE**   *Matteucia, Onoclea*
> - **OSMUNDACEAE**   *Osmunda*
> - **THELYPTERIDACEAE** *Thelypteris*

6   Blade divided into sterile and fertile portions; sterile pinnae located below terminal fertile pinnae, the sterile pinnules 30-70 mm long and 8-23 mm wide, finely toothed, tip rounded to somewhat pointed; fertile pinnae greatly reduced in size, the fertile pinnules 7-11 mm long and 2-3 mm wide   **OSMUNDACEAE, page 284**
Osmunda regalis, page 290

Osmunda regalis

6   Blade not divided into sterile and fertile portions, the pinnules bearing sporangia only slightly if at all reduced in size, both fertile and sterile pinnules mostly 4-20 mm long and 2-10 mm wide. **7**

7   Rhizomes long-creeping, fronds scattered, forming patches; vascular bundles in the stipe 1, U-shaped (even in the lower stipe); sori very small, marginal in sinuses, the indusium cuplike, 2-parted, the outer part a modified tooth of the leaf blade; leaf blades conspicuously finely hairy
   DENNSTAEDIACEAE, page 128
Dennstaedtia punctilobula, page 130

Dennstaedtia punctilobula

7 Rhizomes short-creeping, the fronds clustered, not forming patches; vascular bundles in stipe 2-7 (sometimes uniting to 1 in upper stipe); sori mostly larger, mostly not marginal, the indusium not as above (though cuplike in *Woodsia obtusa*); leaf blades either smooth, with flattened scales, or finely glandular hairy **8**

8 Vascular bundles (3-) 5 (-7) in the stipe
  DRYOPTERIDACEAE, page 136
    Dryopteris, page 138

Dryopteris

8 Vascular bundles 2 in the stipe (or joining upwards near leaf blade into 1) **9**

9 Fronds more than 10 cm wide, the tip acute to acuminate; indusium flaplike; pealike bulblets absent
  ATHYRIACEAE, page 80
    Athyrium filix-femina, page 82

9 Fronds 4-9 cm wide, the tip long-tapering; indusium pocketlike or hoodlike; bulblets often present on upper portion of blade
  CYSTOPTERIDACEAE, page 102
    Cystopteris bulbifera, page 106

Athyrium filix-femina   *Cystopteris bulbifera*

10 Fronds 1-pinnatifid, most of the pinnae not fully divided from one another (rachis winged by leaf tissue for most or all of its length); fronds dimorphic, the fertile much modified, stiff and/or woody **11**

10 Fronds 1-pinnate or 1-pinnate-pinnatifid, the pinnae fully divided from one another (rachis naked for most of its length, but often winged in upper portion); fronds dimorphic or not **12**

11 Fertile leaf woody, brown, with bead-like segments; pinnae margins entire, often wavy or the lowermost even somewhat lobed; pinnae mostly opposite along rachis
  ONOCLEACEAE, page 218
    Onoclea sensibilis, page 219

Onoclea

11 Fertile frond stiff but herbaceous, green, the pinnae linear, not at all bead-like; pinnae margins finely serrulate, otherwise slightly wavy or straight; pinnae mostly alternate along rachis

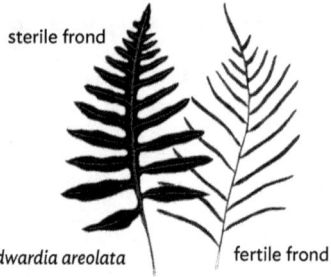

sterile frond

fertile frond

Woodwardia areolata

> **BLECHNACEAE**, page 94
> *Woodwardia areolata*, page 96

12 Rhizomes long-creeping, fronds scattered, forming patches **13**

12 Rhizomes short-creeping, the fronds clustered, not forming patches (or rhizomes of both long and short, but fronds borne only in clusters on the short erect rhizomes in *Matteucia*) **14**

13 Sori roundish, away from the main veins; pinna lobes of sterile fronds with the lateral veins free and pinnately arranged (the lowermost lateral vein sometimes joining that of the adjacent pinna lobe just below the sinus, but the remainder of the lateral veins all free

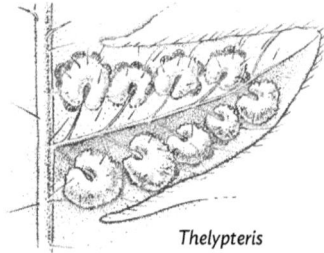

Thelypteris

> **THELYPTERIDACEAE**, page 334
> *Thelypteris*, page 346

13 Sori elongate, end to end along either side of the main veins; pinna lobes of sterile fronds with netted, chain-like venation along the central vein

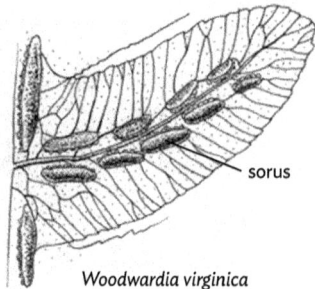

sorus

> **BLECHNACEAE** page 94
> *Woodwardia virginica*, page 98

Woodwardia virginica

14 Plants medium to large, fronds typically 60-300 cm tall; fronds either strongly dimorphic, the fertile fronds very unlike the sterile, brown at maturity (*Matteucia* and *Osmunda cinnamomea*) or fertile pinnae very unlike the sterile, brown at maturity, borne as an interruption in the blade, with normal green pinnae above and below (*Osmunda claytoniana*); rachises scaleless, stipes scaleless (except at the base in *Matteucia*) **15**

14 Plants mostly smaller, the fronds 30-100 cm tall (except *Dryopteris celsa* and *D. goldiana* to 15 dm); fronds not at all or only slightly dimorphic, the fertile differing in various ways, such as having narrower pinnae (as in *Polystichum acrostichoides*, *Diplazium*, and *Thelypteris palustris*) or fertile fronds taller and more deciduous (as in *Asplenium platyneuron* and *Dryopteris cristata*), but not as described in the first lead; rachises and stipes variously scaly or scaleless, but at least the stipe and often also the rachis scaly if the plants over 1 m tall. **16**

15 Fronds strongly tapering to the base from the broadest point (well beyond the midpoint of the blade), lowermost pinnae much less than 1/2 as long as the largest pinnae
ONOCLEACEAE, page 218
Matteucia struthiopteris, page 219

15 Fronds slightly if at all tapering to the base, about equally broad through much of their length, lowermost pinnae much more than 1/2 as long as the largest pinnae
OSMUNDACEAE, page 284
Osmunda, page 284

Matteucia

Osmunda

16 Sori elongate, the indusium elongate, attached along one side as a flap **17**
16 Sori roundish; the indusium kidney-shaped or nearly round, attached by a central stalk, or sometimes absent **19**

17 Stipe and rachis lustrous brownish black; fertile fronds 2-8 (-12) cm wide
ASPLENIACEAE, page 54
Asplenium platyneuron, page 64

Asplenium platyneuron

17 Stipe and rachis green; fertile fronds 10-20 (-30) cm wide **18**

18 Fronds 1-pinnate-pinnatifid (the pinnae pinnatifid)
ATHYRIACEAE, page 80
Deparia acrostichoides, page 88

fertile
frond

18 Fronds 1-pinnate (the pinnae entire)
ATHYRIACEAE, page 80
Diplazium pycnocarpon, page 92

sterile
frond

Deparia acrostichoides

Diplazium pycnocarpon

19 Fronds 1-pinnate, the pinnae toothed
and each with a slight to prominent
lobe near the base on the side towards
the leaf tip, dark green, subcoriaceous
to coriaceous; indusia peltate (round,
stalk attached to the center)
DRYOPTERIDACEAE, page 136
Polystichum acrostichoides, page 170

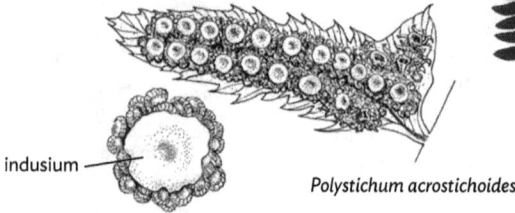

indusium

Polystichum acrostichoides

19 Fronds 1-pinnate-pinnatifid, the pinnae pinnatifid, generally without prominent
basal lobe, light green to dark green, herbaceous to somewhat leathery;
indusium kidney-shaped. **20**

20 Vascular bundles in the stipe 4-7
DRYOPTERIDACEAE, page 136
Dryopteris, page 138

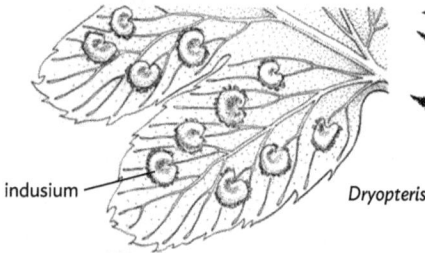

indusium

Dryopteris

20 Vascular bundles in the stipe 2,
uniting upwards
THELYPTERIDACEAE page 334
Thelypteris, page 346

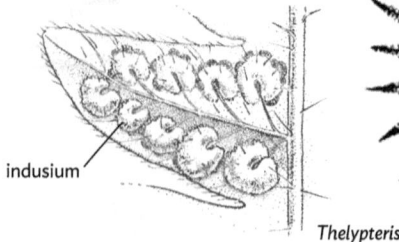

indusium

Thelypteris

True Ferns

Colony of **hay-scented fern** (*Dennstaedtia punctilobula*).

ASPLENIACEAE | Spleenwort Family
Asplenium
SPLEENWORT

Small, delicate ferns with short-creeping to erect rootstocks and slender stipes. Worldwide, the family includes 24 genera and over 400 species, most of which occur in tropical and subtropical regions; many species of *Asplenium* are of hybrid origin. North America is home to 28 species; 10 species in our flora. In the tropics, spleenworts can be found on the forest floor and also growing in trees (epiphytes). In temperate regions, *Asplenium* are more commonly found on rock in crevices and cracks; a number of our species are restricted to limestone.

KEY CHARACTERS
- **Fronds** 1-pinnate, all alike or the fertile slightly smaller.
- Linear **sori** arranged in herringbone fashion along veins.
- Single indusial flap of tissue opening in direction of pinna midrib (except in *Asplenium scolopendrium* with its two-sided, double opening sori).
- **Two vascular stipe bundles** that unite upwards to form an X-shaped bundle (magnification needed to see this).
- **Clathrate scales on rhizome** (present on all members of genus, illus. page 56). Clathrate, or lattice-like, refers to the scale cells having thick, dark outer walls, surrounding a clear, thin, inner area, similar to the leading between panes of stained glass. However, these can be difficult to see without digging up the plant.

NAME
Greek: *a*, not, *splen,* the spleen; refers to early use of a European spleenwort to treat spleen and liver problems.

ADDITIONAL MIDWEST SPECIES
- **Black-stem spleenwort** (*Asplenium resiliens* Kunze), more common south of midwest region, is known from s Ill (endangered) and s Ind (endangered), with a historical record from s Ohio. Found in semi-shade to full sun on moist to dry outcrops of calcareous rocks, often in cedar glades or on limestone cliffs.

Plants evergreen, clumped, from a short, upright rootstalk; **fronds** to 30 cm long; **stipe** and **rachis** shiny black. Similar to **ebony spleenwort** (*A. platyneuron*), but in that species, fronds dimorphic, pinnae margins sharp-toothed, and rachis dark brown; in *A. resiliens,* fronds not dimorphic, pinnae nearly entire, and rachis glossy black.

Asplenium resiliens
BLACK-STEM SPLEENWORT

KEY TO ASPLENIUM | SPLEENWORT

1   Leaf blade entire **2**

1   Leaf blade pinnately compound **3**

2   Tip of blade long tapering to a gradual point, often with plantlets at the tip, these sometimes rooting; Ill, Ind, Iowa, Mich, se Minn, Ohio, Wisc   **1. Asplenium rhizophyllum**
WALKING FERN, page 66

2   Tip of blade abruptly pointed to ± rounded, not producing plantlets at the tip; rare in Mich Upper Peninsula
2. Asplenium scolopendrium
HART'S-TONGUE FERN, page 70

3   Stipe brown below, green above (same color as rachis) **4**

3   Stipe and part or all of rachis brown to black. **7**

4   Blade long-triangular, lobed, the tip long and slender; Ill, Ind, Ohio, s Wisc   **3. Asplenium pinnatifidum**
LOBED SPLEENWORT, page 62

4   Blade short-triangular or oblong, 1- to 3-pinnate **5**

5   Blade 1-pinnate; tissue delicate, tardily deciduous; n Mich, n Wisc
4. Asplenium viride
BRIGHT-GREEN SPLEENWORT, page 74

Asplenium rhizophyllum
WALKING FERN

Asplenium scolopendrium
HART'S-TONGUE FERN

Asplenium pinnatifidum
LOBED SPLEENWORT

Asplenium viride
BRIGHT-GREEN SPLEENWORT

KEY TO ASPLENIUM, CONTINUED

5 Blade 2- or 3-pinnate; tissue firm, evergreen **6**

6 Divisions few, pinnae only 3–5 per side, alternate; plants of limestone; Ill, Ind, Iowa, n Mich, Ohio
5. Asplenium ruta-muraria
WALL-RUE, page 68

6 Divisions many, opposite; plants of acidic rocks; Ind, Ohio
6. Asplenium montanum
MOUNTAIN SPLEENWORT, page 60

7 Rachis dark only 1/3 its length; pinnae lobed and sharp-toothed, with an auricle on upper side; rare in s Ill, s Ind, s Ohio
7. Asplenium bradleyi
BRADLEY'S SPLEENWORT, page 58

7 Rachis dark throughout **8**

8 Sterile and fertile fronds different, the fertile tall and erect, much taller than the spreading sterile fronds; pinnae mostly alternate, the lower much reduced in size; plants usually on soil or sometimes on calcareous rock outcrops; Ill, Ind, Iowa, Mich, se Minn, Ohio, Wisc
8. Asplenium platyneuron
EBONY SPLEENWORT, page 64

8 Fronds uniform; pinnae opposite, the lower pinnae little reduced in size; plants usually on rock. **9**

9 Stipe black; pinnae blue-green, oblong, with a small auricle on upper side of pinnae; margins undulate; rare in s Ill and s Ind
9. Asplenium resiliens
BLACK-STEM SPLEENWORT, page 54

9 Stipe dark brown; pinnae bright green, oval, not auricled; margins round-toothed; Ill, Ind, Mich, Minn, Ohio, Wisc
10. Asplenium trichomanes
MAIDENHAIR SPLEENWORT, page 72

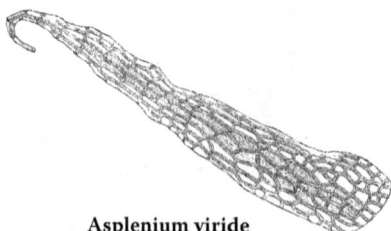

Asplenium viride
clathrate rhizome scale

5

Asplenium ruta-muraria
WALL-RUE

6

Asplenium montanum
MOUNTAIN SPLEENWORT

7

Asplenium bradleyi
BRADLEY'S SPLEENWORT

8

Asplenium platyneuron
EBONY SPLEENWORT

9

Asplenium resiliens
BLACK-STEM SPLEENWORT

10

Asplenium trichomanes
MAIDENHAIR
SPLEENWORT

Asplenium bradleyi D.C. Eaton
BRADLEY'S SPLEENWORT

FIELD TIPS
• small, clumped, evergreen fern
• pinnae dark green, on very short stalks
• stipes dark red-brown, satiny
• crevices of rock cliffs of sandstone (and other non-calcareous rock)
• rare, s Ill, s Ind, s Ohio

MIDWEST RANGE
Rare in s Ill (endangered), s Ind (endangered), Ohio (endangered); becoming more common in the Ozark and Ouachita Mountains of Missouri and Arkansas.

HABITAT
Typically tightly rooted in crevices on bare cliffs of sandstone, granite, chert or other acidic rock; sites usually dry and in full sun. Sometimes found with other spleenworts including *Asplenium montanum*, *A. pinnatifidum* and *A. platyneuron*, plus lichens and mosses.

DESCRIPTION
Rootstock Short-creeping to ascending, occasionally branched.

Frond Evergreen; sterile and fertile fronds alike, 5-30 cm long.

Stipe About half as long as blade, wiry, lustrous dark brown; scales present at base of leaf cluster, these dark reddish to brown, narrowly lance-shaped, grading upwards into hairs.

Rachis Shiny, dark reddish-brown, the coloration extending to middle of rachis, then fading to green.

Blade Narrowly lance-shaped in outline, shiny green when young, 1-pinnate-pinnatifid to 2-pinnate, base truncate (not tapered), basal pinnae (or blade divisions) often somewhat smaller.

Pinnae (5-) 8-15 (-20) pairs, alternate or nearly opposite, very short-stalked; upper basal lobe often slightly enlarged and forming an auricle; margins sharp-toothed to jagged-lobed; veins barely evident.

Sori Between margin and midrib, becoming blackish brown, 1-2 mm long; **indusium** opaque, entire.

SYNONYMS
• *Asplenium × stotleri* Wherry

SIMILAR SPECIES
Often in same habitat as the more common **mountain spleenwort** (*Asplenium montanum*) and **lobed spleenwort** (*A. pinnatifidum*).

NORTH AMERICA

MIDWEST

- In *Asplenium montanum,* fronds broader, pinnae on longer stalks, and upper stipe and rachis green and flattened.
- In *Asplenium pinnatifidum,* pinnae only pinnatifid, in contrast to distinctly pinnate pinnae of *A. bradleyi.*

NOTES

Originated as fertile offspring of cross between **mountain spleenwort** *(Asplenium montanum)* and **ebony spleenwort** *(Asplenium platyneuron).*

NAME

Daniel Eaton named this species in 1873 in honor of Frank H. Bradley, who discovered it in e Tennessee.

LEFT Fronds of **Bradley's spleenwort** *(Asplenium bradleyi)* in typical rock cliff habitat.

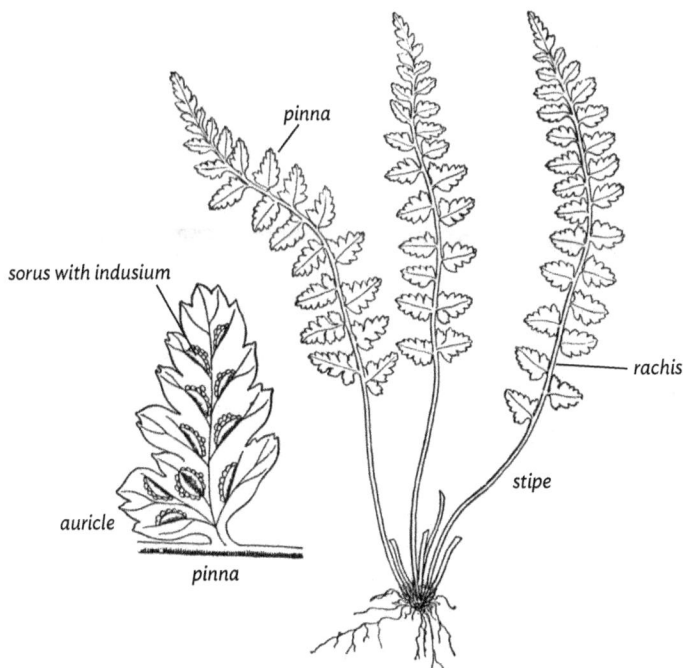

pinna

sorus with indusium

auricle

pinna

rachis

stipe

Asplenium montanum Willd.

MOUNTAIN SPLEENWORT

FIELD TIPS

- small, clumped, evergreen fern
- plants leathery, bluish green
- pinnae widely spaced, stalked; margins indented
- stipe dark purple-brown near base, green above
- rachis green, flattened and winged
- crevices and ledges of rock cliffs, absent from calcareous rocks
- uncommon, s Ind, e Ohio

MIDWEST RANGE

Ohio, s Ind (endangered).

HABITAT

Moist, shaded crevices and ledges of sandstone, gneiss, granite, and shale; often receiving water from seepage from the surrounding rock; absent from calcareous rocks.

DESCRIPTION

Rootstock Short-creeping, erect, tightly clumped to produce a cluster of stems; scales dark brown, strongly clathrate (having a lattice-like pattern).

Frond Evergreen; sterile and fertile fronds alike, gradually tapered toward tip and somewhat arched, to 15 cm long and 5 cm wide.

Stipe Shorter than or about same langth as blade; brown at base, fading to dull green above; scales at base dark brown, grading into hairs upwards on stipe; with 2 oval bundles at extreme base, united just above base to form one nearly circular bundle.

Rachis Dull green, winged, flattened; sparsely hairy.

Blade Oblong-triangular, 2-pinnate at base, always less divided or merely lobed upwards; leathery, dull, often dark blue-green, often with tiny glandular hairs and sparse linear scales.

Pinnae 6 to 9 pairs, opposite to alternate, mostly 3-6 lobed, variable in shape; blunt-tipped; veins forked or simple, never reaching margin, margins coarsely and irregularly indented.

Sori Brown, scattered along veins of pinnules; **indusium** tan, translucent, jagged on margin, hidden by sporangia at maturity, on one side of the sorus, opening toward the middle of the segment.

SIMILAR SPECIES

The dark blue-green color and widely spaced, deeply parted pinnae distinguish *Asplenium montanum* from most other spleenworts.

NORTH AMERICA

MIDWEST

• The pinnae of **Bradley's spleenwort** *(A. bradleyi)* are toothed and less deeply cut, and the dark color of the stipe continues partway up the rachis in that species.

• **Wall-rue** *(A. ruta-muraria)* has a green stipe, its pinnae have longer petioles and the pinnae are widest near the tip.

• **Brittle bladder fern** *(Cystopteris fragilis)* is more dissected (pinnules present) and oval rather than linear sori.

NOTES

Asplenium montanum readily forms hybrids with a number of other *Asplenium* species.

NAME

Sent to Willdenow from the North Carolina mountains and named by him in 1810.

From Latin, *montanus*, growing on mountains, referring to its preferred habitat.

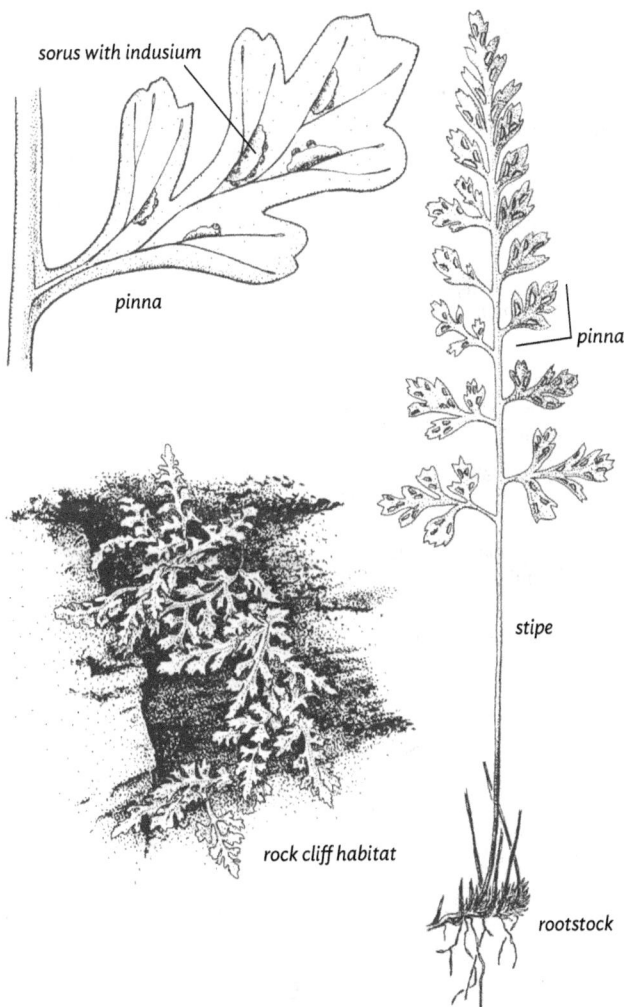

sorus with indusium

pinna

pinna

stipe

rock cliff habitat

rootstock

Asplenium pinnatifidum Nutt.
LOBED SPLEENWORT

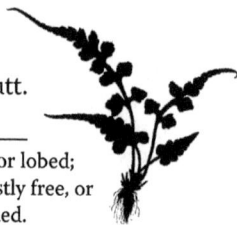

FIELD TIPS

• small, clumped, evergreen fern
• fronds bright green, crinkled, deeply lobed, the tip tail-like and usually pointing downward
• stipe purple-brown at base; rachis green, flat
• crevices in acidic rock cliffs
• mostly southern Midwest

MIDWEST RANGE

Ill, Ind, Ohio; threatened in Wisc.

HABITAT

Dry to moist, shaded crevices of acidic (rarely circumneutral) rocks, especially sandstone and gneiss.

DESCRIPTION

Rootstock Short, very chaffy.

Frond Evergreen, rather tough, sterile and fertile fronds alike, broadly lance-shaped, pinnatifid, deeply cut into pinna-like lobes, long-tapering at tip.

Stipe Dark purple-brown near base, green upwards, 2/3 as long as blade.

Rachis Flat, green.

Blade Elongate-triangular in outline, bilaterally somewhat unsymmetrical, tapered to a tail-like tip (and rarely rooting at tip); pinnatifid (deeply lobed, but not fully pinnate).

Pinnae Segments vary greatly in size and shape; generally ovate, margins rounded or lobed; veins mostly free, or a few joined.

Sori Numerous, straight or slightly curved, running together with age; **indusium** membranous, attached at one side.

SYNONYMS

• *Asplenosorus pinnatifidus* (Nutt.) Mickel

SIMILAR SPECIES

Somewhat similar to its parent *Asplenium rhizophyllum*, but *A. pinnatifidum* distinctly lobed when mature, tends to have longer stipes in proportion to its leaf size, and has a more upright habit; also, lobed spleenwort grows in acid soil and is typically a solitary plant; *Asplenium rhizophyllum* prefers limestone and forms colonies.

NOTES

Lobed spleenwort originated as a hybrid between *Asplenium montanum* and *A. rhizophyllum*.

NAME

Discovered by Muhlenberg in Lancaster Co., Pennsylvania, and named *A. rhizophyllum B, pinnatifidum* in 1813; raised to species rank by Nuttall 5 years later.

NORTH AMERICA

MIDWEST

Asplenium pinnatifidum growing on a moist rock ledge.

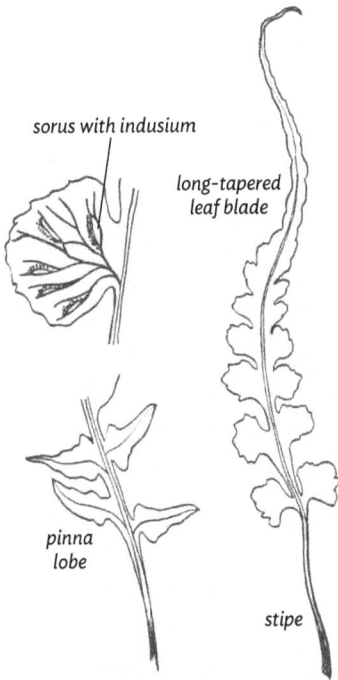

sorus with indusium

long-tapered leaf blade

pinna lobe

stipe

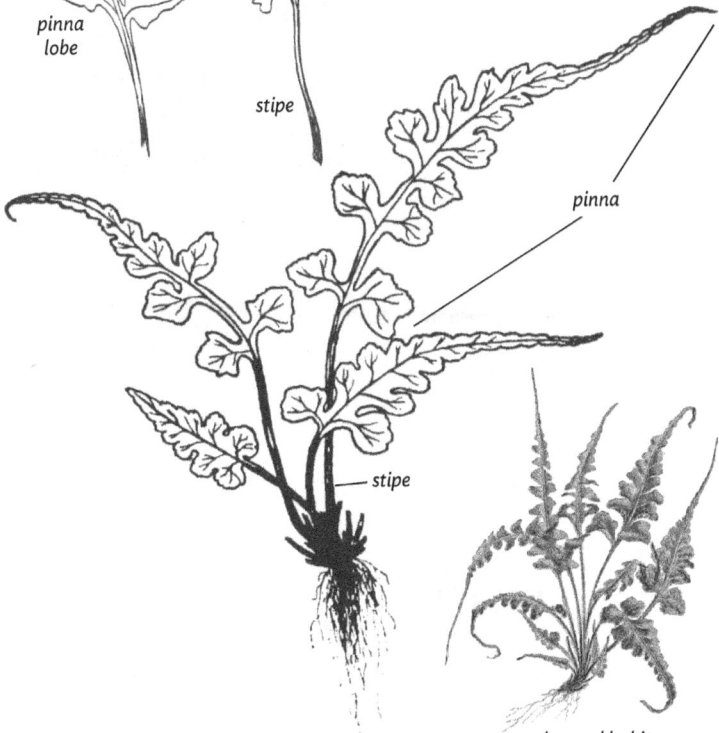

pinna

stipe

clumped habit

Asplenium platyneuron (L.) B.S.P.
EBONY SPLEENWORT

FIELD TIPS

- densely clumped fern growing on both soil and rock
- fronds of two types, the sterile fronds forming rosette at base of larger, darker, upright fertile fronds
- stipe and rachis dark brown, shiny
- pinnae alternate, with ear on upper side
- usually where acidic and at least partially shaded

MIDWEST RANGE

Ill, Ind, Iowa, Mich, se Minn, Ohio, Wisc.

HABITAT

A wide variety of habitats: in partial shade in open woods, old fields and clearings (especially in sandy or loamy soils), less commonly in moss or very shallow soil over rocks, or on limestone or sandstone cliffs and ledges; may also colonize older building foundations and mortared joints.

DESCRIPTION

Rootstock Erect, short, bearing the fronds in dense tufts; scales sparse, dark brown to black, clathrate (with lattice-like pattern).

Sterile fronds Evergreen, prostrate and spreading, to 10 cm long, in a rosette near the ground; more numerous than fertile fronds.

Fertile fronds Semi-evergreen, dark green, longer and more upright than sterile frond; to 40 cm tall and 4 cm wide.

Stipe Much shorter than blade, erect; at first green, then reddish-brown, shiny all the way to end of rachis; threadlike scales present near base, becoming smooth upwards.

Rachis Smooth, red- or purple-brown, shiny.

Blade (fertile) linear lance-shaped, widest above the middle, tapering to either end, 1-pinnate, shiny, commonly with tiny glandular hairs and a few linear scales.

Pinnae 25 to 45 pinna pairs, alternate along the rachis; with Christmas-stocking-like appearance due to an upward (sometimes also downward) pointing auricle (projection) at base of pinna; margins finely toothed; veins forked, never reaching margin; sterile fronds with fewer pinna pairs; lowest pinnae short, triangular.

Sori Positioned along veins of leaflets; aligned diagonally to costa (pinna midrib); **indusium** linear, 2 mm long, on one side of the sorus, silvery, thin,

NORTH AMERICA

MIDWEST

margins slightly toothed; **sporangia** brown, maturing in late summer.

SYNONYMS
• *Acrostichum platyneuron* L.
• *Asplenium ebenum* Ait.

SIMILAR SPECIES
• **Black-stem spleenwort** (*Asplenium resiliens*, rare in s Ill and s Ind) similar but fronds not dimorphic, pinnae nearly entire, and rachis glossy black. In *A. platyneuron*, fronds dimorphic, pinnae margins somewhat toothed, and rachis dark brown.

NOTES
Ebony spleenwort forms hybrids with a number of other spleenworts. The hybrid with **mountain spleenwort** (*A. montanum*) gives rise to the fertile species **Bradley's spleenwort** (*A. bradleyi*).

NAME
Linnaeus in 1753 proposed the name *Acrostichum platyneuros* for material from Virginia, comprising this plant and a *Polypodium*. Oakes combined the Linnaean species name with the correct genus, as recorded by Eaton in 1878.

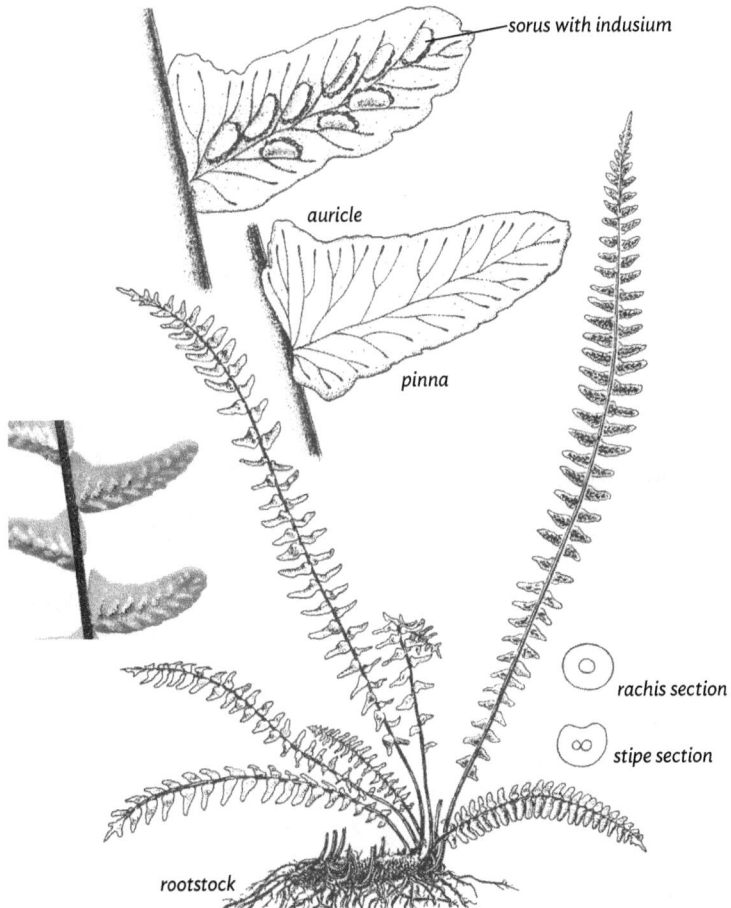

sorus with indusium

auricle

pinna

rachis section

stipe section

rootstock

Asplenium rhizophyllum L.
WALKING FERN

FIELD TIPS
- small evergreen fern
- leaf blades undivided, long-tapering to tip, which can root to form new plants
- found on moist, mossy limestone

MIDWEST RANGE
Ill, Ind, Iowa, Mich (threatened), se Minn, Ohio, Wisc.

HABITAT
Moist, shaded, mossy limestone or dolomite rocks, boulders, ledges and crevices, generally found on northerly exposures; also in moist deciduous woods.

DESCRIPTION
Rootstock Erect or ascending, scales dark brown, narrowly triangular, clathrate.

Frond Evergreen, sterile and fertile fronds similar but fertile fronds typically larger than sterile fronds; to 30 cm long by 1-5 cm wide.

Stipe Variable in length; dark brown and scaly at base, becoming green and without scales above; upper stipe with tiny club-shaped hairs; vascular bundles 2 at base, these united upwards.

Rachis Green, dull, nearly smooth, grooved above.

Blade Simple, leathery, variable in size and shape, narrowly triangular to linear-lance-shaped; basal lobes usually rounded, but sometimes pointed; tip rounded to very long-tapering and, if tapering, often forming plantlet and rooting at tip; sparsely hairy; margins entire to wavy, rarely irregularly lobed; veins obscure, netlike.

Pinnae Absent.

Sori Linear along the veins, scattered irregularly on underside of leaf blade, often joined at vein junction; **indusium** attached on one side of the sorus, small, green, **sporangia** gold-yellow when young, becoming red-brown.

SYNONYMS
- *Camptosorus rhizophyllus* (L.) Link

SIMILAR SPECIES
- **Lobed spleenwort** (*Asplenium pinnatifidum*) similar but is distinctly lobed, tends to have longer stipes in proportion to its leaf size, and has a more upright habit; also, lobed spleenwort grows in acid soil and is typically a solitary plant; *Asplenium rhizophyllum* prefers limestone and forms colonies.

NORTH AMERICA

MIDWEST

NAME

Linnaeus received this fern from Virginia and Canada and named it *Asplenium rhizophylla* in 1753, the ending *a* being a misprint for *um*. Link founded a genus *Camptosorus* to include this species and a related Asiatic fern in 1833; modern treatments include the species in *Asplenium*.

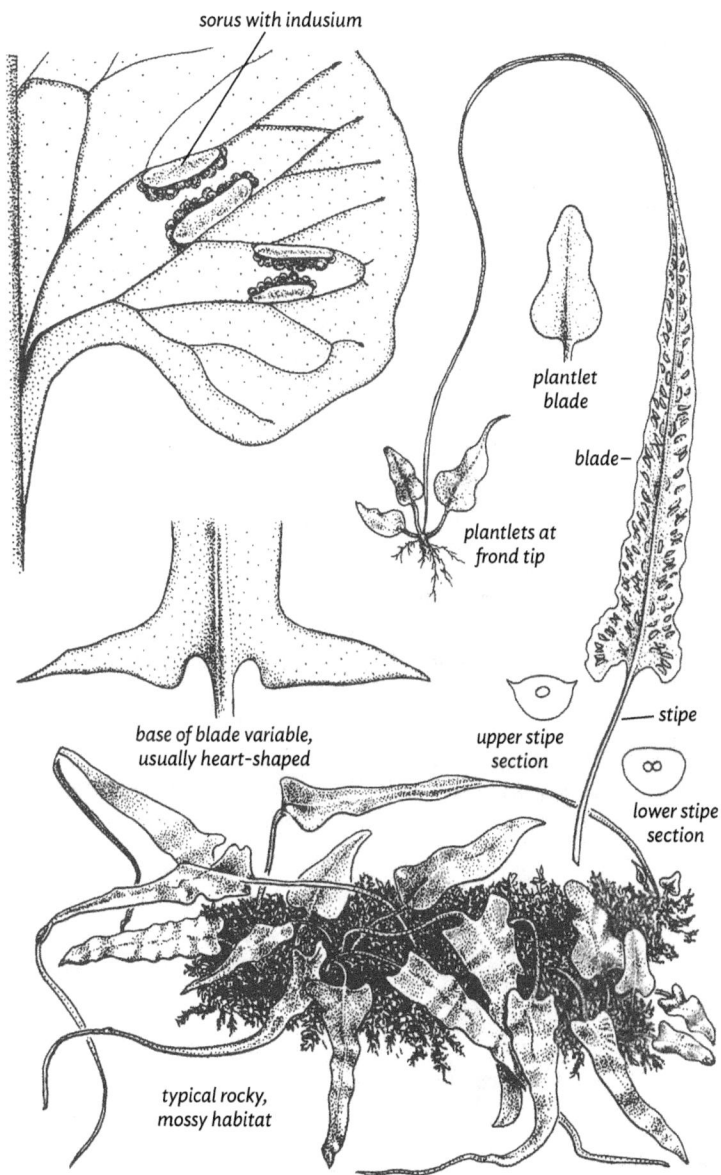

sorus with indusium

plantlet blade

blade−

plantlets at frond tip

stipe

upper stipe section

lower stipe section

base of blade variable, usually heart-shaped

typical rocky, mossy habitat

Asplenium ruta-muraria L.

WALL-RUE

FIELD TIPS

• small evergreen fern
• pinnae widely spaced, stalked, divided into small rounded or fan-shaped segments
• shaded limestone cliffs and rocky slopes
• uncommon in Midwest

MIDWEST RANGE

s Ind (threatened), e Iowa, Mich Upper Peninsula (Drummond Island only, endangered), s Ohio (threatened); reported for Illinois.

HABITAT

Crevices in limestone cliffs and mossy, calcareous talus slopes; in partial to full shade, rarely in full sun; may colonize masonry walls. Best growth made in partial shade on neutral to alkaline soil.

DESCRIPTION

Rootstock Erect, sparsely branched, tipped with crowded remains of old stipes; scaly, the scales hidden in the roots, dark brown, clathrate (with lattice-like pattern).

Frond Evergreen, clumped, fertile and sterile fronds alike, bluish green, arching and delicate, to 15 cm long by 5 cm wide, somewhat resembling flat-leaved parsley or common rue (*Ruta graveolons*), leading to its scientific name.

Stipe Generally longer than blade, 1.5-7 cm long, green throughout or the base purple-brown; with a few scales at base, upwards becoming multicellular hairs; with two circular or oval bundles at base, united above to form one bundle.

Rachis Green; smooth, grooved, with sparse hairs

Blade Oblong-triangular in outline, 2-pinnate at base, always less divided upwards, leathery, dull, usually with tiny glandular hairs and a few linear scales.

Pinnae 2 to 5 pairs, these widely spaced and distinctly stalked, alternate, variable in shape, rounded to fan-shaped; margins facing rachis and costa entire; opposite margin with blunt teeth; serrate or crenate; veins forked, not extending to margin.

Sori Linear, along veins, often covering entire underside; **indusium** translucent, pale tan, margin fringed with hairs, hidden by sporangia at maturity, on one side of the sorus, opening toward middle of the segment; **sporangia** brown.

SYNONYMS

• *Asplenium cryptolepis* Fernald
• *Amesium ruta-muraria* (L.) Newman

NORTH AMERICA

MIDWEST

NOTES

Our smallest spleenwort and only spleenwort of the region with a triangular-shaped blade (in outline); easily overlooked due to small size and sometimes inaccessible rock cliff habitat; the clumped fronds also have a habit of curling up in dry weather.

A common fern on stone walls in Europe.

NAME

The European wall-rue was named *Asplenium ruta-muraria* by Linnaeus in 1753.

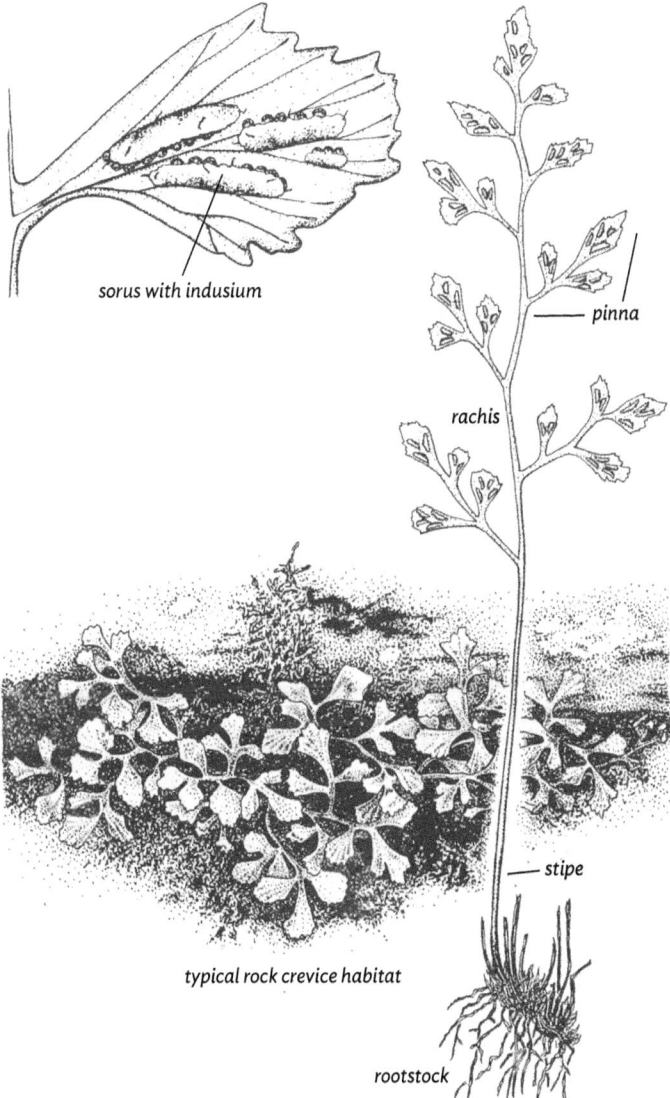

sorus with indusium

pinna

rachis

stipe

typical rock crevice habitat

rootstock

Asplenium scolopendrium L.

HART'S-TONGUE FERN

FIELD TIPS

- distinctive, very rare, evergreen fern
- leaf blades entire, strap-shaped, with smooth to wavy margins
- sori elongated, of varying length
- shaded dolomite in Michigan's eastern Upper Peninsula

MIDWEST RANGE

Eastern Upper Peninsula of Mich (endangered); additional North American populations located in ne USA, se Ontario, and Alabama and Tennessee

HABITAT

Cool, moist, shaded dolomite boulders and crevices, talus slopes, and rocky woodlands underlain by dolomite.

DESCRIPTION

Rootstock Erect, short, some tipped with older, dead stipe bases.

Frond Evergreen, fertile and sterile fronds alike, arching, glossy bright green, somewhat leathery, to 35 cm long by 5 cm wide.

Stipe Shorter than blade, 4-12 cm long, 1-2 mm wide, light brown, grooved, with scales when young; vascular bundles 2, c-shaped, joined upwards.

Rachis Light brown at base, yellowish upwards, smooth.

Blade Simple, strap-shaped or lance-shaped, uniform in width, base lobed or heart-shaped, heart shaped to lobed at base of blade; tip with blunt point; margins entire becoming wavy with age; veins forked, stopping short of margin.

Pinnae Absent.

Sori Linear, paired along a vein so that diagonal to rachis, in upper half of blade, **indusium** thin, flap-like, opening toward vein, whitish, on one side of the sorus; **sporangia** brown.

SYNONYMS

- *Phyllitis japonica* Komarov subsp. *americana* (Fernald) A. Löve & D. Löve
- *Phyllitis scolopendrium* (L.) Newman var. *americana* Fernald

NOTES

Common in Europe; rare in North America where populations widely scattered, and given varietal status *(A. scolopendrium* var. *americanum)* because of the smaller fronds, narrower scales, and blades tending to bear sori in the upper half rather than on the entire length.

NORTH AMERICA

MIDWEST

Popular as an ornamental plant, and cultivars available with varying frond form, including frilled margins, forked fronds and cristate forms. The American variety is considered to be difficult to cultivate, and cultivated plants are derived from European forms.

NAME

The European Harts-tongue was known to the ancients as *Scolopendrium* (from the Greek *scolopendra*, centipede, alluding to the two rows of sori which resemble feet of a centipede; Linnaeus made this the species name under *Asplenium* in 1753.

The tongue-shaped leaves lead to the common name; *hart* is an old word for deer.

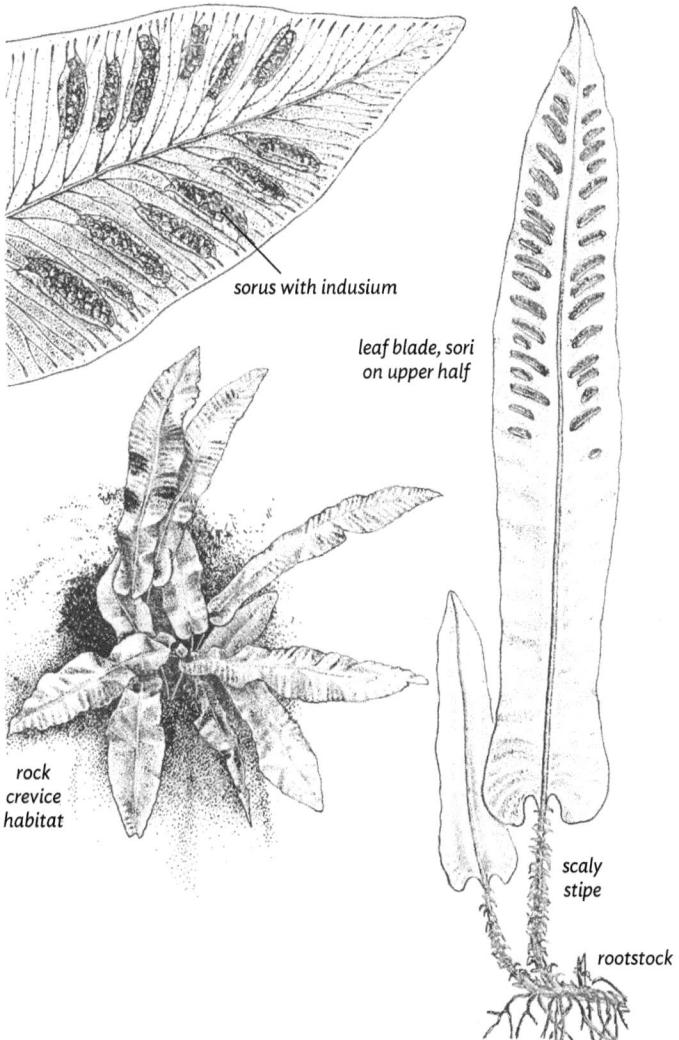

sorus with indusium

leaf blade, sori on upper half

rock crevice habitat

scaly stipe

rootstock

Asplenium trichomanes L.
MAIDENHAIR SPLEENWORT

FIELD TIPS

- small, densely clumped, evergreen fern
- sterile fronds small and spreading, fertile fronds longer and upright
- stipe and rachis shiny purple-brown
- pinnae opposite, rounded, not eared
- rock cliffs and boulders, usually where shaded

MIDWEST RANGE

Southern Ill, s Ind, n Mich, e Minn (threatened), Ohio, Wisc (endangered).

HABITAT

Damp, shaded, mossy limestone cliffs and boulders, limestone crevices; also found on non-calcareous rocks such as sandstone.

DESCRIPTION

Rootstock Short-creeping, often branched, scaly near apex, the scales nearly black or sometimes with brown margin, lance-shaped, clathrate.

Frond Evergreen, 20 cm long by 1.5 cm wide, fertile and sterile fronds alike or nearly so, but the sterile fronds developing earlier and oriented along ground surface.

Stipe Much shorter than blade, wiry, dark purple-brown, shiny, sometimes with a few scales; base with 1-2 oval or round bundles. A diagnostic feature (visible with 10x hand lens) is a narrow wing along entire length of the stipe and rachis.

Rachis Purple-brown, smooth, grooved; old rachises persistent.

Blade Linear in outline, 1-pinnate, widest above middle, tapering to either end, dark green, smooth or minutely hairy.

Pinnae 20 to 35 pairs, opposite or nearly so or sometimes alternate, oval, the base unequal and sometimes wedge-shaped, slightly toothed on sides and the rounded apex; veins evident.

Sori Oblong to linear, 2-5 pairs per pinna, situated on the veins between the midrib and the margin; **indusium** translucent, pale tan, hidden by sporangia at maturity, on one side of the sorus, **sporangia** brown.

SIMILAR SPECIES

- **Ebony spleenwort** (*Asplenium platyneron*) has shiny, purple-brown stipes and dark green pinnae as in *A. trichomanes.* However maidenhair spleenwort is smaller overall, with

NORTH AMERICA

MIDWEST

once-cut fronds, and very small, oval-shaped pinnae.

• Distinguished from **rock polypod**y *(Polypodium virginianu*m) by its prominent, dark rachis and narrow fronds of nearly constant width.

NOTES

Two subspecies, are defined: subsp. *quadrivalens* D. E. Mey., which prefers calcareous substrates, and the more

common subsp. *trichomanes*, which prefers acidic substrates.

NAME

Named by Linnaeus from European specimens in 1753. Subsequently found to be widespread and essentially identical in North America.

From Greek, *thrix*, hair, referring to persistent stipes and rachises without pinnae.

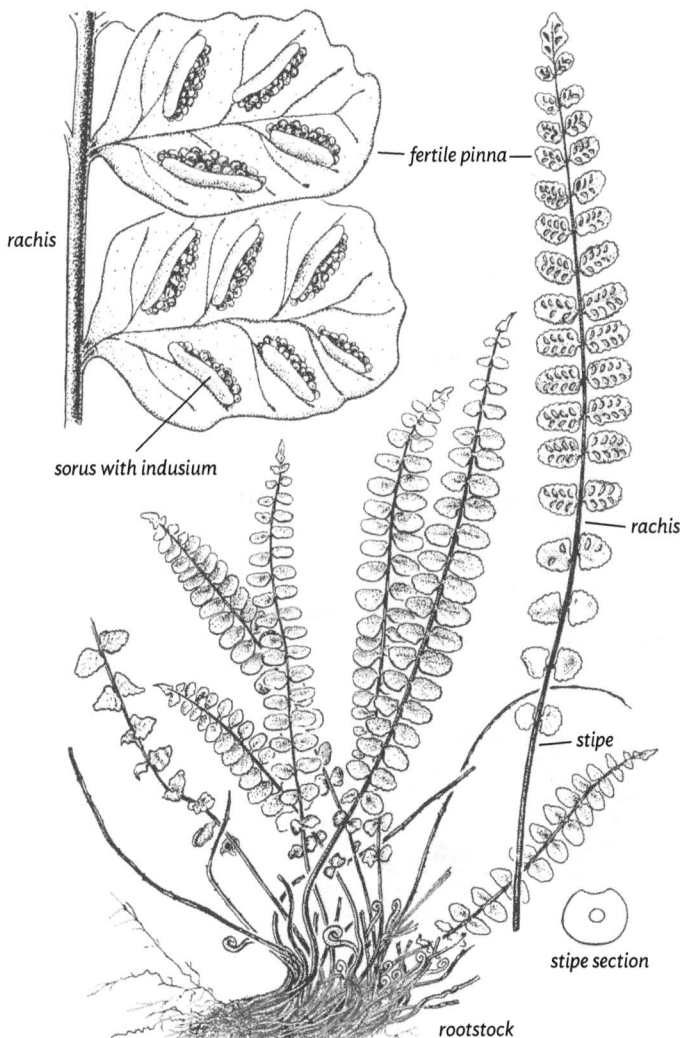

rachis

fertile pinna

sorus with indusium

rachis

stipe

stipe section

rootstock

Asplenium viride Huds.
BRIGHT-GREEN SPLEENWORT

FIELD TIPS

- delicate, clumped, nearly evergreen fern
- lower and middle pinnae widely spaced
- rachis bright green
- shaded limestone habitats
- n Michigan (mainly Upper Peninsula) and n Wisconsin

MIDWEST RANGE

Northern Mich, n Wisc.

HABITAT

Shaded crevices and ledges of limestone; talus slopes and boulders.

DESCRIPTION

Rootstock Short-creeping to erect, branching; scales few, dark brown to black, with lattice-like pattern.

Frond Nearly evergreen, clumped from the short rhizome, 5-15 cm long, slightly dimorphic, the fertile fronds stiff, erect; sterile fronds smaller, prostrate.

Stipe Shorter than blade, red-brown at base, green above near rachis, lustrous; usually with a few scales near base, grading into gland-tipped hairs upwards; vascular bundles 1 or 2.

Blade linear, widest above middle, tapering to either end, 1-pinnate, thin, pale green, smooth or with sparse tiny hairs.

Pinnae 6 to 21 pairs; rhombic or ovate, rounded to acute at tip, subopposite; midrib indistinct; margins round-toothed or entire; veins sometimes forked.

Sori Linear, 1-2 pairs per pinna, paired across the midrib and often joined when mature; **indusium** white or translucent, entire, often deciduous, attached on one side of the sorus, **sporangia** brown.

SYNONYMS

- *Asplenium ramosum* L.
- *Asplenium trichomanes-ramosum* L.

SIMILAR SPECIES

This fern with its bright green rachis and upper stipe is usually easily identified.

- Sometimes growing with **maidenhair spleenwort** (*Asplenium trichomanes*) but rachis in that species is purple-brown.
- Small specimens may resemble **smooth cliff fern** (*Woodsia glabella*), but *Asplenium viride* has elongate sori (vs. round in *Woodsia*), with the

NORTH AMERICA

MIDWEST

indusium attached on one side, typical for the genus.

Hudson's later name *Asplenium viride*. *Viride* means green.

NAME

Described by Linnaeus in his 1753 Species Plantarum, under the name *Asplenium trich. ramosum*. That name was later rejected in favour of William

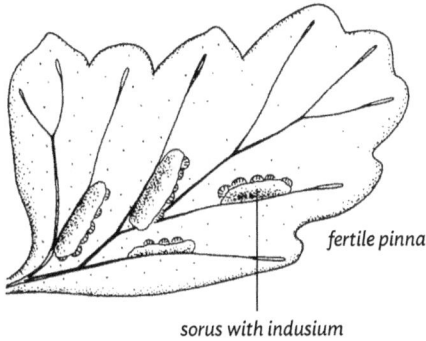

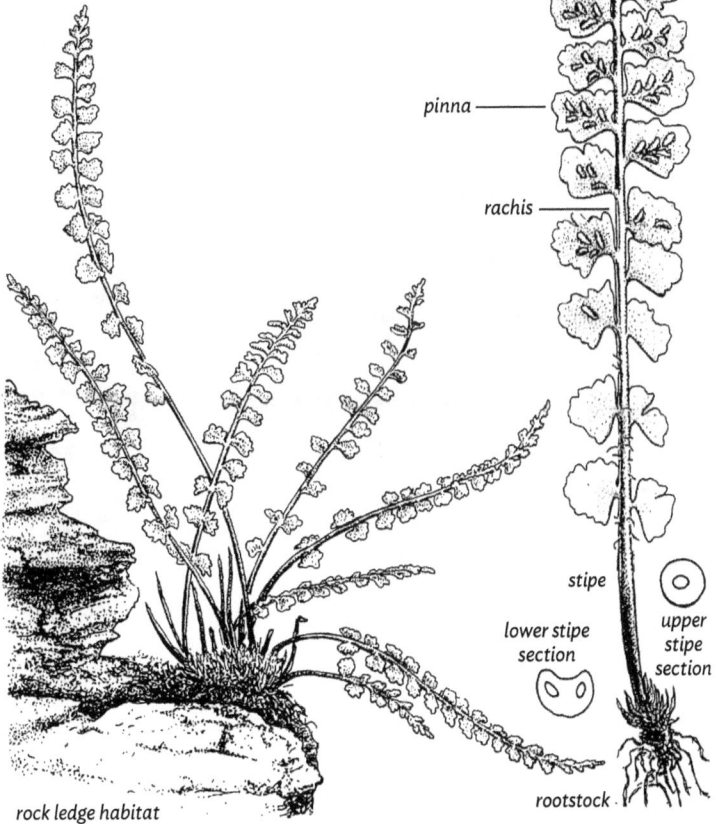

fertile pinna

sorus with indusium

pinna ——

rachis ——

stipe

lower stipe section

upper stipe section

rootstock

rock ledge habitat

Bradley's spleenwort (*Asplenium bradleyi*).   CHRIS HOESS

Mountain spleenwort (*Asplenium montanum*).

Ebony spleenwort (*Asplenium platyneuron*).

Walking fern (*Asplenium rhizophyllum*).

Hart's-tongue fern (*Asplenium scolopendrium*).

Maidenhair spleenwort (*Asplenium trichomanes*).

Bright-green spleenwort (*Asplenium viride*).

ATHYRIACEAE
Lady Fern Family

Medium to large ferns of worldwide distribution; family includes 5 genera and ca. 600 species (previously mostly included in Aspleniaceae, Dryopteridaceae or Woodsiaceae); 3 genera in our flora, each with a single species.

NOTES

Until 1800, *Athyrium* was included in genus *Asplenium*, as the indusia are similar. *Deparia* and *Diplazium*, formerly considered members of *Athyrium*, are now treated as separate species, although the differences are not great. In contrast to *Deparia*, *Athyrium* has a more deeply grooved rachis, which is continuous from rachis to costa (vs. discontinuous in *Deparia*); in *Diplazium*, the leaf blade is merely 1-pinnate.

KEY TO ATHYRIACEAE | LADY FERN FAMILY

1   Pinnae margins finely toothed, but otherwise undivided; Ill, Ind, e Iowa, Mich, s Minn, Ohio, Wisc   **1. Diplazium pycnocarpon**
GLADE FERN, page 92

1   Pinnae deeply lobed or divided, margins sometimes finely toothed **2**

2   Blades 2-pinnate (with the pinnae again divided), the pinnules also sometimes deeply lobed; common, regionwide
2. Athyrium filix-femina
LADY FERN, page 82

2   Blades with deeply lobed pinnae; Ill, Ind, e Iowa, Mich, se Minn, Ohio, Wisc   **3. Deparia acrostichoides**
SILVERY-SPLEENWORT, page 88

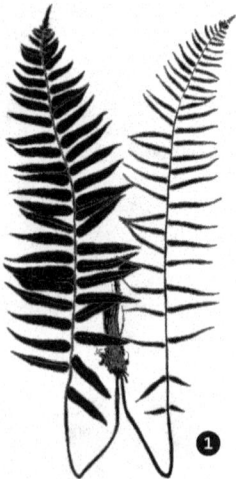

Diplazium pycnocarpon
GLADE FERN

Athyrium filix-femina
LADY FERN

Deparia acrostichoides
SILVERY-SPLEENWORT

Athyrium
LADY FERN

Medium to large ferns, usually in woodlands. *Athyrium* includes ca. 220 species worldwide, mostly in temperate regions. There are 2 (or more) species in North America, 1 in our flora.

KEY CHARACTERS

- Clumped deciduous fern.
- Sterile and fertile **fronds** alike.
- **Blade** 2-pinnate-pinnatifid (but variable and less and more divided occur).
- **Vascular bundles** in stipe 2 only, these uniting upwards to form a U-shape.
- **Sori** elongate, near base of pinnules, usually hooked at one end or crescent-shaped, covered by persistent **indusium**.

NAME

From Greek, *athyrium,* meaning a small door, the indusium is hinged on one side.

NOTES

Athyrium filix-femina is one of the region's more common ferns. Circumboreal in distribution, lady fern is found in North to South America, Europe, and Asia.

Lady ferns were historically used to treat a variety of medical conditions. Powdered extracts from the rhizome of *Athyrium filix-femina*, like similar preparations from male fern (*Dryopteris filix-mas*), were used for many ailments, especially for purging worms from the digestive systems of humans and animals.

A large number of cultivated varieties are used for landscaping; best growth is made in partial shade on moist, acidic soil.

Lady fern (*Athyrium filix-femina*), upper portion of frond.

Athyrium filix-femina (L.) Roth

LADY FERN

FIELD TIPS

- common medium to large fern
- fronds clumped, deciduous
- sori elongate; hooked at one end, horseshoe-shaped, or sometimes straight
- usually in woodlands

MIDWEST RANGE

Common throughout region.

HABITAT

Moist to dry woods, swamps, shores, meadows, thickets, ravines.

DESCRIPTION

Rootstock Short-creeping to nearly erect, scaly.

Crozier Round-oblong, 1-2 cm wide, densely covered with linear, dark brown to purple scales.

Frond Deciduous, forming an irregular clump; new fronds produced all summer; sterile and fertile fronds alike, to 120 cm long by 30 cm wide.

Stipe Straw-colored above, base dark red-brown or black, swollen; in some forms very red, old stipes persistent; scales light to dark brown, lance-shaped; vascular bundles 2, curved, back-to-back, uniting to a U-shape above.

Rachis Pale, slightly grooved to flat; smooth or with glands or hairs.

Blade Elliptic to lance-shaped, 2-pinnate-pinnatifid (but variable, less and more divided blades also occur), yellow-green to bright green.

Pinnae 20 to 30 pairs, mostly alternate, stalk very short or absent, oblong lance-shaped, lower pinnae longer or shorter than middle pinnae; costae grooved above, the groove continuous from rachis to costae to costules; margins toothed or lobed; veins free, forking, reaching the margin.

Sori At base of leaf segments, straight, or more often hooked at one end or horseshoe-shaped; **indusium** elongate, laterally attached, finely toothed; **sporangia** brownish gray to dark brown or yellow.

SYNONYMS

- *Athyrium angustum* (Willd.) C. Presl
- *Athyrium asplenioides* (Michx.) A.A. Eaton

SIMILAR SPECIES

- **Hay-scented fern** (*Dennstaedtia punctilobula*), not in clumps, fronds with white, sticky gland-tipped hairs.

NORTH AMERICA

MIDWEST

• **Silvery-spleenwort** (*Deparia acrostichoides*) has blades 1-pinnate-pinnatifid vs. 2-pinnnate-pinnatifid in *Athyrium*; also groove in rachis and costa not continuous in *Deparia*.

• **New York fern** (*Thelypteris novaboracensis*) margins not toothed.

• **Wood ferns** (*Dryopteris* spp.) have round sori, rather than the elongate or curved sori of *Athyrium*.

NAME

Linnaeus first named the European lady-fern *Polypodium filix femina*. Michaux altered this to *Nephrodium filix femina* for a related Canadian plant. Other specimens from Canada were named *Aspidium angustum* by Willdenow in 1810; his species was classified as *Athyrium* by Presl in 1825.

From the Latin *filix*, fern; *femina*, female or woman.

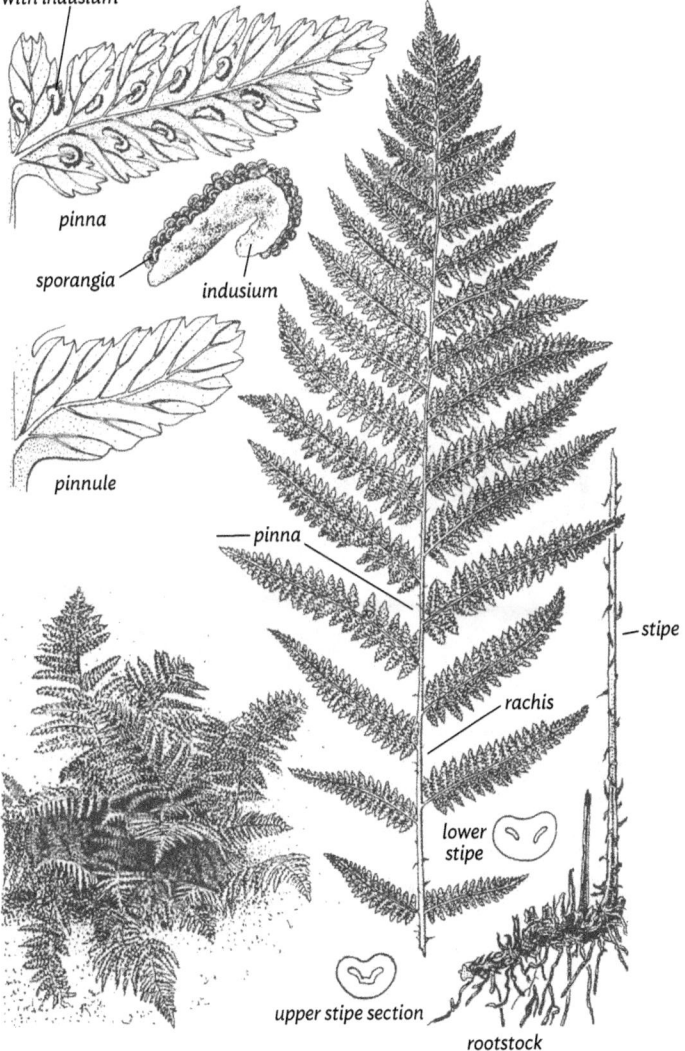

sorus with indusium

pinna

sporangia

indusium

pinnule

pinna

stipe

rachis

lower stipe

upper stipe section

rootstock

Lady fern (*Athyrium filix-femina*), habit (l) and crozier (r).

Lady fern (*Athyrium filix-femina*), frond.

Lady fern (*Athyrium filix-femina*), upper side of pinnules.

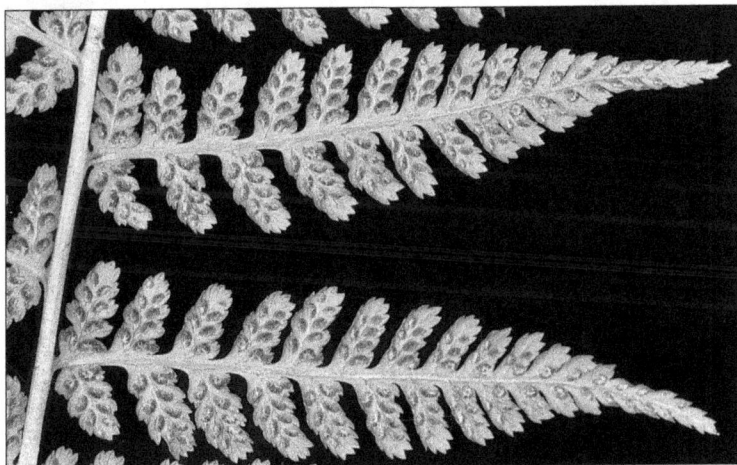

Lady fern (*Athyrium filix-femina*), underside of fertile pinnae.

Deparia

SILVERY SPLEENWORT

Ferns of terrestrial habitats; about 70 species worldwide, primarily in tropical regions of Asia, Africa, Australia and the Pacific Islands; 2 species in North America; 1 in our flora.

KEY CHARACTERS

• Medium to large deciduous fern, forming loose clusters from a creeping rhizome.
• Sterile and fertile **fronds** slightly different, fertile fronds longer, developing earlier than sterile.
• **Pinnae** widely separated, deeply lobed but not cut into pinnules.
• **Sori** in herringbone pattern.
• **Indusia** linear, whitish when young.

SIMILAR GENERA

Distinguished from **lady fern** (*Athyrium*) and **glade fern** (*Diplazium*) by discontinuity in the grooves from rachis to costae, and by the hairs on the rachis and costae. The linear sori are very similar to *Diplazium*, but are never back-to-back along a vein.

NAME

Previously included in *Asplenium*, *Athyrium* and *Diplazium*.

From the Greek *depas*, cup or a beaker, thought to describe shape of the indusium.

Silvery-spleenwort (*Deparia acrostichoides*), clustered fronds.

Deparia acrostichoides • **ATHYRIACEAE | LADY FERN FAMILY**

Silvery-spleenwort (*Deparia acrostichoides*), fronds.

Silvery-spleenwort (*Deparia acrostichoides*), underside of fertile pinnae.

Deparia acrostichoides (Sw.) M. Kato
SILVERY SPLEENWORT

FIELD TIPS

- medium to large deciduous fern, sterile and fertile fronds slightly different
- fronds single but close together from rhizome
- blade tapered at both ends; pinnae widely separated and deeply lobed; lowest pair of pinnae point downwards
- sori in herringbone pattern
- indusia linear, whitish when young

MIDWEST RANGE

Ill, Ind, e Iowa, Mich, se Minn, Ohio, Wisc.

HABITAT

Moist, shaded deciduous or mixed conifer-hardwood forests, ravines and slopes, shaded streambanks and seeps, swamp edges; soils typically acidic.

DESCRIPTION

Rootstock Short-creeping, black, scaly; stipes about 1 cm apart, forming an asymmetric clump.

Frond Deciduous, arising separately (but close together) from the rhizome; sterile and fertile fronds somewhat different; fertile fronds longer, more erect, developing earlier than sterile; 110 cm long by 20 cm wide.

Stipe 1/2 to 2/3 as long as blade, base dark red-brown, scaly and swollen, straw-colored then green above, the scales changing to long white hairs upwards on stipe; shallowly grooved above; vascular bundles 2, curved or peanut-shaped, uniting to a U-shape above.

Rachis Light green, very hairy, with a few scales; grooved.

Blade Elliptic to ovate-lance-shaped, tapering to both ends, 1-pinnate-pinnatifid.

Pinnae 20 to 25 pairs, mostly alternate, oblong, truncate at base, long-tapering; lowest pinnae shorter than middle pinnae and usually pointing downward; costae shallowly grooved above, discontinuous from rachis to costae, hairs on costae and costules; margins entire or finely toothed; veins free, rarely forked, reaching the margin.

Sori Elongate, ± straight, or slightly curved, in a herringbone pattern, originating on midvein of pinnule; **indusium** linear, laterally attached, persistent, whitish when young (leading to the 'silvery' name), becoming dark brown, along the vein

NORTH AMERICA

MIDWEST

and opening outward; **sporangia** brownish early, later blue-gray.

SYNONYMS
• *Athyrium thelypterioides* (Michx.) Desv.
• *Diplazium acrostichoides* (Sw.) Butters

SIMILAR SPECIES
• **Lady fern** (*Athyrium filix-femina*) has dissected pinnules (not merely lobed) with toothed margins.

• Sterile fronds of **cinnamon fern** (*Osmunda cinnamomea*) and **interrupted fern** (*Osmunda claytoniana*) are wider, the pinnae closer together, and the stipes scaleless.

NAME
First named *Asplenium acrostichoides* by Swartz in 1801, and two years later *A. thelypteroides* by Michaux, with reference to occurrences in Virginia and North Carolina.

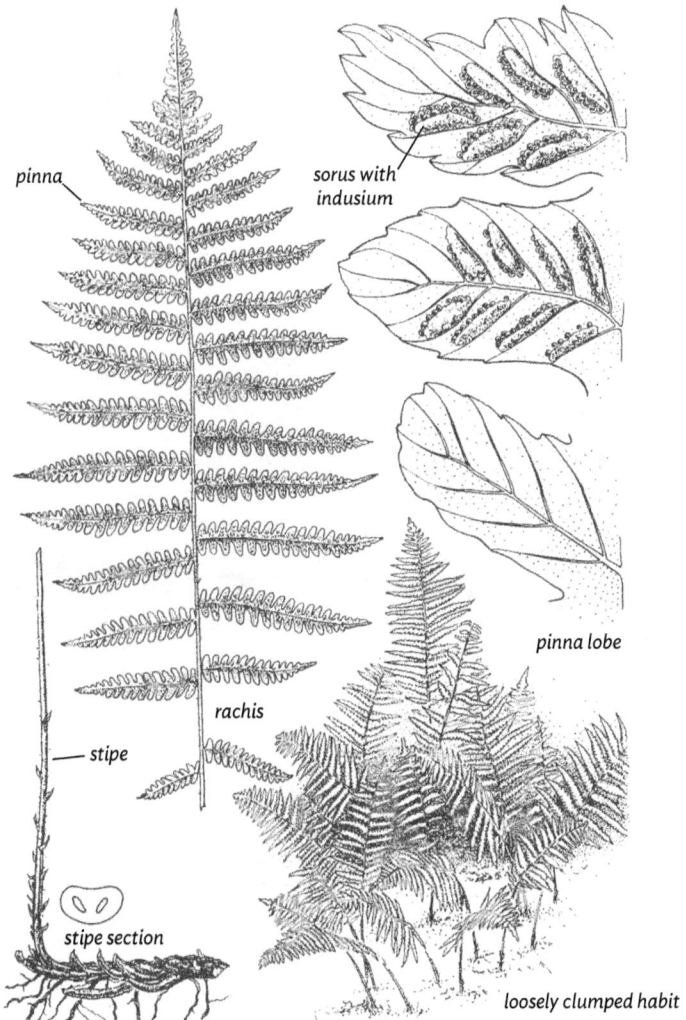

pinna

sorus with indusium

pinna lobe

rachis

stipe

stipe section

loosely clumped habit

Diplazium
GLADE FERN

Large genus of about 300-400 species worldwide, found mostly in tropical regions; only 3 species in North America, and 1 in our flora.

KEY CHARACTERS

• Large deciduous fern, fronds clumped or solitary.

• **Blade** 1-pinnate; pinna margins wavy, not sharp-toothed.

• **Sori** in herringbone pattern, silvery green when young.

• **Indusium** elongate, may appear raised due to expanding sporangia beneath.

SIMILAR GENERA

Diplazium distinguised from other members of the family by the 1-pinnate blades. Similar to **lady fern** (*Athyrium*), but sori in *Diplazium* never hook over the veins, and are sometimes paired back-to-back; also, grooves in the rachis are U-shaped vs. V-shaped in *Athyrium*. Glade fern differs from **silvery spleenwort** (*Deparia*) in having the grooves continuous from costa to rachis.

NAME

Previously included in *Asplenium* and *Athyrium*.

From Greek, *diplazios*, double, *plasion*, oblong; referring to the 2-valved indusium of some members of the genus that splits open on both sides of a vein.

Glade fern (*Diplazium pycnocarpon*), sterile fronds.

Glade fern (*Diplazium pycnocarpon*), blade 1-pinnate, the pinnae alternate.

Glade fern (*Diplazium pycnocarpon*), underside of fertile pinnae, sori aligned in herringbone pattern.

Diplazium pycnocarpon (Spreng.) Broun
GLADE FERN

FIELD TIPS

- large deciduous fern, fronds clumped or solitary
- blade 1-pinnate; pinna margins wavy, not sharp-toothed
- sori in herringbone pattern, silvery green when young

MIDWEST RANGE

Ill, Ind, e Iowa, Mich, s Minn (threatened), Ohio, Wisc.

HABITAT

Moist, open to partially shaded woods, meadows and slopes; soils rich and circumneutral.

DESCRIPTION

Rootstock Short-creeping, stipes 1-2 cm apart; scales brown or tan, broadly lance-shaped.

Frond Deciduous, solitary or clustered; sterile and fertile fronds somewhat different; fertile fronds taller, erect; sterile fronds arching; 100 cm long by 20 cm wide.

Stipe About 1/2 as long as blade (stipe of fertile frond longer and stiffer than those of sterile fronds), hairy (at least when young), reddish-brown and scaly at base, green upwards, deeply grooved above; vascular bundles 2, uniting to a V-shape above.

Rachis Green to straw colored, underside slightly hairy.

Blade Oblong-lance-shaped, ± narrowed to base, 1-pinnate.

Pinnae 20 to 40 pairs, narrow, to 1 cm wide, tapered to long sharp tip, base rounded to heart-shaped, lowest pinnae with very short stalk; costae grooved above, continuous from rachis to costae; smooth; margins entire to crenate; veins free, forking. Sterile pinnae often somewhat twisted, not stiff; fertile pinnae narrowed, straight and stiff.

Sori Linear or slightly curved , running from midvein halfway or almost to margin in a herringbone pattern; **indusium** linear, persistent, translucent, laterally attached; **sporangia** dark brown when mature.

SYNONYMS

- *Asplenium pycnocarpon* Spreng.
- *Athyrium pycnocarpon* (Spreng.) Tidestr.

SIMILAR SPECIES

The 1-pinnate leaf blade and linear sori in a herringbone pattern are distinctive.

NORTH AMERICA

MIDWEST

NAME

Discovered by Michaux "on the banks of the Ohio River" and named by him *Asplenium angustifolium* in 1803. However, this name was already in use for a tropical fern, and Sprengel proposed *Asplenium pycnocarpum* in 1804; Tidestrom proposed *Athyrium pycnocarpum* in 1906. Modern floras place the species within *Diplazium*.

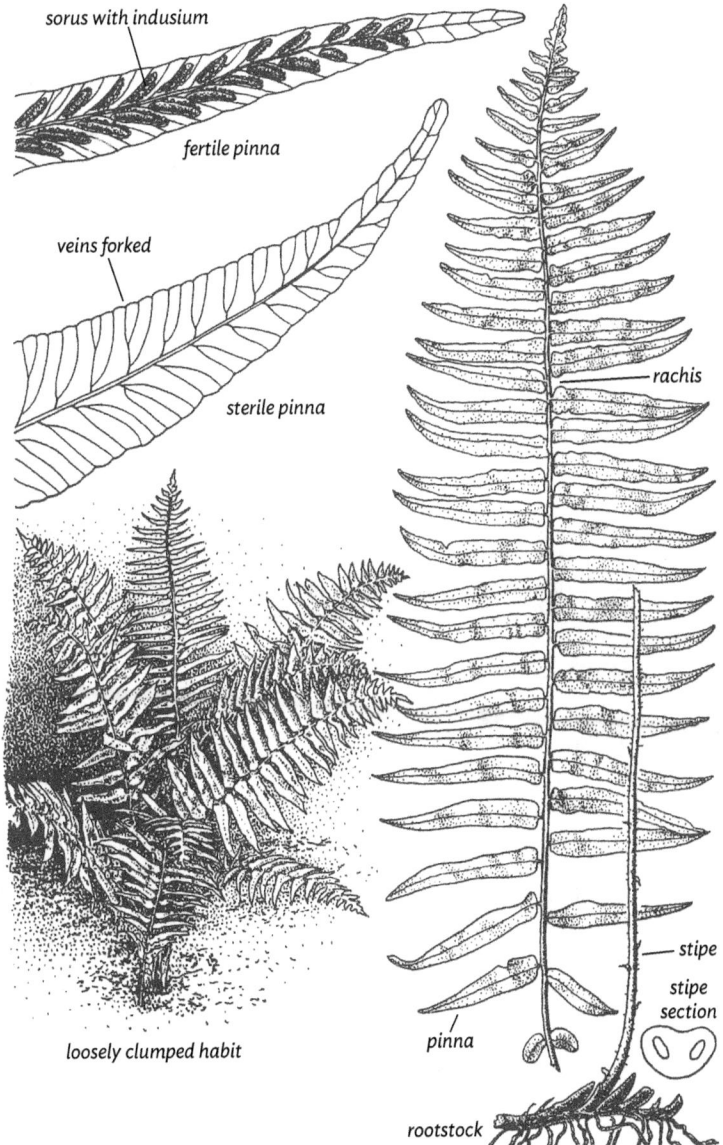

sorus with indusium

fertile pinna

veins forked

sterile pinna

rachis

loosely clumped habit

pinna

stipe

stipe section

rootstock

BLECHNACEAE | Chain Fern Family
Woodwardia
CHAIN FERN

Our sole members of this family are the chain ferns *(Woodwardia),* a small genus found in North and Central America, eastern Asia and the Mediterranean region of Europe. There are 3 species in North America; 2 in the Midwest. Our *Woodwardia* species are medium to large deciduous ferns of shaded wet woods and acidic bogs. Sterile and fertile fronds similar (*Woodwardia virginica*) or different (*Woodwardia areolata*), pinnatifid or 2-pinnatifid; veins joining to form small enclosed areas (areolae), then free to the margin or forming additional areolae. **Sori** linear or oblong, parallel to the midveins, borne along the veinlets, which form the outer side of the first row of areolae. **Indusium** persistent, opening on the side adjacent to the midrib.

KEY CHARACTERS

• Coarse deciduous ferns of wet acidic places, in shade (or sun where very wet).

• Deciduous, sterile and fertile **blades** alike (*Woodwardia virginica*) or markedly different (*Woodwardia areolata*).

• Colony-forming from long-creeping rootstock.

• **Linear sori** in chainlike rows along each side of midvein of pinna or pinnule.

• **Indusia** attached by their outer margin, opening towards midvein.

• **Netted veins** along the costae and midveins form areoles (enclosed spaces).

NOTES

Apart from the distinctive sori, our two species of chain fern have little in common, and the two species were previously placed in separate genera. *Woodwardia areolata* is dimorphic, *W. virginica* monomorphic; *W. areolata* is net-veined throughout, *W. virginica* is net-veined only surrounding the sori; *W. areolata* has 2 vascular bundles; *W. virginica* has 7-9 bundles at base of stipe; *W. areolata* has scaly stipes, *W. virginica* is glabrous.

NAME

Named for Thomas Jenkinson Woodward, (1745-1820), an English botanist.

CULTURE

Woodwardia areolata is readily grown in a shady garden where the soil is wet to moist and acidic. Leaf blades are an attractive pinkish color when young, becoming a glossy bronze-green when mature. *Woodwardia virginica* tends to be invasive but can be used to fill a soggy place or grown in a container.

KEY TO WOODWARDIA | CHAIN FERN

1 Sterile and fertile fronds similar; fronds 1-pinnate-pinnatifid (the pinnae deeply lobed), mostly 13–25 per side; Ind, Mich, Ohio
1. Woodwardia virginica
VIRGINIA CHAIN FERN, page 98

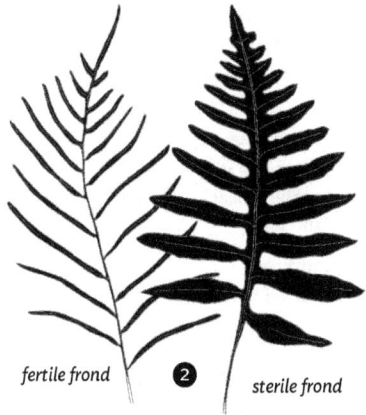

1 Sterile and fertile fronds very different; sterile fronds merely deeply lobed (the lobes connected at the axis by a narrow wing of tissue), the lobes entire to somewhat undulate, mostly 6–14 per side; fertile fronds with very narrow linear pinnae; uncommon in s Ill, Ind, Ohio
2. Woodwardia areolata
NETTED CHAIN FERN, page 96

fertile frond   *sterile frond*

Woodwardia virginica
VIRGINIA CHAIN FERN

Woodwardia areolata
NETTED CHAIN FERN

Colony of **Virginia chain fern** *(Woodwardia virginica)*.

Woodwardia areolata (L.) T. Moore
NETTED CHAIN FERN

FIELD TIPS

- large deciduous fern; sterile and fertile fronds different; fertile pinnae linear
- fronds glossy green single from rhizome but may form dense colonies
- lowest pinnae pair not winged at rachis
- veins joined to form a network
- sori elongate, in chainlike rows on narrow fertile pinnae

MIDWEST RANGE

s Ill, Ind, s Mich, Ohio.

HABITAT

Open to shaded wet woods (where often rooted in mud), acidic bogs, seeps, cobbly beaches, rarely on rock cliffs and ledges.

DESCRIPTION

Rootstock Long-creeping, slender, dark-brown to black; scales many, satiny brown.

Crozier Densely covered with tan scales.

Frond Deciduous, to 70 cm long, arising singly from the rhizome but may appear massed due to extensive rhizomes; sterile and fertile fronds different, sterile fronds slightly shorter, emerging first, fertile fronds emerging in midsummer, with darker stipes and narrower pinnae. Divisions of fertile frond narrowly linear and almost distinct.

Stipe About same length as blade, with a few scales; sterile fronds reddish brown below, straw-colored above. Fertile fronds darker, to purple-black; vascular bundles 2 at base, merging above to a 3-sided structure.

Rachis With a few scales on underside, olive-green to straw-colored.

Blade Lance-shaped, fertile blade 1-pinnate, sterile blade less divided above where it is winged along the rachis; thin-textured, reddish on emergence, bright green later, scaly-glandular when young but soon smooth.

Pinnae 7 to 12 pairs; sterile pinnae alternate, lance-shaped; the veins raised, net-like throughout; fertile pinnae subopposite, thin and contracted; margins wavy and finely toothed.

Sori In chainlike rows on each side of costae; **indusium** flap-like, tucked under sporangia, disintegrating with age, opening towards the costa;

NORTH AMERICA

MIDWEST

sporangia reddish-brown.

SYNONYMS
• *Lorinseria areolata* (L.) C. Presl

SIMILAR SPECIES
• Superficial resemblance to **sensitive fern** (*Onoclea sensibilis*), but *Woodwardia areolata* has very different fertile fronds (linear sori vs. beads in *Onoclea*), finely toothed rather than entire margins, and by its basal pinnae, which are alternate rather than subopposite as in *Onoclea*.

NOTE
Dimorphism sometimes incomplete, resulting in fronds fertile at top and sterile below.

NAME
This fern reached Linnaeus from Maryland and Virginia, and he named it *Acrostichum areolatum* in 1753. Presl founded the genus *Lorinseria* on it at the same time.

From Latin, *areolatus*, a small place, referring to areas enclosed by the netted veins.

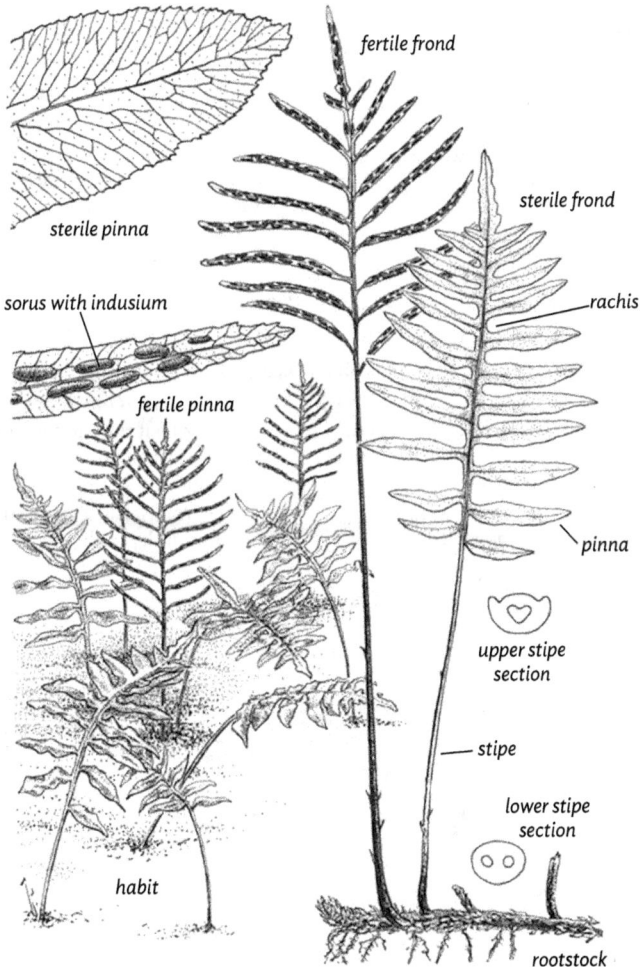

fertile frond

sterile pinna

sterile frond

sorus with indusium

rachis

fertile pinna

pinna

upper stipe section

stipe

lower stipe section

habit

rootstock

Woodwardia virginica (L.) Sm.

VIRGINIA CHAIN FERN

FIELD TIPS

- large deciduous fern; sterile and fertile fronds alike
- fronds upright, single from the rhizome, may form colonies with fronds facing one direction
- stipe long, swollen and spongy at base, shiny dark purple-brown
- blade 1-pinnate-pinnatifid
- veins both netted and free
- sori linear, in chainlike rows

MIDWEST RANGE

n Ind, Mich, Ohio; historical record from ne Ill.

HABITAT

Open to shaded acidic swamps, bogs, wet places in woods, sandy or peaty lake shores.

DESCRIPTION

Rootstock Long-creeping and branching, ropelike; scales few near tips, dark brown, triangular.

Crozier Reddish brown, scale and hairs absent.

Frond Deciduous, arising singly from the rhizome; sterile and fertile fronds alike, 100 cm long by 30 cm wide.

Stipe About same length as blade; base swollen, spongy, dark purple to black, smooth, with 2 large vascular bundles and 3-7 smaller ones, fewer at top of stipe.

Rachis Dark purple-brown becoming green upwards; smooth and shiny, grooved.

Blade Narrowly ovate, widest at middle, 1-pinnate-pinnatifid, thin-textured, with glands and scales when young, the glands persisting.

Pinnae 12 to 23 pairs; alternate, middle ones to 15 cm; lower pinnae bending forward and towards the base; margins finely dentate; veins netted next to the rachis and costae, free otherwise.

Sori In chainlike rows, but distinct along costae; **indusium** flap-like, opening towards costa; **sporangia** purple-brown.

SYNONYMS

- *Anchistea virginica* (L.) C. Presl

SIMILAR SPECIES

- **Cinnamon fern** (*Osmunda cinnamomea*) but that species grows in clumps rather than in a line from creeping rhizome, has different sterile and fertile fronds, and a green stipe (rather than dark purple-brown).

NORTH AMERICA

MIDWEST

NOTES

The chains of sori on the areolae adjacent to the midrib are distinctive.

NAME

Collected in Virginia in colonial days, and named *Blechnum virginicum* by Linnaeus. J. E. Smith placed it in genus *Woodwarclia* in 1793, but Presl classified it as *Anchistea* in about 1849.

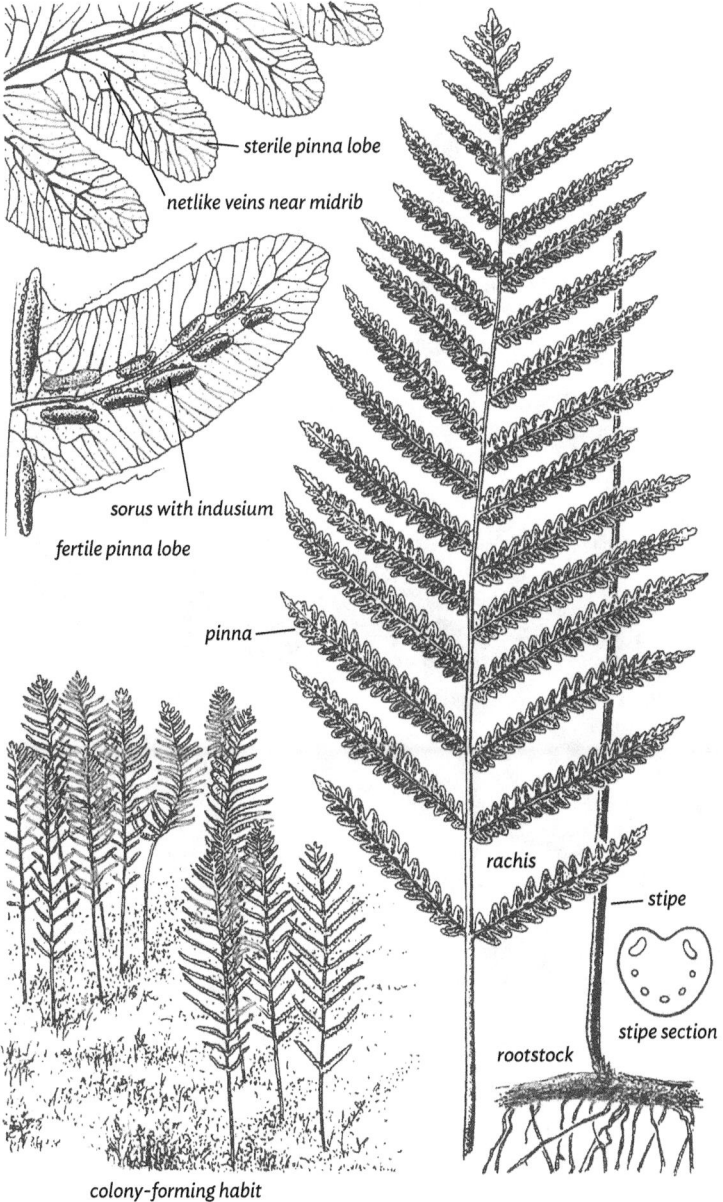

sterile pinna lobe

netlike veins near midrib

sorus with indusium

fertile pinna lobe

pinna

rachis

stipe

stipe section

rootstock

colony-forming habit

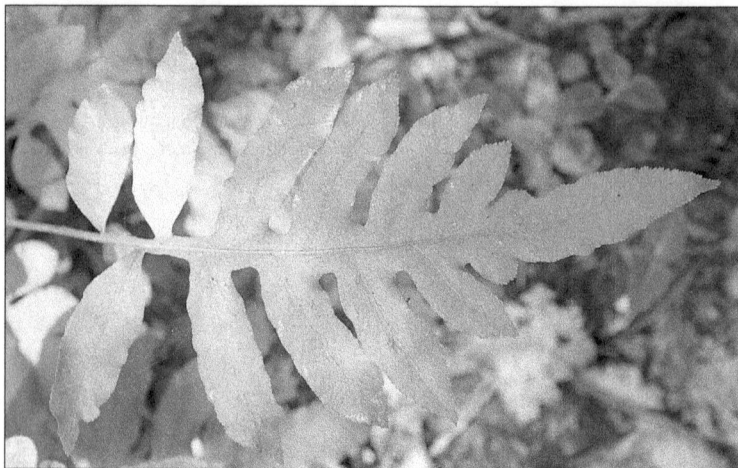

Sterile frond, **Netted chain fern** *(Woodwardia areolata).*

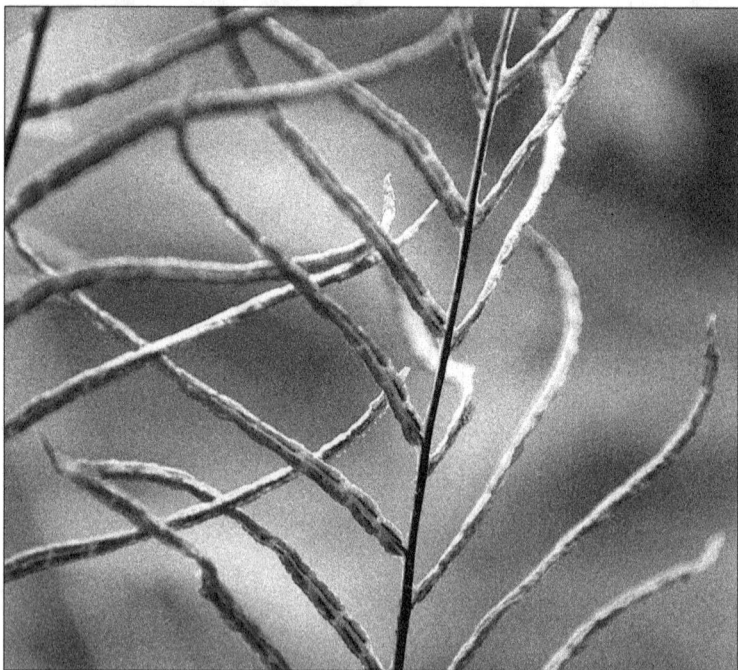

Fertile fronds, **Netted chain fern** *(Woodwardia areolata).* These develop in lates
ummer or early fall and may eventually overtop the sterile fronds.

Sterile frond, **Virginia chain fern** *(Woodwardia virginica).*

Underside of fertile frond, **Virginia chain fern** *(Woodwardia virginica).*

CYSTOPTERIDACEAE
Bladder Fern Family

A small family of 4 genera and about 60 species, mostly in temperate regions. Two genera in our flora, *Cystopteris* and *Gymnocarpium*, common ferns throughout the north temperate zone. *Cystopteris* (ca. 23 species, 9 in North America, 6 in our flora) also ranges southward in montane habitats of the Andes and Himalayas, and to Australia, New Zealand, Hawaii, and southern Africa. *Gymnocarpium* species number 7 worldwide; 5 in North America, 3 in our flora. Characteristic are a swollen knob at the base of each pinna, the indusium is always absent, and 3-parted leaf-blades.

KEY TO CYSTOPTERIDACEAE | BLADDER FERN FAMILY

1 Blades ternate (divided into 3 more or less equal parts); indusium absent
 1. Gymnocarpium
OAK FERN, page 120

1 Blades 1-pinnate; indusium present
 2. Cystopteris
BLADDER FERN, page 103

Gymnocarpium
OAK FERN

Cystopteris
BLADDER FERN

Cystopteris
BLADDER FERN

Cystopteris species are delicate, monomorphic to slightly dimorphic ferns found in terrestrial habitats or growing on rocks, or in some regions as epiphytes. Species of *Cystopteris* commonly hybridize making field identification difficult. One species, ***Cystopteris bulbifera***, produces vegetative reproductive structures called bulblets, which can germinate to form new plants (bulblets also rarely formed on plants of *C. laurentiana* and *C. tennesseensis,* the bulblets often misshapen).

KEY CHARACTERS
• **Stipe** grooved, hairy or smooth, vascular bundles 2, round or oblong.
• **Leaf blade** 1-3 pinnate-pinnatifid, ovate lance-shaped.
• Costae grooves continuous from rachis to costae, margins toothed, veins free, simple or forked, ending at margin.
• **Sori** round, in 1 row between midrib and margin, **indusium** hood-shaped, often shed at maturity, beneath sorus on midrib side, sporangia brown.

SIMILAR GENERA
Cystopteris and *Woodsia* are small ferns having thin-textured leaves and growing mostly on or near rock outcrops; they also frequently occur together and are often confused. Distinguishing characters are:
• *Cystopteris* has an undivided indusium, pocket-like or hood-like, attached around one side of the sorus; *Woodsia* has an indusium divided into a series of scale-like or hair-like structures, attached below sorus.
• In *Cystopteris,* stipe bases deciduous; *Woodsia* has persistent, dark stipe bases.
• In *Cystopteris,* veins reach pinnule margin; final veinlets in *Woodsia* do not extend to margin.

NAME
Greek: *kystos,* bladder, and *pteris,* a fern (the ancient Greeks used *pteris* to describe all ferns); the indusium is inflated or bladder-like.

CYSTOPTERIS HYBRIDS
• **Cystopteris × illinoensis** R.C. Moran (*C. bulbifera* × *C. tenuis*); known from several places in s Wisc, n Ill and e Ohio.
• **Cystopteris × wagneri** R.C. Moran (*C. tennesseensis* × *C. tenuis*); in the USA, known only from several locations in Ohio.

Cystopteris xillinoensis
C. bulbifera x C. tenuis

Cystopteris xwagneri
C. tennesseensis x C. tenuis

KEY TO CYSTOPTERIS | BLADDER FERN

1 Blade broadest at or near base; rachis and midribs of pinnules with small gland-tipped hairs; bulblets sometimes present on underside of rachis. **2**

1 Blade broadest near middle; rachis and midribs of pinnules without glandular hairs; bulblets not present **4**

2 Gland-tipped hairs usually dense on the various axes; blade triangular, widest at base and long-tapering to tip; bulblets often present on rachis of mature fronds; stipe conspicuously red in the earliest leaves; regionwide   **1. Cystopteris bulbifera**
BULBLET BLADDER FERN, page 106

2 Gland-tipped hairs sparse; blade more abruptly tapered at tip; bulblets few, often misshapened ; stipe not red in earliest fronds (this pair of species can be difficult to distinguish, and both occupy similar rock habitats) **3**

3 Blade usually widest at base, triangular in outline; Ill, Ind, Iowa, se Minn, Ohio, Wisc   **2. Cystopteris tennesseensis**
TENNESSEE BLADDER FERN, page 114

3 Blade usually widest just a little above base, the outline more ovate or lance-shaped; n Ill, e Iowa, n Mich, Minn, Wisc
3. Cystopteris laurentiana
ST. LAWRENCE BLADDER FERN, page 110

4 Stem usually extends past the point of attachment of the current year's fronds (that is, fronds emerging slighly behind rhizome tip); rhizome with tan to golden hairs (seen by carefully removing leaf litter and soil near outermost fronds); plants typically in forest soil, less commonly on rocks; regionwide   **4. Cystopteris protrusa**
LOWLAND BLADDER FERN, page 112

4 Stem not extending past the current fronds (fronds emerging at rhizome tip), rhizome hairs absent; usually growing on rocks **5**

5 Pinnae usually perpendicular to rachis and not curving toward tip of frond; margins with sharp teeth; plants on rock outcrops; ne Ill, Iowa, Mich, Minn, Wisc   **5. Cystopteris fragilis**
BRITTLE BLADDER FERN, page 108

5 Pinnae often angled toward tip of frond and/or curved toward tip of frond; margins usually with low and rounded teeth; plants often on rocks, less commonly on forest floor; regionwide
6. Cystopteris tenuis
UPLAND BRITTLE BLADDER FERN, page 116

NOTE Glands in *Cystopteris* are small and may be shed during growing season, and by midsummer, glands usually absent; this is especially noticeable on *C. laurentiana* except for persistent glands on the indusia.

Cystopteris bulbifera
BULBLET BLADDER FERN

Cystopteris tennesseensis
TENNESSEE BLADDER FERN

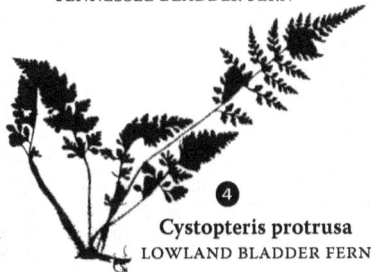

Cystopteris laurentiana
ST. LAWRENCE BLADDER FERN

Cystopteris protrusa
LOWLAND BLADDER FERN

Cystopteris fragilis
BRITTLE BLADDER FERN

Cystopteris tenuis
UPLAND BRITTLE
BLADDER FERN

Cystopteris bulbifera (L.) Bernh.
BULBLET BLADDER FERN

FIELD TIPS
- loosely clumped deciduous fern; sterile and fertile fronds alike
- fronds long-tapered to tip
- pea-like bulblets on upper portion of blade
- rachis and costae with many gland-tipped hairs (small, but visible with hand lens)

MIDWEST RANGE
Regionwide.

HABITAT
Moist crevices and ledges on cliffs, shaded talus slopes, hummocks in cedar swamps, usually where calcareous, rarely on soil; occasional near seeps and in moist alluvial woods.

DESCRIPTION

Rootstock Short-creeping, black, covered with old stipe bases, with a few brown scales.

Frond Deciduous, usually loosely clustered; sterile and fertile fronds alike, but early fronds smaller, sterile, less divided; to 75 cm long by 15 cm wide.

Stipe In groups of 3-8 from stem apex, grooved, maroon or reddish when young, straw-colored to green above, sparsely scaly at base, vascular bundles 2, round or oblong.

Rachis Shiny, yellow, delicate, with gland-tipped hairs.

Blade Narrowly triangular, widest at base, arching, 2-pinnate-pinnatifid, bearing bulblets on the rachis (or sometimes the costa) in upper 1/3 of blade; light green or yellowish-green.

Pinnae 18 to 30 pairs, perpendicular to the rachis, nearly opposite at base, becoming alternate upward, lance-shaped, basal pinnae slightly longer than next pair above; costae grooves continuous from rachis to costae, costae usually densely covered by gland-tipped hairs; margins notched; veins free, simple or forked, directed to notches between teeth.

Sori Round, in 1 row between midrib and margin; **indusium** on a vein, cup-shaped, beneath sorus on midrib side; **sporangia** brown to black.

SYNONYMS
- *Filix bulbifera* (L.) Underw.

SIMILAR SPECIES
The **bulblets**, common on mature fronds, distinguish *Cystopteris bulbifera* from all other *Cystopteris* except *C.*

NORTH AMERICA

MIDWEST

laurentiana and *C. tennesseensis*, where bulblets are only rarely formed.

• **Brittle fragile fern** (*Cystopteris fragilis*) but blade much shorter and without glandular hairs.

NAME

This North American fern was collected in Canada and sent to Linnaeus, who named it *Polypodium bulbiferum* in 1753; it was included in the new genus *Cystopteris* by Bernhardi in 1806.

From Greek, *bolbos*, bulb, and Latin *fero*, to bear; referring to the bulblets.

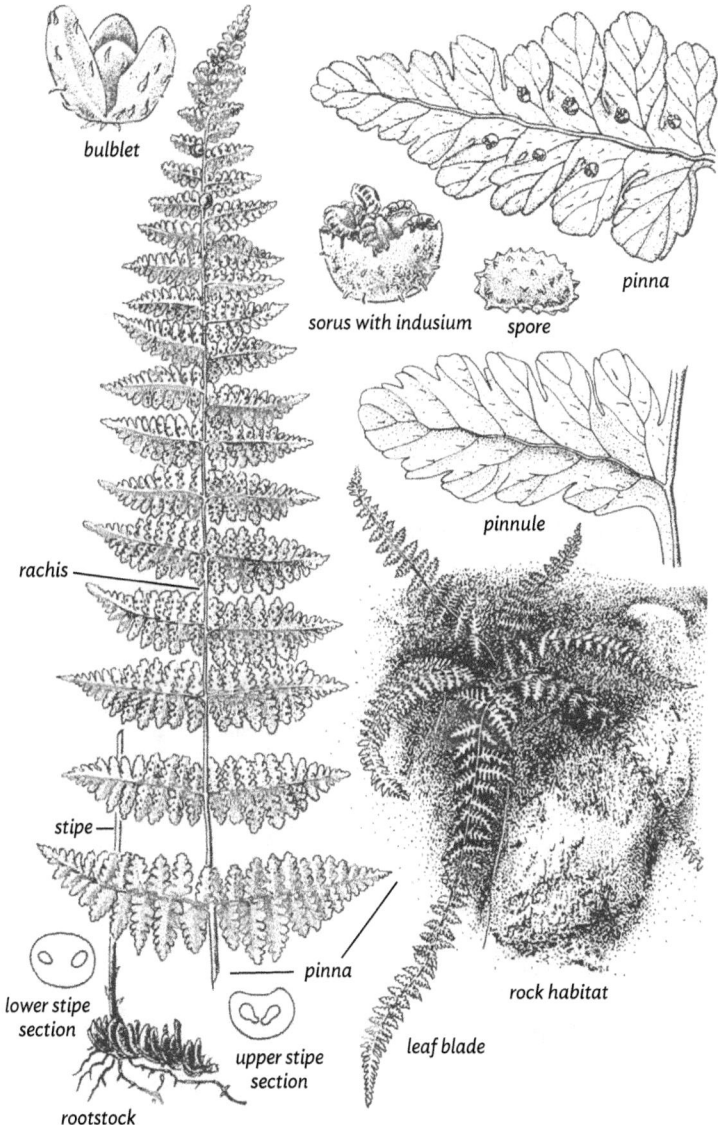

bulblet

sorus with indusium spore

pinna

pinnule

rachis

stipe

pinna

lower stipe section

upper stipe section

rootstock

leaf blade

rock habitat

Cystopteris fragilis (L.) Bernh.
BRITTLE BLADDER FERN

FIELD TIPS
- small, deciduous, clumped fern; sterile and fertile fronds alike; fronds may dry up during summer drought
- lower pinnae widely spaced
- veins simple or forked, usually ending at teeth
- indusium hoodlike over the sorus (but may be shed as plant matures)

MIDWEST RANGE
ne Ill, Iowa, Mich, Minn, Wisc; reported for Ind, historical records from e Ohio.

HABITAT
Calcareous or acidic cliff faces, moist talus slopes, usually where partially shaded; also sometimes in moist woods and atop decaying tree stumps.

DESCRIPTION
Rootstock Short-creeping, shorter than curent year's fronds, hairy and covered with old stipe bases; scales few, tan to light brown, lance-shaped.

Frond Deciduous, clustered; sterile and fertile fronds alike, to 40 cm long.

Stipe 3- to 8-clustered at stem apex, grooved, dark at base, straw-colored to green above, sparsely scaly, vascular bundles 2, round or oblong.

Rachis Smooth, without scales.

Blade Ovate lance-shaped, 2-pinnate-pinnatifid, smooth.

Pinnae 9 to 15 pairs; opposite or nearly so, lance-shaped, widest just below the middle, perpendicular to the rachis, lowest pinnae pairs bending forward and down; costae grooves continuous from rachis to costae; margins variably lobed, toothed to smooth; veins free, simple or forked, usually running to teeth.

Sori Round, in 1 row between midrib and margin; **indusium** forming a hood over the sori, but shriveling with maturity, attached beneath sori on midrib side; **sporangia** brown to black.

SYNONYMS
- Cystopteris dickieana Sim
- Filix fragilis (L.) Underw.

SIMILAR SPECIES
- **Lowland bladder fern** (Cystopteris protrusa) but blade of C. protrusa wider (less than 2.5 times longer than wide); blade of C. fragilis narrower (more than 2.5 times longer than wide).
- **Bulblet bladder fern** (Cystopteris bulbifera), but C. fragilis never producing bulblets.

NORTH AMERICA

MIDWEST

• **Upland brittle bladder fern**
(*Cystopteris tenuis*), but blade of this
species larger and pinnae angled
toward frond tip (perpendicular in
fragile fern).

• **Blunt-lobe cliff fern** (*Woodsia
obtusa*), which has a few scales on
rachis, these are absent in *Cystopteris*.

NOTES

In earlier classifications, *Cystopteris
fragilis* included *C. laurentiana, C.*

tennesseensis, C. protrusa, and *C. tenuis*
as varieties.

NAME

Named *Polypodium fragile* by Linnaeus
in 1753, and referred to a new genus,
Cystopteris, by Bernhardi in 1806.

From Latin, *fragilis*, easily broken;
referring to the fragile stipes that
are easily broken when bent.

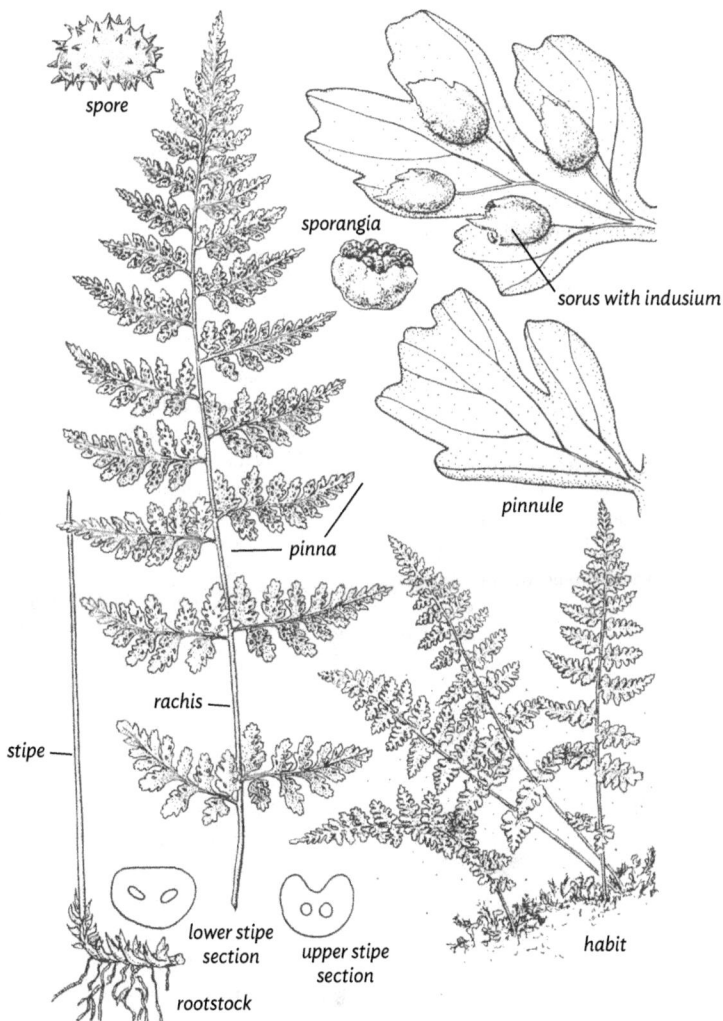

spore

sporangia

sorus with indusium

pinnule

pinna

rachis

stipe

lower stipe section

upper stipe section

rootstock

habit

Cystopteris laurentiana (Weatherby) Blasdell
ST. LAWRENCE BLADDER FERN

FIELD TIPS
- hybrid between *Cystopteris bulbifera* and *C. fragilis*
- blade ovate in outline, with sparse gland-tipped hairs
- sometimes with a few small, scaly, misshapen bulblets
- moist calcareous rock habitats

MIDWEST RANGE
n Ill, e Iowa, Mich Upper Peninsula, se Minn, Wisc.

HABITAT
Cracks and ledges on calcareous rock outcrops, usually in partial or full shade.

DESCRIPTION
Rootstock Short-creeping, covered with old stipe bases; scales tan to light brown, lance-shaped.

Frond Deciduous, loosely clustered; sterile and fertile fronds alike, nearly all fronds fertile; to 50 cm long by 12 cm wide.

Stipe Clustered at tip of rootstock, grooved, brown at base to straw-colored above, sparsely scaly at base; vascular bundles 2, round or oblong.

Blade Ovate, usually widest above base, rarely with bulblets on the rachis, 2-pinnate-pinnatifid, light green or yellowish-green; sparse, gland-tipped hairs present on rachis and costa.

Pinnae 12 to 16 pairs; perpendicular to the rachis, ± opposite, lance-shaped; costae grooves above continuous from rachis to costae; margins serrate; veins free, simple or forked, running to teeth and notches.

Sori round, in 1 row between midrib and margin; **indusium** cup-shaped, beneath sorus on midrib side, on a vein; **sporangia** brown to black.

SYNONYMS
- *Cystopteris fragilis* (L.) Bernh. var. *huteri* (Hausman) Luerss.\
- *Cystopteris fragilis* (L.) Bernh. var. *laurentiana* Weath.

SIMILAR SPECIES
- Similar to **bulblet bladder fern** (*Cystopteris bulbifera*), but plants smaller, blade widest above base, with only sparse glandular hairs, bulblets rare or absent, stipe brown, not maroon or reddish, and veins running to tips of teeth and notches.

- *Cystopteris laurentiana* larger than **brittle bladder fern** (*C. fragilis*), with

NORTH AMERICA

MIDWEST

more pinnae, sometimes with a few small, reddish, misshapen bulblets; and veins sometimes running to notches; in *C. fragilis* veins mostly to teeth-tips only.

NOTES
Believed to have originated as a hybrid between *Cystopteris bulbifera* and *C. fragilis*.

NAME
Laurentiana, for the St. Lawrence River and surrounding region.

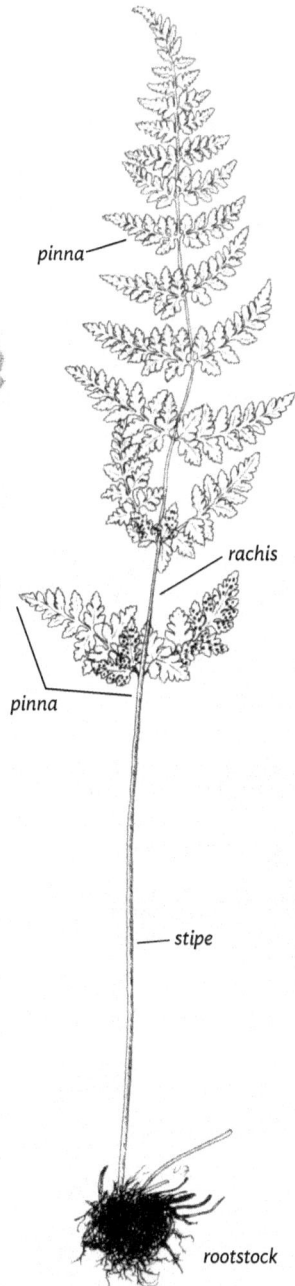

pinna

rachis

sorus with indusium

pinna

stipe

rootstock

Cystopteris protrusa (Weatherby) Blasdell

LOWLAND BLADDER FERN

FIELD TIPS

- deciduous fern of moist woods
- usually in soil rather than rock
- blade lance-shaped; pinnules of lower pinnae distinctly stalked
- rhizome tip extends past emergent fronds

MIDWEST RANGE

Ill, Ind, Iowa, Mich, Minn, Ohio, Wisc; common in s portions of region.

HABITAT

Soil of rich, moist forests and alluvial woods; rarely on rock; can form colonies from its creeping rhizomes.

DESCRIPTION

Rootstock Creeping, extending about 2-3 cm beyond current year's fronds, covered with old stipe bases and with tan to golden hairs, and with light brown scales, these mostly near rhizome tip.

Frond Deciduous, loosely clustered; sterile and fertile fronds alike, but early fronds smaller, sterile, less divided; 40 cm long by 10 cm wide.

Stipe Rising 1-several cm behind the rhizome tip, grooved, dark at base, straw-colored to green upwards; scales at base; vascular bundles 2, round or oblong.

Rachis Smooth.

Blade Ovate lance-shaped, 2-pinnate-pinnatifid, smooth.

Pinnae 6 to 12 pairs; lance-shaped, widest just below the middle, lowest pinnae pair usually bending forward and down; costae grooves continuous from rachis to costae; veins free, simple or forked, usually ending at tooth.

Sori Round, in 1 row between midrib and margin; **indusium** ovate, forming hood over the sorus, but shriveling with maturity, beneath sorus on midrib side; **sporangia** brown to black.

SYNONYMS

- *Cystopteris fragilis* (L.) Bernh. var. *protrusa* Weath.

SIMILAR SPECIES

- Distinguished from other **bladder ferns** (*Cystopteris*) by growing in soil rather than on rocks, and by the protruding, golden-haired rhizome (rhizome tip can be felt with fingertips just beyond outermost emerging fronds).

NORTH AMERICA

MIDWEST

NOTES

Separated from *Cystopteris fragilis* in 1935, when C. A. Weatherby named it variety *protrusa*; now considered a distinct species.

sorus with indusium

sorus with indusium

pinna

rachis

stipe

rootstock

habit

Cystopteris tennesseensis Shaver
TENNESSEE BLADDER FERN

FIELD TIPS

- deciduous fern of rock cliffs
- hybrid between *Cystopteris bulbifera* and *C. protrusa*, appearance intermediate between the two parents
- bulblets present or not, often misshapen, dark, scaly

MIDWEST RANGE

Ill, s Ind, Iowa, se Minn, Ohio, Wisc.

HABITAT

Cracks and ledges of limestone and sandstone rocks, with best growth in calcium-rich habitats; rarely in soil.

DESCRIPTION

Rootstock Creeping, not cordlike, with many old stipe bases; not hairy, scales lance-shaped, light brown

Fronds Crowded near tip of rootstock, sterile and fertile stems alike; to 45 cm long, nearly all fronds with sori.

Stipe Mostly dark brown at base, gradually becoming straw-colored upwards, sparsely scaly at base.

Rachis With or without bulblets (if present usually misshapen); axils of pinnae sometimes with gland-tipped hairs.

Blade Triangular in outline, short-tapered to tip, 2-pinnate-pinnatifid, usually widest at or near base.

Pinnae Usually perpendicular to rachis, not curving toward blade tip; margins serrate; veins running to sinuses and teeth.

Sori Round; **indusium** cup-shaped, with scattered gland-tipped hairs.

SYNONYMS

- *Cystopteris fragilis* (L.) Bernh. var. *simulans* (Weath.) McGregor
- *Cystopteris fragilis* (L.) Bernh. var. *tennesseensis* (Shaver) McGregor
- *Cystopteris utahensis* Windham & Haufler (apparently the same species and known from sw USA)

SIMILAR SPECIES

- Similar to **bulblet bladder fern** (*Cystopteris bulbifera*) but differs in having only a few scattered glandular hairs, in having fewer bulblets, the bulblets misshapen, scaly and dark (vs. smooth and green in *C. bulbifera*), the sori larger (2-5 mm wide vs. 1-2 mm in *C. bulbifera*) and in lacking red stipes in young plants. Also, in *C. tennesseensis*, veins run to sinuses and teeth whereas in *C. bulbifera*, veins run to sinuses.

Also similar to our other species of

NORTH AMERICA

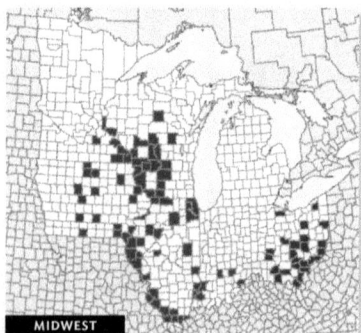

MIDWEST

Cystopteris, see key, page 104.

NOTES

Believed to have originated from an ancient cross between *Cystopteris bulbifera* and *C. protrusa. C. tennesseensis* may also hybridize with *C. tenuis* to form *C.* × *wagneri. C. tennesseensis* may grow together with other members of the genus in mixed populations.

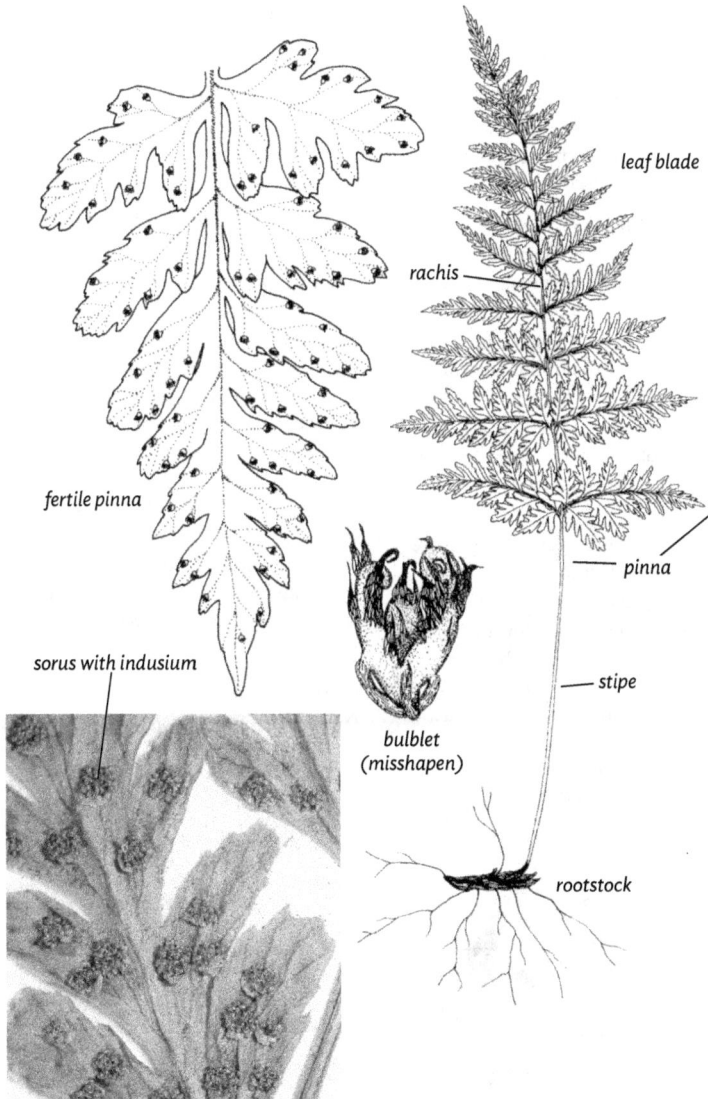

leaf blade

rachis

fertile pinna

pinna

sorus with indusium

bulblet (misshapen)

stipe

rootstock

Cystopteris tenuis (Michx.) Desv.
UPLAND BRITTLE BLADDER FERN

FIELD TIPS
- deciduous, bright green, loosely clumped fern
- usually found on rock but also in soil

MIDWEST RANGE
Regionwide.

HABITAT
Shaded, mossy rock outcrops, sandstone cliffs; less commonly in soil in moist forests; forming small patches but not extensive colonies.

DESCRIPTION

Rootstock Short-creeping, covered with old stipe bases; scales tan to light brown, lance-shaped.

Frond Deciduous, loosely clustered; sterile and fertile fronds alike; to 40 cm long.

Stipe Grooved, dark brown at base, straw-colored above; sparsely scaly at base; vascular bundles 2, round or oblong.

Blade Ovate lance-shaped, 2-pinnate-pinnatifid, smooth.

Pinnae ca. 12 pairs, ± opposite, lance-shaped, widest just below the middle, angled upward from rachis and curving towards tip; costae grooves continuous from rachis to costae; margins lobed, with rounded teeth; veins free, simple or forked, directed into teeth and notches.

Sori Round, in 1 row between midrib and margin; **indusium** ovate to lance-shaped, forming a hood over the sorus, but shriveling with maturity, attached beneath sorus on midrib side; **sporangia** brown to black.

SYNONYMS
- *Cystopteris fragilis* (L.) Bernh. var. *mackayi* G. Lawson

SIMILAR SPECIES
- *Cystopteris tenuis* previously considered a variety of **brittle bladder fern** (*C. fragilis*) and is similar in appearance to that species and **lowland bladder fern** (*C. protrusa*); the pinnae of *Cystopteris tenuis* tend to be more curved in shape and angled toward tip of frond, and margins with rounded teeth rather than sharp teeth.

Also differs from *Cystopteris protrusa* in having fronds clustered at end of short, rather than long-creeping rhizome.

NOTES
Named by Linnaeus *Polypodium fragile* in 1753 (original name for *Cystopteris fragilis*), and referred to new genus

NORTH AMERICA

MIDWEST

Cystopteris by Bernhardi in 1806. The varietal name *mackayi* was applied by Lawson in 1889.

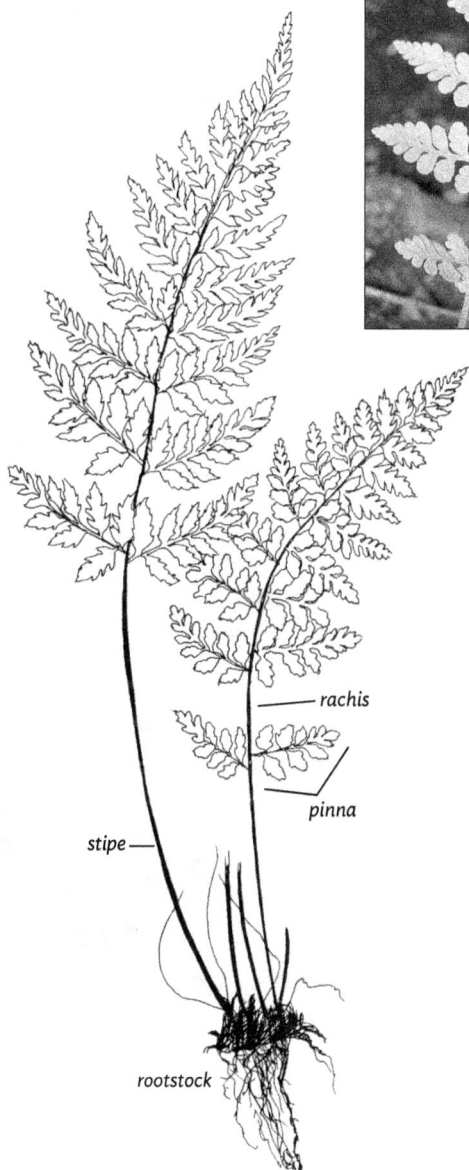

rachis

pinna

stipe

rootstock

Bulblet bladder fern (*Cystopteris bulbifera*), note 2 bulblets along rachis.

Brittle bladder fern (*Cystopteris fragilis*).

Lowland bladder fern (*Cystopteris protrusa*).

Upland brittle bladder fern (*Cystopteris tenuis*).

Gymnocarpium
OAK FERN

Gymnocarpium are small to medium, delicate ferns of cool, shaded places, either in soil or on rock. *Gymnocarpium* includes 7 species, with 5 occurring in temperate regions of North America; 3 in our flora.

KEY CHARACTERS
• Plants forming small colonies from long-creeping, smooth, dark rhizomes.
• **Frond** deciduous, sterile and fertile fronds alike.
• **Stipe** grooved, smooth; vascular bundles 2, oblong.
• **Leaf blade** nearly horizontal, pinnatifid to 3-pinnate-pinnatifid, triangular to pentagonal in outline.
• **Pinnules** on lower side of pinnae longer than those on upper side, costae grooved above, not continuous from rachis to costae, segments entire or crenate, veins free, simple or forked.
• **Sori** small, round, in 1 row between midrib and margin, **indusium** absent, sporangia brownish.

NAME
From the Greek, *gymnos,* naked, *karpos,* fruit, because there are no indusia covering the sori.

ADDITIONAL MIDWEST SPECIES
• **Asian Oak Fern** [*Gymnocarpium jessoense* (Koidzumi) Koidzumi]; local in ne Iowa, Mich Upper Peninsula, ne Minn, and Wisc; usually on or near non-calcareous rock or talus, in shade or partial sun. **Fronds** deciduous, arising singly from the creeping rhizomes; fronds usually less than 40 cm long; **blade** broadly triangular, 2 pinnate pinnatifid and ternate (divided into three roughly equal branches at the base); **sori** round, located on underside of blade; **indusium** absent.

NOTES
The region's 3 species of *Gymnocarpium* may be distinguished as follows:
• In *G. jessoense,* underside of rachis and blade clearly glandular and upperside smooth; lowest pinna pair curved toward tip of blade.
• *G. dryopteris* smooth (without hairs or glands) on both sides.
• *G. robertiana* glandular on both surfaces.

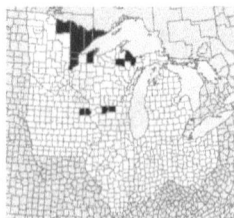

Gymnocarpium jessoense
Asian oak fern

KEY TO GYMNOCARPIUM | OAK FERN

1 Rachis of leaf blade and midveins on the underside of the pinnae smooth; blades with the two basal segments nearly as large as the rest of the blade, the blade thus about as wide as long; Ill, Iowa, Mich, Minn, Ohio, Wisc **1. Gymnocarpium dryopteris**
NORTHERN OAK FERN, page 122

1 Rachis of blade and larger midveins on the underside of the pinnae finely glandular with short-stalked glands; the two basal segments clearly smaller than the rest of the blade, the blade thus usually longer than wide **2**

2 Blades glandular on the upper surface, especially on the midveins of the pinnae and adjacent to the rachis; lowermost (first) pinnae pair diverging at a right angle, not upward pointing; the second pinnae pair usually with a petiole; nw Ill, e Iowa, Mich, Minn, Wisc **2. Gymnocarpium robertianum**
LIMESTONE OAK FERN, page 124

2 Blades smooth on the upper surface; lowermost (first) pinnae pair usually clearly upwardly pointing; the second pinnae pair sessile; ne Iowa, n Mich, Minn, Wisc **3. Gymnocarpium jessoense**
ASIAN OAK FERN, page 120

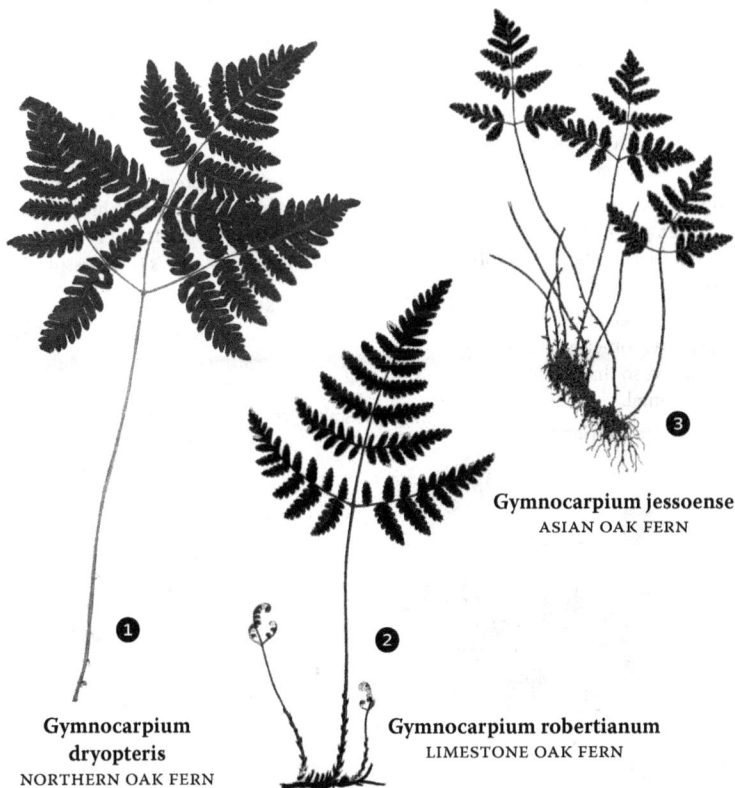

Gymnocarpium jessoense
ASIAN OAK FERN

Gymnocarpium dryopteris
NORTHERN OAK FERN

Gymnocarpium robertianum
LIMESTONE OAK FERN

Gymnocarpium dryopteris (L.) Newman
NORTHERN OAK FERN

FIELD TIPS
- delicate deciduous fern of moist woods, sometimes on rock
- frond 3-parted, bright green, blades triangular, tilted to near horizontal position
- indusium absent

MIDWEST RANGE
Ill (endangered), Iowa (threatened), Mich, Minn, Ohio (endangered), Wisc.

HABITAT
Moist woods of all types, sometimes on cliffs and boulders.

DESCRIPTION
Rootstock Long-creeping and branching, slender, dark brown to black, scaly.

Crozier Delicate, green, 3-parted, one for each developing pinna; produced all summer.

Frond Deciduous, singly from the rhizome; sterile and fertile fronds alike; to 40 cm long.

Stipe Grooved, wiry, straw-colored, darker, purplish at the base, scaly at base, vascular bundles 2, oblong at stipe base, becoming round above.

Rachis Green to dark purplish-green, delicate, sometimes glandular, bending so that blade more or less horizontal.

Blade Broadly triangular, appearing 3-parted in outline, 2-pinnate-pinnatifid, vivid green; uppermost pinna arched to more or less horizontal position; lower 2 pinna nearly parallel to ground; smooth, or underside with a few glandular hairs.

Pinnae 6 to 10 pairs; the lowest pair, exactly opposite, attached at a swollen junction, each similar to the remainder of the blade, only slightly smaller; the lower, basal pinnule of the basal pinna most divided; costae grooved, continuous from rachis to costae; margins entire to crenate; veins free, simple or forked.

Sori Round, in rows near the margin, often merging at maturity; **indusium** absent; **sporangia** brownish.

SYNONYMS
- *Dryopteris dryopteris* (L.) Britton
- *Dryopteris linnaeana* C. Chr.
- *Phegopteris dryopteris* (L.) Fée
- *Thelypteris dryopteris* (L.) Slosson

SIMILAR SPECIES
- The less common **limestone oak fern** (*Gymnocarpium robertianum*) differs in having the lowest 2 pairs of pinnae with stalks (vs. lowest pair only in *G.*

NORTH AMERICA

MIDWEST

dryopteris) and the blade with many short glandular hairs.

NOTES
Believed to have originated as hybrid between *Gymnocarpium appalachianum* and *G. disjunctum*. In our region, may hybridize with *G. jessoense* and *G. robertianum*.

NAME
Known to Linnaeus in Europe, and named by him *Polypodium dryopteris* in 1753; included in *Phegopteris* by Fée nearly 100 years later.

From Greek, *drys*, oak + *pteron*, a wing, describing shape of the pinnae; *pteris* was used by the ancient Greeks for all ferns.

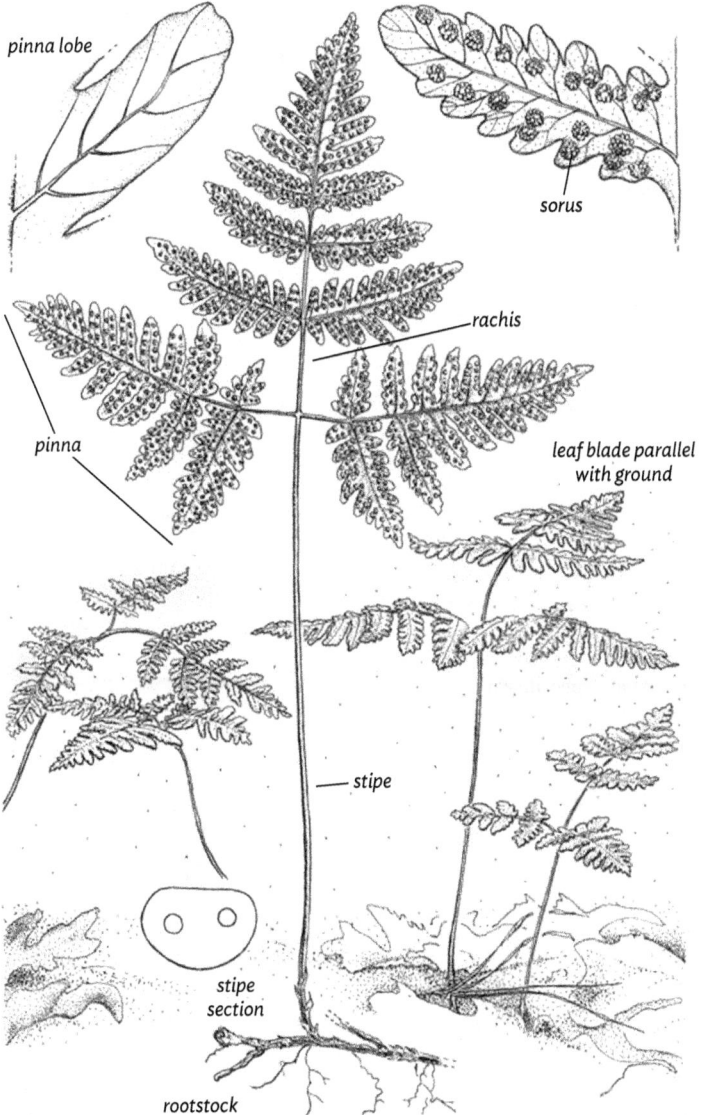

pinna lobe

sorus

rachis

pinna

leaf blade parallel with ground

stipe

stipe section

rootstock

Gymnocarpium robertianum
(Hoffmann) Newman

LIMESTONE OAK FERN

FIELD TIPS
- deciduous fern of calcareous soil
- blade triangular, covered with glandular hairs
- less common than *G. dryopteris*

MIDWEST RANGE
nw Ill (endangered), e Iowa, Mich (threatened), Minn, Wisc.

HABITAT
In shade of calcareous woods and on hummocks in swamps of northern white cedar (*Thuja occidentalis*), sometimes on shallow soil over calcareous rock (often with *Thuja*).

DESCRIPTION
Rootstock Long-creeping, branching, dark brown to black.

Frond Deciduous, single along rhizome; sterile and fertile fronds alike; to 40 cm long.

Stipe Grooved, dull green; scaly at base, the scales soon deciduous; vascular bundles 2, oblong.

Blade Triangular, stiff, oriented somewhat horizontally (but not as much as *G. dryopteris*), 2-pinnate-pinnatifid, surface dull, very glandular, especially on underside and rachis.

Pinnae 9 to 12 pair, lowest pair similar in form (not size) to the next pair (unlike *G. dryopteris*); costae grooved above, continuous from rachis to costae; margins entire to slightly crenate; veins free, simple or forked.

Sori Round, closer to the margin, on both edges of the segments; **indusium** absent; **sporangia** brownish.

SYNONYMS
- *Dryopteris robertiana* (Hoffm.) C. Chr.
- *Gymnocarpium dryopteris* (L.) Newman var. *pumilum* (DC.) B. Boivin
- *Phegopteris robertiana* (Hoffm.) A. Braun ex Asch.

SIMILAR SPECIES
Differs from the region's two other *Gymnocarpium* species by similarity in form (not size) of lowest pinnae pair to the second pair, the central leaflet clearly larger than two lower leaflets, the rank-smelling, very glandular blade (visible with a hand lens), and generally larger size.

NORTH AMERICA

MIDWEST

NAME

From Latin, 'of Robert,' referring to similarity of leaves to St. Robert's geranium.

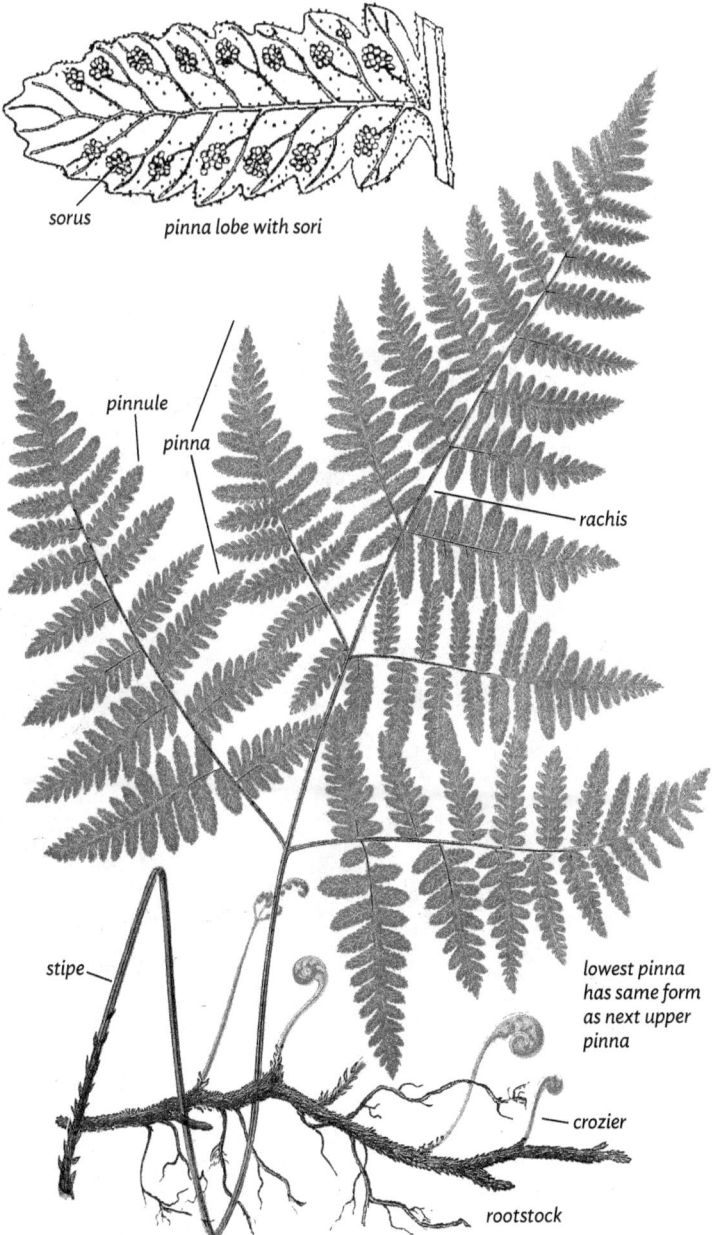

sorus · *pinna lobe with sori*

pinnule

pinna

rachis

stipe

lowest pinna has same form as next upper pinna

crozier

rootstock

Northern oak fern (*Gymnocarpium dryopteris*).

Northern oak fern (*Gymnocarpium dryopteris*).

Limestone oak fern (*Gymnocarpium robertianum*).

Limestone oak fern (*Gymnocarpium robertianum*).

DENNSTAEDTIACEAE
Bracken Fern Family

Large, deciduous, colony-forming ferns. **Sori** on blade margins, either in round separate clusters covered by a cup-shaped indusium (*Dennstaedtia*); or in *Pteridium*, more or less continuous along margin of frond and covered by recurved border of the leaf segment (forming an outer false indusium).

Worldwide, 37 genera, about 360 species, mostly of tropical regions; 4 genera in North America, 2 genera in our flora.

KEY TO DENNSTAEDTIACEAE | BRACKEN FERN FAMILY

1   Leaf blades elongate in outline, at least 4x as long as broad, not leathery; sori globular, separate; Ill, Ind, Mich, Ohio, s Wisc
1. Dennstaedtia punctilobula
HAY-SCENTED FERN, page 130

1   Leaf blades broadly triangular in outline, about as broad as long, somewhat leathery; sori linear, continuous along margins as a marginal band; regionwide   **2. Pteridium aquilinum**
NORTHERN BRACKEN FERN, page 132

Dennstaedtia punctilobula
HAY-SCENTED FERN

Pteridium aquilinum
NORTHERN BRACKEN FERN

Dennstaedtia
HAY-SCENTED FERN

Large deciduous ferns. Worldwide, ca. 35 species, mostly in the tropics; 3 species in North America, 1 species in our flora. Rare in our region apart from e Ohio, becoming common and sometimes weedy in ne USA where may form large, crowded colonies.

KEY CHARACTERS
- Colony-forming from slender creeping rhizomes.
- Sterile and fertile **fronds** alike; **blade** covered with small, white, gland-tipped hairs, these hay-scented (especially as drying).
- **Indusia** cup-shaped; **sori** at tip of vein on margin of pinnule lobes.

pinnule underside and cup-shaped indusium

NAME
Named for German botanist, August Wilhelm Dennstaedt (1776-1826).

Pteridium
BRACKEN FERN

Coarse deciduous ferns, spreading by rhizomes and sometimes covering large areas. Previously considered to contain a single species but now divided into ca. 11 species of worldwide distribution, 1 described in this Flora.

KEY CHARACTERS
- Large coarse, deciduous ferns, sterile and fertile **fronds** alike, with long, creeping, hairy rhizomes.
- **Blades** generally triangular-shaped, mostly 3-pinnate.
- **Sori** continuous, along the margin, **indusium** double, the outer false and formed by the recurved border of the leaf segment, the inner true indusium tiny or absent.

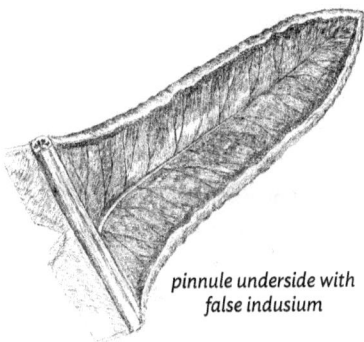

pinnule underside with false indusium

NOTES
Pteridium is the world's most widely distributed genus of fern, sometimes becoming a serious weed and very difficult to eradicate due to the deeply seated rhizomes. Bracken ferns also release allelopathic chemicals, which in combination with its shady canopy and thick litter, inhibit other plant species from establishing.

NAME
From Greek, *pteridion*, a small fern. *Bracken* is of Old Norse origin, related to the Swedish word *bräken*, meaning fern.

Dennstaedtia punctilobula

(Michx.) T. Moore

HAY-SCENTED FERN

FIELD TIPS

• large deciduous fern
• fronds single from the rhizome, may form dense colonies
• fronds with glandular hairs on upper and lower surface, margins round-toothed
• indusium cup-shaped, sorus at end of vein on blade margin

MIDWEST RANGE

Ill (endangered), Ind, Mich (threatened), Ohio, s Wisc; becoming common in ne and se USA.

HABITAT

Partially shaded to sunny woods, roadsides, fields, clearings, rocky slopes.

DESCRIPTION

Rootstock Long-creeping, slender, with dark, red-brown, jointed hairs near new growth; scales few or absent.

Crozier Covered with silver-white gland-tipped hairs.

Frond Deciduous, arising singly along the rhizome and forming colonies; sterile and fertile fronds alike.

Stipe Shorter than blade, straw-colored to brown, darker to nearly black at base, grooved above, glandular hairy, vascular bundles 1, arranged in a U-shape.

Rachis Pale to straw-colored, slender, hairy.

Blade Lance-shaped, lacy, papery, 2-pinnate-pinnatifid (sometimes less or more divided), yellow-green to pale green, with silver-gray, jointed hairs on both surfaces and gland-tipped hairs containing a fragrant wax below.

Pinnae Broadest at base, segments ovate to lance-shaped, margins with rounded teeth, veins free, pinnately branched.

Sori Globose to almost cylindric, marginal at vein tips; **indusium** a circular cup.

SIMILAR SPECIES

• **Lady fern** (*Athyrium filix-femina*) is clumped, not spreading by rhizomes; stipe with dark brown scales, not hairs.

• Blade of **New York fern** (*Parathelypteris noveboracensis*) gradually tapered to base, lowest pinnae very small.

NAME

Michaux discovered this fern in Canada and named it *Nephrodium punctilobulum* in 1803. It was later mistakenly transferred to the tropical

NORTH AMERICA

MIDWEST

genus *Dicksonia*. In 1857, however, Moore showed it to belong in Bernhardi's genus *Dennstaedtia*.

From Latin, *punctum*, small spot, *lobulus*, small lobe; referring to the sori appearing as small spots on the pinnule lobes.

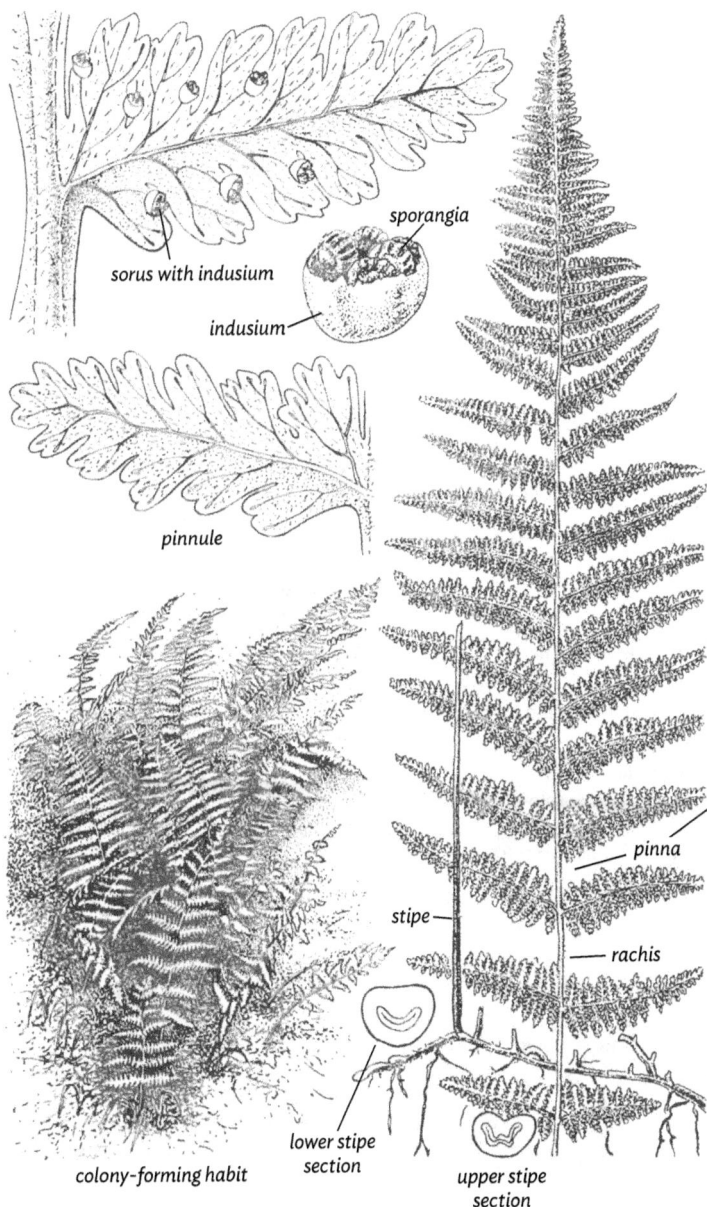

sporangia

sorus with indusium

indusium

pinnule

colony-forming habit

lower stipe section

stipe

rachis

pinna

upper stipe section

Pteridium aquilinum (L.) Kuhn
NORTHERN BRACKEN FERN

FIELD TIPS
- large deciduous fern, very common
- fronds in rows from rhizomes, often forming large colonies
- blade broadly triangular in outline, held almost parallel to the ground

MIDWEST RANGE
Common across nearly all of region.

HABITAT
Dry, acidic, sandy woods, clearings, abandoned fields, waste places, burned areas, in sun to partial shade; often forming large colonies from extensive and long-creeping rhizomes; becoming a serious, difficult to eradicate weed in some places.

DESCRIPTION
Rootstock Long-creeping and branching, cordlike, to 2 cm wide, deep-seated (30 cm or more deep); scales absent.

Crozier With 3 sections, uncoiling like opening of an eagle's claw, covered with silvery gray hairs

Frond Deciduous, scattered on the rhizome; sterile and fertile fronds alike, to ca. 1 m tall.

Stipe Variable in length, somewhat woody, dark purple-brown at base, straw-colored above; vascular bundles numerous (often more than 10), of various sizes and shapes.

Rachis Green, grooved, slightly hairy.

Blade Large, broadly triangular, 2- to 4-pinnate, waxy.

Pinnae Lowest very large; pinnules variable; costae rachis and costae grooved above, veins free, forked, except for a marginal strand.

Sori Continuous, along the margin; **indusium** double, the outer false, reflexed, the inner distinct or absent.

NOTES
Bracken fern fiddleheads contain the carcinogenic terpene *ptaquiloside* and should not be eaten.

NAME
Ours nearly all *P. aquilinum* var. *latiusculum* (Desv.) Underwood ex Heller

Linnaeus named the European bracken *Pteris aquilina* in 1753. Seven years later Scopoli placed it in genus *Pteridium*, on the basis of the special indusium.

Aquilinum, from Latin *aquilus*, eagle-like; perhaps referring to talon-like shape of the unfurling crozier.

NORTH AMERICA

MIDWEST

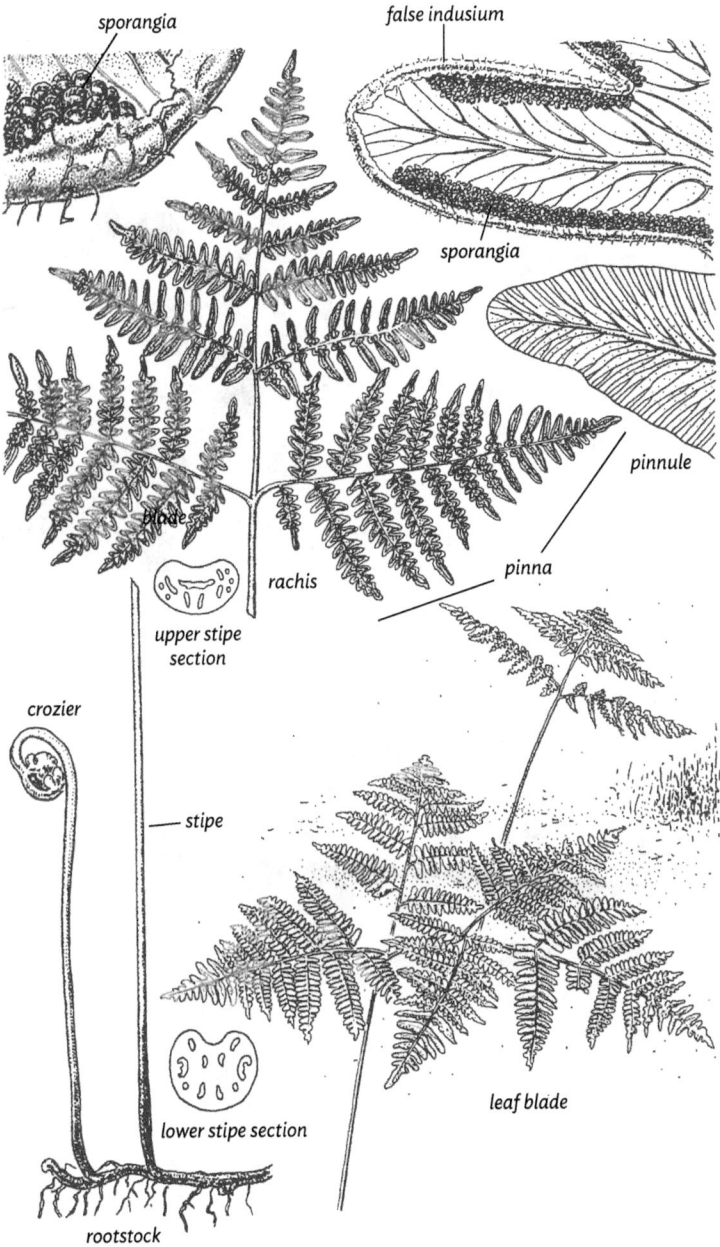

sporangia

false indusium

sporangia

pinnule

Blade

rachis

pinna

upper stipe section

crozier

stipe

lower stipe section

leaf blade

rootstock

Hay-scented fern (*Dennstaedtia punctilobula*).

Hay-scented fern (*Dennstaedtia punctilobula*), sori at ends of veins along margins.

Northern bracken fern (*Pteridium aquilinum*), large leaf blade nearly parallel with ground.

Northern bracken fern (*Pteridium aquilinum*), fertile pinnules with sori near margins.

DRYOPTERIDACEAE
Wood Fern Family

Small to large ferns growing in mineral or organic soils, or in rock crevices; plants usually clumped, sterile and fertile **fronds** generally more or less alike. Worldwide ca. 66 genera, more than 1400 species, especially in tropical regions and in forests; 3 genera in our flora.

KEY CHARACTERS

• Usually clumped ferns of wet to dry habitats.

• **Blade** 1- to 4-pinnate, often with scales, hair-like scales, and/or hairs (except clear, needle-like hairs generally absent, these are typical in Thelypteridaceae); ultimate leaf segments (smallest subdivision of blade) not entire; veins free.

• **Sori** round, in most species on veins or vein tips of lower blade surface (usually not marginal); **indusia** peltate (umbrella-like) or kidney-shaped.

NOTES

The family previously included a number of other genera: *Athyrium, Cystopteris, Deparia, Diplazium, Matteuccia, Onoclea* and *Woodsia*. These are now separated into several new families as treated in this Flora.

NAME

From Greek, *drys*, oak, and *pteris*, fern.

ADDITIONAL MIDWEST SPECIES

• **Asian net-vein holly fern** (*Cyrtomium fortunei* J. Sm.); evergreen clumped fern, introduced from Asia and known from several shaded forest locations in s Ind and occasional in s USA.

Rootstock erect, with large tan scales. **Frond** evergreen, sterile and fertile fronds alike; stipe scaly, the scales covering much of the stipe, vascular bundles 4 or more in an arc. **Leaf blade** 1-pinnate, terminated by a pinna similar to the others ('imparipinnate'), leathery, with needlelike scales on underside, smooth above. Pinnae holly-like, margins crenate to spinulose, veins netted. **Sori** round, in 2 or more rows between midrib and margin, **indusium** peltate (round), persistent or not, sporangia brown.

Cyrtomium shares the peltate indusium with *Polystichum* and also sometimes has an upward ear or auricle at base of pinnae; differs in having a terminal pinna similar to the lateral ones; it shares the grooves continuous from rachis to costae with *Dryopteris*; differs from both genera in having netlike rather than free veins.

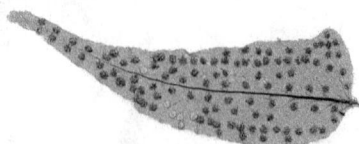

Underside of pinna with mature sori; most indusia already shed.

Cyrtomium fortunei
ASIAN NET-VEIN HOLLY FERN

KEY TO DRYOPTERIDACEAE | WOOD FERN FAMILY

1   Fronds 1-pinnate-pinnatifid to more divided, the pinnae pinnatifid or themselves fully divided, lacking a prominent basal lobe, light green to dark green, herbaceous to nearly leathery; indusia kidney-shaped   **1. Dryopteris**
WOOD FERN, page 138

1   Fronds 1-pinnate, the pinnae toothed and each with a slight to prominent lobe near the base on the side towards the leaf tip, dark green, leathery or nearly so; indusia peltate (umbrella-like) 2

2   Uncommon introduced fern, in Midwest region known only from several locations in southern Indiana; pinnae 4-25 pairs per frond, veins rejoining to form a netlike pattern   **3. Cyrtomium**
NET-VEIN HOLLY FERN, page 136

2   Native, mostly widespread species; pinnae many, 25-50 pairs on larger leaves; veins branching, free and not rejoining in a netlike pattern   **2. Polystichum**
HOLLY FERN, SWORD FERN, page 168

Cyrtomium
NET-VEIN HOLLY FERN

Polystichum
HOLLY FERN, SWORD FERN

Dryopteris
WOOD FERN

Dryopteris
WOOD FERN

Mostly medium to large ferns, with scaly, stout stipes and variably dissected blades. Worldwide, about 250 species of *Dryopteris*, most commonly in temperate regions of Asia. North America is home to 14 species, 10 species in our flora. Wood ferns, as the name implies, are often found in forests, but our species are found in habitats ranging from wetlands to moist or dry forests to talus and rock outcrops. Some species commonly hybridize, complicating field identification; 3 of the region's most common hybrids are listed below. *Dryopteris* similar in many respects to *Polystichum*, the latter differing in the peltate (umbella-shaped) indusium and lack of continuity in grooves between rachis and costa.

KEY CHARACTERS

- **Fronds** range from 1-pinnate-pinnatifid to 3-pinnate-pinnatifid; we have no simple or pinnatifid *Dryopteris*.
- **Stipes** scaly, without hairs (except in *D. intermedia*), the **vascular bundles** in a C-shape.
- Upperside of stipe, rachis, and costa with a continuous groove.
- **Indusium** kidney-shaped over a round **sorus**.

NAME

From Greek, *drys*, oak, referring to the many species that are found in oak woodlands, and *pteron*, a wing, which describes the shape of the pinnae (*pteris* was used by the ancient Greeks for all ferns). The dryads of Greek mythology were the nymphs that inhabited oaks. In Britain and Europe, known as **shield ferns** or **buckler ferns**, a buckler being a small, round shield, in reference to the shape of the indusium. Previously, wood ferns were classified as *Aspidium* or *Lastrea*.

DRYOPTERIS HYBRIDS

Three *Dryopteris* hybrids are fairly common in the midwest region:

- **Dryopteris × boottii** (Tuckerman) Underwood: *D. cristata × D. intermedia* Fronds more dissected than *D. cristata*.
- **Dryopteris × triploidea** Wherry: *D. carthusiana × D. intermedia* Fronds similar to parents but often somewhat larger.
- **Dryopteris × uliginosa** (A. Braun ex Dowell) Druce: *D. carthusiana × D. cristata* Usually in swamps and wet woods.

Hybrids may be recognized by an appearance intermediate between the parent species, and the presence of abortive spores.

Dryopteris xbootii
D. cristata x D. intermedia

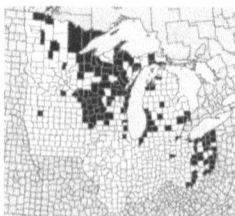

Dryopteris xtriploidea
D. carthusiana x D. intermedia

Dryopteris xuliginosa
D. carthusiana x D. intermedia

KEY TO DRYOPTERIS | WOOD FERN

1 Leaf blades small, to 25 cm long, densely scaly on underside; plants of granitic rock cliffs and talus; Mich Upper Peninsula, n Minn, Wisc **1. Dryopteris fragrans**
FRAGRANT WOOD FERN, page 154

1 Leaf blades larger, 25-120 cm long, underside scales sparse or absent ; plants of various habitats but not normally on steep cliffs **2**

2 Blades 2-pinnate with the pinnules deeply cut (more than halfway to the base), or 3-pinnate **3**

2 Blades 1-pinnate (with the pinnae deeply lobed), to 2-pinnate with the pinnules shallowly cut (halfway or less to the midrib) **5**

3 Innermost pinnules (next to rachis) on lower side of lowest pinnae clearly shorter than adjacent pinnule; midribs of pinnae, indusia, and rachis with tiny glands; leaf blades fully evergreen; regionwide **2. Dryopteris intermedia**
EVERGREEN WOOD FERN, page 158

3 Innermost pinnules (next to rachis) on lower side of lowest pinnae slightly to clearly longer than adjacent lower pinnule; midribs of pinnae, indusia, and rachis without glands; leaf blades not evergreen, dying back in winter **4**

4 Blades lance-shaped triangular; innermost pinnules (next to rachis) on lower side of lowest pinnae 1.5-5 cm long, about 2 times as long as the innermost pinnule on upper side of the pinna; regionwide **3. Dryopteris carthusiana**
SPINULOSE WOOD FERN, page 142

Dryopteris fragrans
FRAGRANT WOOD FERN

Dryopteris intermedia
EVERGREEN WOOD FERN

Dryopteris carthusiana
SPINULOSE WOOD FERN

KEY TO DRYOPTERIS | WOOD FERN, CONTINUED

4   Blades ovate-triangular; innermost pinnules (next to rachis) on lower side of lowest pinnae large, (4-) 6-12 cm long, 2-3.5 times as long as innermost pinnule on upper side of the pinna; Mich Upper Peninsula, Minn, n Wisc   **4. Dryopteris expansa**
                                         SPREADING WOOD FERN, page 150

5   Sori very close to margins of blade; base of stipe with many pale brown scales; Ill, Ind, Iowa, Mich, se Minn, Ohio, Wisc
                                         5. Dryopteris marginalis
                                         MARGINAL WOOD FERN, page 160

pinnule underside
Dryopteris marginalis
Marginal wood fern

5   Sori well away from margin of blade, ± in the middle of the pinnules or lobes of the pinnae (or even closer to midvein); stipes with scales either few or dark brown or both **6**

6   Stipes very short, 1/5–2/5 length of leaf blade; rachis and costa (midvein of pinna) scaly; plants of dry, rocky forests; n Mich, Minn, n Wisc (and reported for ne Ill and Ohio)
                                         6. Dryopteris filix-mas
                                         MALE FERN, page 152

6   Stipes 2/5–3/5 length of blade; scales mostly only on stipes; plants of moist forests and swamps. **7**

7   Lower pinnae triangular, widest near the rachis **8**
7   Lower pinnae ovate, the widest point well away from the rachis, pinnae narrowing both to tip and to base **9**

8   Larger leaf blades to 15 cm wide; pinnae of fertile leaf blades tilted horizontally (like the slats of a venetian blind); regionwide
                                         7. Dryopteris cristata
                                         CRESTED WOOD FERN, page 148

8   Larger leaf blades 11-22 (-25) cm wide; pinnae of fertile leaf blades not tilted, the frond essentially flat; local in s Ill, Mich, Ohio
                                         8. Dryopteris clintoniana
                                         CLINTON'S WOOD FERN, page 146

9   Leaf blades abruptly narrowed to tip; sori near costa; occasional regionwide   **9. Dryopteris goldiana**
                                         GOLDIE'S WOOD FERN, page 156

9   Leaf blades gradually tapered to tip; sori midway between costa and margin; rare in s Ill, n Ind, sw Mich and ne Ohio
                                         10. Dryopteris celsa
                                         LOG FERN, page 144

Dryopteris expansa
SPREADING WOOD FERN

Dryopteris marginalis
MARGINAL WOOD FERN

Dryopteris filix-mas
MALE FERN

Dryopteris cristata
CRESTED
WOOD FERN

Dryopteris celsa
LOG FERN

Dryopteris clintoniana
CLINTON'S WOOD FERN

Dryopteris goldiana
GOLDIE'S WOOD FERN

Dryopteris carthusiana (Vill.) H.P. Fuchs
SPINULOSE WOOD FERN

FIELD TIPS
- deciduous clumped fern of moist to wet woods
- blade 2-pinnate-pinnatifid, hairs usually absent
- innermost pinnule of basal pinnae longer than next lower pinnules
- indusia kidney-shaped

MIDWEST RANGE
Common across most of Midwest region.

HABITAT
Moist to wet, deciduous or conifer woods, streambanks, swamp hummocks.

DESCRIPTION
Rootstock Short-creeping to erect, densely scaly, covered with old stipe bases, forming an irregular clump.

Frond Deciduous, clustered; sterile and fertile fronds nearly alike; sterile fronds tending to be somewhat shorter, arching; to 60 cm long by 15 cm wide.

Stipe Shorter than blade, straw-colored, grooved, base enlarged, light brown scales at base, fewer above; vascular bundles mostly 3-7 in a C-shaped pattern.

Rachis With scattered scales.

Blade Ovate lance-shaped, 2-pinnate-pinnatifid; upper side smooth, lower side smooth or with a few scales or glands.

Pinnae 12 to 14 pairs, held almost horizontally, tips pointing upward; the lowest one not reduced or only slightly reduced in length compared to second pair; first lower pinnule of basal pinna twice as long as upper one, and also larger than second lower pinnule; costae grooved above, continuous from rachis to costae; margins margins serrate, teeth spine-tipped; veins free, forked.

Sori Round, in 1 row between midrib and margin, at the tips of veins, usually absent on the lowest pinna; **indusium** reniform, gray-white, at a sinus; **sporangia** dark brown.

SYNONYMS
- *Dryopteris austriaca* var. *spinulosa* (O.F. Müll.) Fisch.
- *Dryopteris spinulosa* (O.F. Müll.) Watt

SIMILAR SPECIES
- **Spreading wood fern** (*Dryopteris expansa* is similar to *D. carthusiana*, but lower basal pinnules of lowest pinnae of *D. expansa* very large relative to upper basal pinnules.

NORTH AMERICA

MIDWEST

• Separated from **evergreen wood fern** (*D. intermedia*) by deciduous rather than evergreen habit, blade generally less dissected, and lower pinnule on the lowest pinna larger than the next one (and that pinnule also attached at or very near point of attachment of first upward pointing pinnule); in *D. intermedia*, this lower inner pinnule on the lowest pinna is somewhat shorter than the next one.

NOTES

An early and apparently effective treatment for intestinal tapeworms.

NAME

First named in Europe *Polypodium spinulosum* by O. F. Mueller in 1767, and placed in the genus *Dryopteris* by Kuntze in 1891.

Carthusiana, named after botanist Johan Friedrich Cartheuser (1704-1777). An alternative origin is the village of Carthusium in Dauphiné, France, where Dominique Villars collected it.

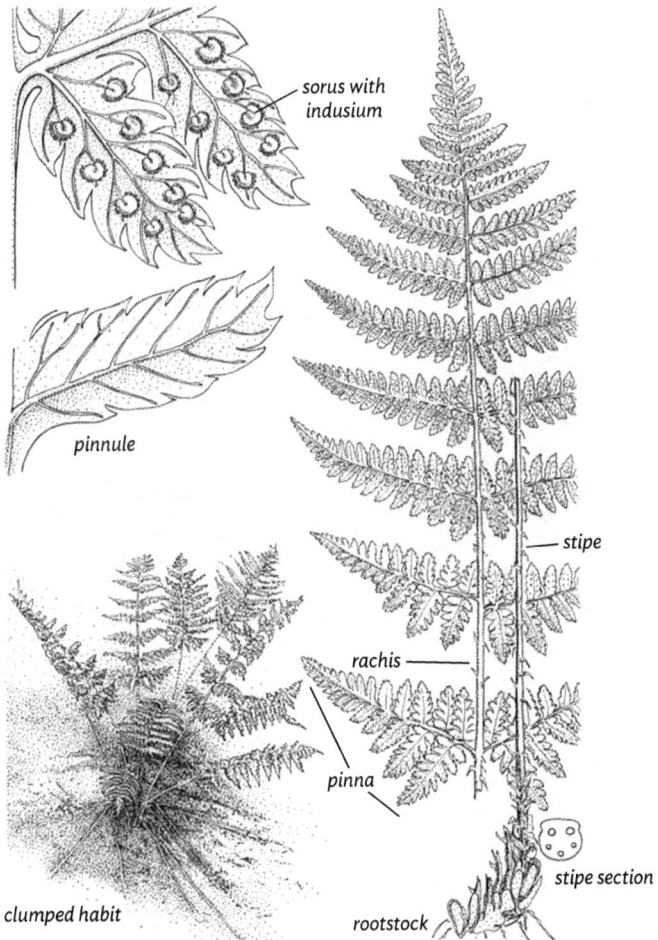

sorus with indusium

pinnule

stipe

rachis

pinna

stipe section

clumped habit

rootstock

Dryopteris celsa
(Wm. Palmer) Knowlt., Palmer & Pollard ex Small
LOG FERN

FIELD TIPS
- rare deciduous fern of wet places
- similar to *Dryopteris goldiana*, but blade narrower, tapering gradually to tip, and uniformly dark green

MIDWEST RANGE
s Ill (endangered), n Ind, sw Mich (threatened), ne Ohio (endangered); more common in se USA.

HABITAT
Seepage slopes, swamp hummocks and downed logs.

DESCRIPTION
Rootstock Short-creeping.

Frond Deciduous, clustered; sterile and fertile fronds alike; 120 cm long by 30 cm wide.

Stipe Half or more as long as blade, grooved, bearing mixed broad and narrow pale brown scales with darker central stripe; vascular bundles 3-7 in C-shaped pattern.

Blade Ovate lance-shaped, gradually tapered to tip, 1-pinnate-pinnatifid; upper side smooth, lower side with linear to ovate scales.

Pinnae 15 to 20 pairs; basal pinnae linear-oblong, much reduced; costae grooved above, continuous from rachis to costae; margins crenately toothed; veins free, forked.

Sori Round, in 1 row between midrib and margin; **indusium** kidney-shaped, at a sinus; **sporangia** brownish.

SYNONYMS
- *Dryopteris goldiana* (Hook. ex Goldie) A. Gray subsp. *celsa* Wm. Palmer
- *Dryopteris wherryi* F.W. Crane

SIMILAR SPECIES
- **Clinton's wood fern** (*Dryopteris clintoniana*) similar but lowest pinnae wider at base than at middle; in *D. celsa*, lower pinnae narrower at base than middle.

- **Goldie's wood fern** (*Dryopteris goldiana*) similar but blade in *D. goldiana* tapers abruptly to a short pointed tip, and has dark and light green coloration; blade of *D. celsa* tapers gradually to tip and is uniformly dark green.

NOTES
Dryopteris celsa is a fertile hybrid between *D. goldiana* and *D. ludoviciana* and is known to hybridize with other species; hybrids can usually be identified by the dark-striped scales.

NORTH AMERICA

MIDWEST

Formerly treated as a subspecies of *Dryopteris goldiana.*

NAME

Celsa means held high, in reference to the plant's growing perched up on logs or hummocks above water level in swamps.

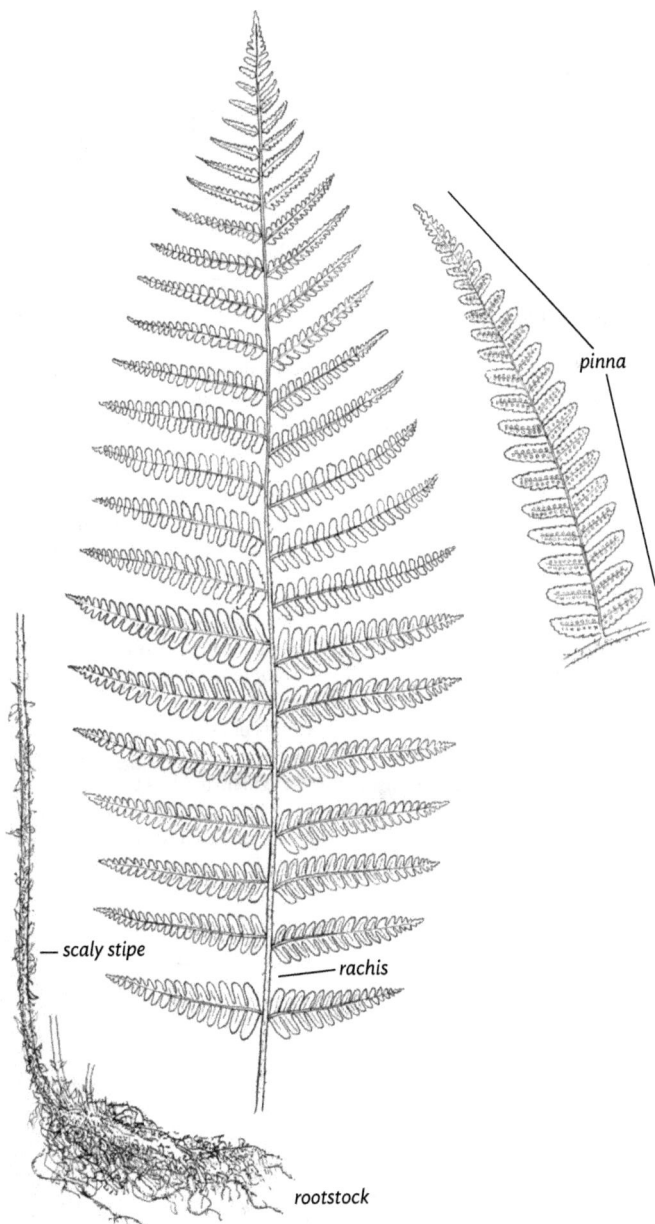

pinna

— scaly stipe

—— rachis

rootstock

Dryopteris clintoniana (D.C. Eat.) Dowell
CLINTON'S WOOD FERN

FIELD TIPS
- loosely clumped fern of wet places
- sterile fronds evergreen, fertile fronds deciduous
- blade with nearly parallel sides, tapering only near tip
- fertile pinnae twisted toward horizontal

MIDWEST RANGE
s Ill, Mich, Ohio (endangered); historical records for n Ind and se Wisc.

HABITAT
Moist to wet woodlands and swamps.

DESCRIPTION
Rootstock Short-creeping to erect, dark brown to black, with old stipe bases, densely scaly.

Frond Usually evergreen (sterile fronds), borne in 1-2 rows along the rhizome; sterile and fertile fronds nearly alike but fertile fronds deciduous, slightly smaller; 35-80 (-120) cm long by 20 cm wide.

Stipe Shorter than blade, grooved, straw-colored, scaly (at least near base) at base, scales scattered, tan; vascular bundles mostly 3-7 in a c-shaped pattern.

Rachis Green to pale green, grooved; scaly, especially at base of pinnae.

Blade In outline, lance-shaped with nearly parallel sides, tapering only in upper 1/4 of blade; 1-pinnate-pinnatifid (nearly more); upper side smooth, lower side with scales.

Pinnae 14 to 16 pairs, more alternate than opposite above the basal pair, less twisted out of plane of blade than in *Dryopteris cristata*; costae grooved above, continuous from rachis to costae; margins sharp-toothed, minutely bristle-tipped; veins free, forked, mostly not reaching margin.

Sori Round, on veins, closer to costa than margin; **indusium** kidney-shaped, at a sinus; **sporangia** brownish.

SYNONYMS
- *Dryopteris cristata* (L.) A. Gray var. *clintoniana* (D.C. Eaton) Underw.
- *Dryopteris ×poyseri* Wherry

SIMILAR SPECIES
- Similar to **crested wood fern** (*Dryopteris cristata*) and previously treated as a variety of that species; fronds of *D. clintoniana* tend to be larger, the pinna longer, more narrowly triangular, and less twisted; blade of sterile fronds more abruptly narrowed at tip, similar to Goldie's wood fern.

NORTH AMERICA

MIDWEST

NOTES

Originated as a hybrid between *Dryopteris cristata* and *D. goldiana*.

NAME

Named by D. C. Eaton *Aspidium cristatum* var. *clintonianum* in 1867.

Transferred to *Dryopteris*, though still as a variety, by Underwood in 1893; raised to species rank by Dowell in 1906.

Named for Judge G.W. Clinton (1807-1885), a naturalist of Buffalo, New York.

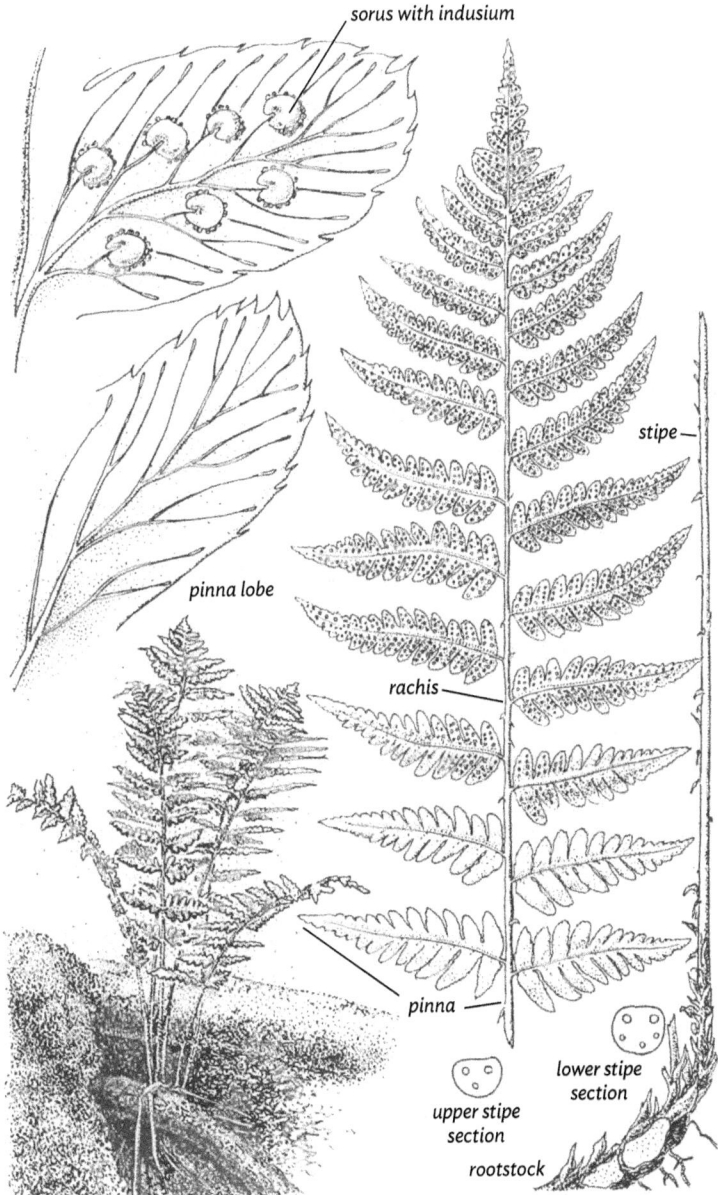

sorus with indusium

pinna lobe

stipe

rachis

pinna

upper stipe section

lower stipe section

rootstock

Dryopteris cristata (L.) Gray
CRESTED WOOD FERN

FIELD TIPS

- medium fern of wet places
- fronds narrow
- pinnae short, widely spaced, twisted horizontally
- lowermost pinnae broadly triangular

MIDWEST RANGE

n Ill, Ind, Iowa, Mich, Minn, Ohio, Wisc.

HABITAT

Hummocks in bogs and sedge meadows, open to semi-shaded places in swamps, wet woods, shrubby wetlands; soils typically acidic.

DESCRIPTION

Rootstock Short-creeping to erect, stout, dark brown to black, covered with old stipe bases, very scaly.

Frond borne in 1-2 rows along the rhizome; sterile fronds evergreen, somewhat waxy and leathery; fertile fronds less so; sterile fronds shorter and narrower at base than the taller (to 80 cm long), erect fertile fronds.

Stipe Usually shorter than blade, grooved, straw-colored, scaly (at least near base); vascular bundles mostly 3-7 in a c-shaped pattern.

Rachis Green, with a few scales.

Blade Narrowly lance-shaped or with parallel sides, 1-pinnate-pinnatifid, often somewhat leathery; upper side smooth, lower side with scales.

Pinnae 10 to 15 pairs, alternate or opposite; fertile pinnae twisted out of plane of blade and perpendicular to it; lower pinnae widely separated, broadly triangular; upwards closer together and narrower; costae grooved above, continuous from rachis to costae; margins spine-toothed; veins free, forked, mostly not reaching margin.

Sori Round, in 1 row between midrib and margin; **indusium** kidney-shaped, shriveling upon ripening, attached at a sinus; **sporangia** dark brown.

SIMILAR SPECIES

- **Dryopteris × boottii** (Tuck.) Underw. (*D. cristata × D. intermedia*), usually found in swamp forests, in appearance a narrow (and less cut) *D. carthusiana*, but noticeably glandular (see page 142).

NAME

This species was named from European specimens *Polypodium cristatum* by Linnaeus in 1753, and made a *Dryopteris* by Gray in 1848. American plants identical with the European.

NORTH AMERICA

MIDWEST

Latin, *cristata,* like a comb; fertile
fronds have the pinnae twisted 90°, and
thus appear comblike in profile.

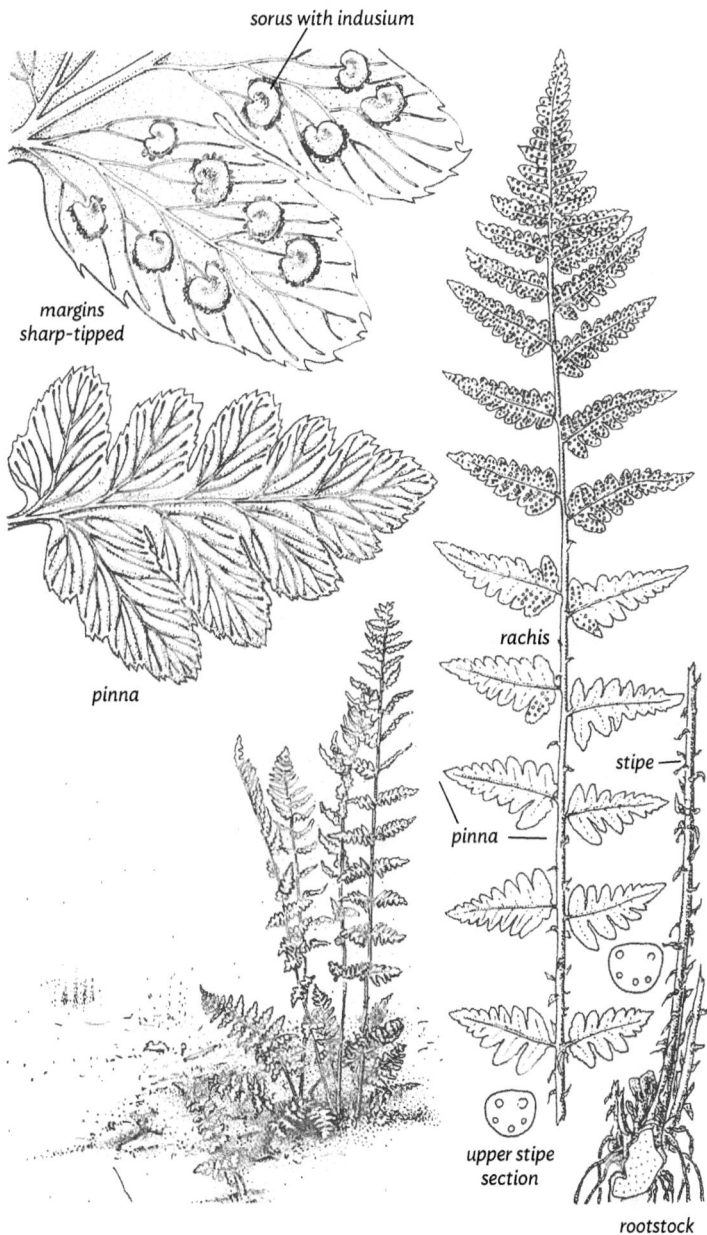

sorus with indusium

*margins
sharp-tipped*

pinna

rachis

stipe

pinna

*upper stipe
section*

rootstock

Dryopteris expansa
(C. Presl) Fraser-Jenkins & Jermy

SPREADING WOOD FERN

FIELD TIPS
- large, very lacy, clumped fern
- sterile and fertile fronds alike
- pinnae tapered to sharp tip
- lower basal pinnule much longer than upper basal pinnule (often 2-3 times as long)

MIDWEST RANGE
e Minn, Mich Upper Peninsula, n Wisc; more common in nw North America and e Canada.

HABITAT
Cool moist hardwood or mixed hardwood-conifer forests, usually on slopes and in ravines.

DESCRIPTION
Rootstock Erect, occasionally branching, covered in old stipe bases.

Frond Tardily deciduous (i.e., slow to die-back in winter), clustered; sterile and fertile fronds alike; to 100 cm long.

Stipe Shorter than blade, grooved, straw-colored, darker at base, scaly, the scales scattered, brown with dark brown central stripe; vascular bundles mostly 3-7 in a c-shaped pattern.

Blade Triangular-ovate, slightly spreading to arching, 3-pinnate-pinnatifid (sometimes more at base), yellowish-green, smooth or with some glands.

Pinnae 12 to 14 pairs, well spaced, long-tapered to tip, the lowest usually curving upwards; first lower pinnules on lowest pinnae equal to or longer than adjacent pinnules, and 2-3x length of first upper pinnule; costae grooved above, continuous from rachis to costae, narrow scales on lower surface; margins serrate, often spine-tipped; veins free, forked, usually not reaching margin.

Sori Round, small, in 1 row between midrib and margin, but sparse throughout the blade; **indusium** kidney-shaped, at a sinus; **sporangia** black.

SYNONYMS
- *Dryopteris assimilis* S. Walker
- *Dryopteris dilatata* auct. non (Hoffm.) A. Gray
- *Dryopteris spinulosa* (O.F. Müll.) Watt var. *dilatata* auct. non (Hoffm.) Underw.

NORTH AMERICA

MIDWEST

NOTES

Very long and wide lower pinnules near the rachis give the frond an asymmetric appearance compared to *Dryopteris carthusiana* and *D. intermedia*.

NAME

From Latin, *expando*, to spread out, in reference to spreading habit of the clumped fronds.

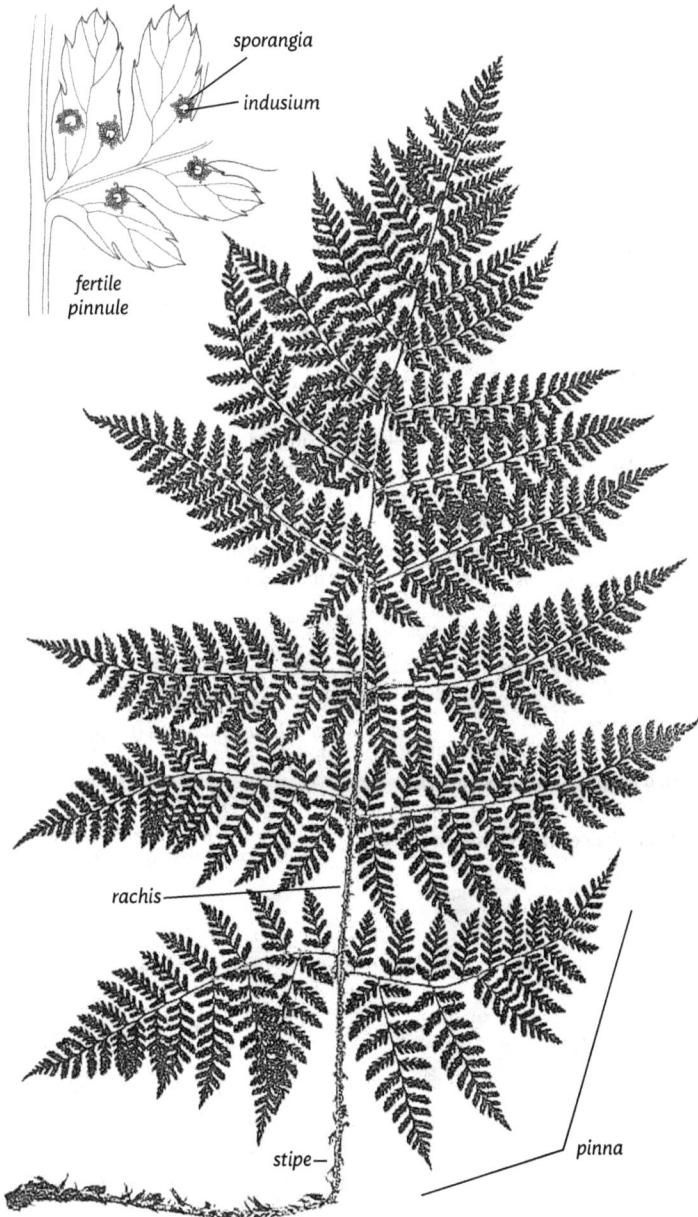

sporangia

indusium

fertile pinnule

rachis

stipe

pinna

Dryopteris filix-mas (L.) Schott
MALE FERN

FIELD TIPS
• large, dark green, clumped fern of calcareous habitats
• fronds deciduous, 1-pinnate-pinnatifid
• stipe with both wide and hairlike scales

MIDWEST RANGE
n Mich, Wisc; perhaps adventive in ne Ill and Ohio (endangered).

HABITAT
Moist, rocky woods, talus slopes and outcrops; typically over limestone and where shaded, rarely where open.

DESCRIPTION
Rootstock Erect, thick, dark brown to black, densely scaly, covered with old stipe bases.

Frond Deciduous (sometimes nearly evergreen), clustered, sterile and fertile fronds alike, erect to somewhat arching; 30-120 cm long and 10-25 cm wide.

Stipe Grooved, straw-brown, scales pale brown, of two types - one broad and one hairlike; vascular bundles 5 or 7 in a C-shaped pattern.

Rachis Green, not grooved or slightly grooved at leaf tip; underside with scales.

Blade Ovate lance-shaped, widest at middle, strongly narrowed at base; 1-pinnate-pinnatifid to nearly 2-pinnate; herbaceous to somewhat leathery.

Pinnae 16 to 24 pairs, lance-shaped, straight to slightly curved upwards; basal pinnae ovate lance-shaped, reduced in size; basal lower pinnule and basal upper pinnule equal, fully attached to costa along the base; costae grooved above, continuous from rachis to costae; margins serrate but not bristle-tipped; veins free, forked.

Sori Round, in 1 row between midrib and margin, on the upper half of the frond; **indusium** kidney-shaped, pale green at first, then whitish gray, then rusty brown, then shriveling, at a sinus; **sporangia** black or dark brown.

SYNONYMS
• *Aspidium filix-mas* (L.) Swartz
• *Lastrea filix-mas* (L.) C. Presl.
• *Polypodium filix-mas* L.
• *Polystichum filix-mas* (L.) Roth
• *Thelypteris filix-mas* Nieuwl.

SIMILAR SPECIES
• Fronds of **marginal wood fern** (*Dryopteris marginalis*) more leathery,

NORTH AMERICA

MIDWEST

the stipe longer, sori nearly marginal.

NOTES

Previously a popular and effective treatment for tapeworms; the root contains an oleoresin that paralyses tapeworms and other internal parasites; since but has been replaced by less toxic and more effective drugs.

NAME

From Latin, *filix*, fern + *mas*, male. Sometimes referred to in ancient texts as worm fern.

Male fern (*Dryopteris filix-mas*), clumped habit (upper left), fertile pinna (upper right), portion of blade (bottom).

Dryopteris fragrans (L.) Schott
FRAGRANT WOOD FERN

FIELD TIPS
- small evergreen fern of dry rock cliffs and boulders
- blade glandular, scented, tapered gradually toward both ends
- curled, dead fronds at base of plant persist for several years

MIDWEST RANGE
Mich Upper Peninsula, Minn, Wisc.

HABITAT
Dry, sunny or partially shaded cliffs, boulders, and talus slopes, on either calcareous or non-calcareous rock.

DESCRIPTION
Rootstock Erect, short, with old stipe bases and old persistent fronds; covered with brown scales.

Frond Evergreen, clustered; persistent old gray or brown fronds surround base of plant; sterile and fertile fronds alike; usually less than 25 cm long and 5 cm wide.

Stipe Much shorter than blade, grooved, straw-colored, densely scaly; vascular bundles mostly 3-7 in a C-shaped pattern.

Rachis Glandular, with many reddish-brown scales.

Blade Narrowly elliptic, gradually tapering toward both ends; pinnate-pinnatifid to 2-pinnate; leathery, with yellow, fragrant glandular hairs; lower side with many brown to reddish scales.

Pinnae Mostly 20 to 30 pairs, opposite, sometimes overlapping, oblong, sessile or short-stalked, lowest pair much shorter than middle pinnae; margins crenately toothed but not bristle-tipped; costae grooved above, continuous from rachis to costae; veins obscure, free, mostly not reaching margin.

Sori Round, large, halfway between midvein and margin; **indusium** kidney-shaped, glandular, whitish, becoming brown, often overlapping; **sporangia** chocolate brown.

SYNONYMS
- *Aspidium fragrans* Sw.
- *Filix fragrans* Farwell
- *Lastrea fragrans* Presl
- *Nephrodium fragrans* Richards
- *Polypodium fragrans* L.
- *Polystichum fragrans* (L.) Roth
- *Thelypteris fragrans* Nieuwl.
- *Woodsia xanthosporangia* Ching

SIMILAR SPECIES
- May be mistaken for **rusty cliff fern** (*Woodsia ilvensis*), but the persistent, curled, dead fronds are characteristic of *Dryopteris fragrans.*

NORTH AMERICA

MIDWEST

NAME

Latin, *fragrans*, fragrant; fronds have a
sweet, fruity or hay-like odor when
crushed.

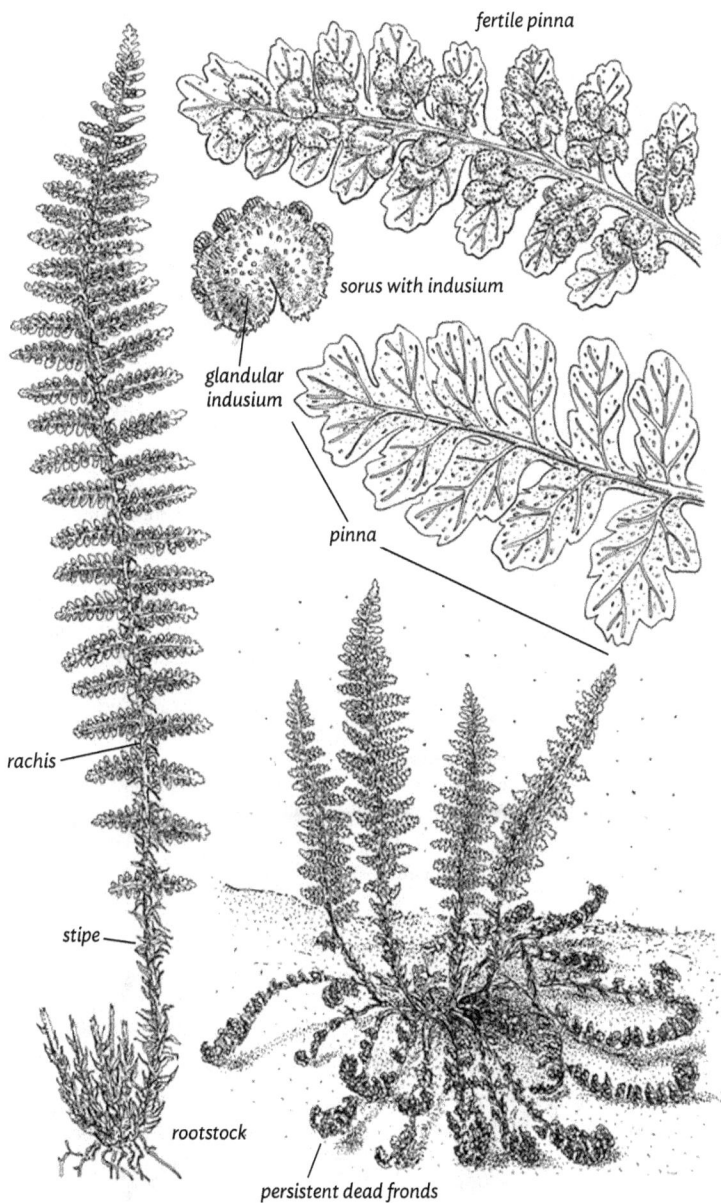

fertile pinna

sorus with indusium

*glandular
indusium*

pinna

rachis

stipe

rootstock

persistent dead fronds

Dryopteris goldiana (Hook. ex Goldie) Gray
GOLDIE'S WOOD FERN

FIELD TIPS
- region's largest wood fern
- fronds clustered, arched
- blades abruptly tapered at tip
- stipe with glossy scales, dark brown with paler edges
- young fronds usually with faint yellow-green highlights on outer edges

MIDWEST RANGE
Ill, Ind, Iowa, Mich, Minn, Ohio, Wisc.

HABITAT
Rich, moist, shaded woods, especially in ravines and near seeps and springs; swamp margins, rocky slopes.

DESCRIPTION
Rootstock Short-creeping to erect, stout, with old stipe bases, densely scaly.

Frond Deciduous, clustered; sterile and fertile fronds alike, to 120 cm long.

Stipe Shorter than blade, grooved, straw-colored, scaly; scales scattered, dark, glossy brown to nearly black, with pale border; vascular bundles mostly 3-7 in a C-shaped pattern.

Rachis Green with tan scales; lower portion not grooved, upper segment slightly grooved, with purplish trough.

Blade Ovate or broadly lance-shaped, tapered abruptly at tip; 1-pinnate -pinnatifid to 2-pinnate-pinnatifid; not glandular, underside scaly, scales absent above.

Pinnae 15 to 20 pairs, lance-shaped, on short stalks, lower pinnae mostly opposite, alternate upwards; lowest pair usually slightly shorter than middle pinnae; costae grooved above, continuous from rachis to costae; margins crenate or serrate, teeth tipped with small bristle; veins free, forked, most not reaching the margin.

Sori Round, in one row near the midvein; **indusium** kidney-shaped, white to transparent when immature, shriveling, attached at a sinus; **sporangia** lead gray, then dark brown or black.

SYNONYMS
- *Aspidium goldianum* Hooker ex Goldie
- *Thelypteris goldiana* (Hooker) Nieuwland

NORTH AMERICA

MIDWEST

- *Nephrodium goldianum* (Hooker) Hooker & Greville
- *Lastrea goldiana* Presl
- *Polystichum goldieanum* Keys.

SIMILAR SPECIES
- Blade of **small log fern** (*Dryopteris celsa*) tapers gradually to tip and is uniformly dark green color; blade in *D. goldiana* tapers abruptly to a short

pointed tip, and outer edges often somewhat yellow-green.

NAME
First collected by John Goldie, a Scottish botanist (1793-1886), at Montreal, Canada; this fern was named *Aspidium goldianum* by Hooker in 1822, and transferred to *Dryopteris* by Gray in 1848.

sorus with indusium

pinnule

rachis

scales

stipe

pinna

upper stipe section

rootstock

Dryopteris intermedia (Muhl. ex Willd.) Gray
EVERGREEN WOOD FERN

FIELD TIPS
- clumped, somewhat arching, evergreen fern
- first downward pointing pinnule on lowest pinna shorter than pinnule next to it
- our only *Dryopteris* with hairs on stipe and rachis (as well as scales)

MIDWEST RANGE
Ill, Ind, Iowa (threatened), Mich, Minn, Ohio, Wisc; common in northern portions of Midwest.

HABITAT
Moist to dry, deciduous, mixed conifer-hardwood, or coniferous forests, often on slopes and in ravines.

DESCRIPTION
Rootstock Nearly erect, with old stipe bases, densely scaly.

Frond Evergreen, clustered; sterile and fertile fronds alike; to 70 cm long.

Stipe Shorter than blade, grooved, green to straw-colored, with hairs and tan scales; vascular bundles mostly 3-7 in an arc.

Rachis With whitish glandular hairs and a few scales.

Blade Ovate in outline; 3-pinnate at base, gradually less above; glandular.

Pinnae 12 to 20 pairs; lance-shaped and more-or-less in plane of blade, opposite or nearly so, lowest pair not reduced in size; inner lower pinnule on the lowest pinna somewhat shorter than the next one; costae grooved above, continuous from rachis to costae; margins toothed, bristle-tipped; veins free, forked, mostly not reaching margin.

Sori Round, in 1 row between midrib and margin; **indusium** kidney-shaped, at a sinus, with small glandular hairs; **sporangia** brown.

SYNONYMS
- *Dryopteris austriaca* (Jacq.) Woynar ex Schinz & Thell. var. *intermedia* (Muhl. ex Willd.) Morton
- *Dryopteris spinulosa* (O.F. Müll.) Watt var. *intermedia* (Muhl. ex Willd.) Underw.

SIMILAR SPECIES
- **Spinulose wood fern** *(Dryopteris carthusiana)* often confused with *D. intermedia*, but in most plants, lower basal pinnule of the lowest pinna is longer than the adjacent pinnules; in *D. intermedia* lower basal pinnule is shorter than the adjacent pinnules. Fronds of *D. intermedia* evergreen and

NORTH AMERICA

MIDWEST

those of *D. carthusiana* are not, helpful especially in winter and spring. Glands on indusia and midveins of pinnules of *D. intermedia* are present, but can be difficult to see, requiring a hand lens, or are shed with age; no glands on *D. carthusiana*.

NOTES

Dryopteris intermedia known to hybridize with 8 species; hybrids all have distinctive glandular hairs on the indusia and usually also on costae.

NAME

Discovered by Muhlenberg in Lancaster Co., Pennsylvania. His manuscript name for it, *Aspidium intermedium*, was published by Willdenow in 1810, and transferred to *Dryopteris* by Gray in 1848.

From Latin *intermedius*, presumably intermediate between *D. carthusiana* and *D. campyloptera* (found in New England).

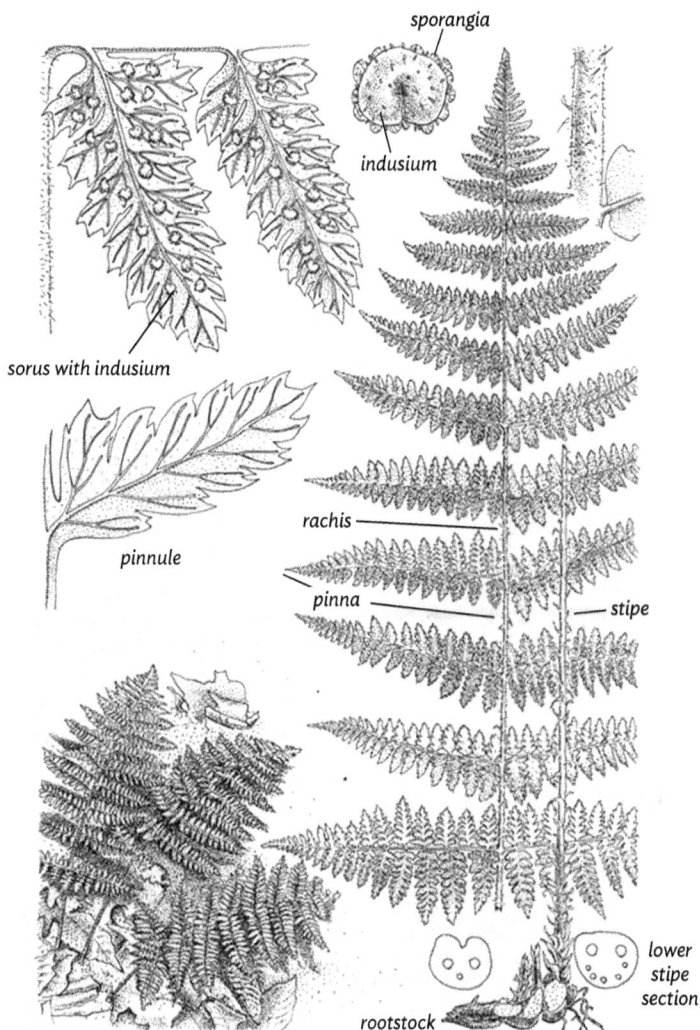

sporangia

indusium

sorus with indusium

pinnule

rachis

pinna

stipe

lower stipe section

rootstock

Dryopteris marginalis (L.) Gray
MARGINAL WOOD FERN

FIELD TIPS
- large evergreen fern, usually where rocky
- fronds leathery, sometimes blue-green; dead fronds persist at base of plant
- pinule margins smooth or round-toothed
- stipes very scaly
- sori very near margins, in upper portion of blade

MIDWEST RANGE
Ill, Ind, Iowa (threatened), Mich, se Minn (threatened), Ohio, Wisc.

HABITAT
Rocky wooded slopes and ravines, streambanks, usually in partial shade; soils alkaline to acidic.

DESCRIPTION
Rootstock Upright, thick, with old stipe bases; densely covered with gold-brown chaffy scales.

Crozier Densely covered with golden brown hairs.

Frond Evergreen, clustered; sterile and fertile fronds alike; to 100 cm long or more.

Stipe Shorter than blade, grooved, reddish brown at swollen base, straw-colored above, densely covered with tan scales; vascular bundles mostly 3- 7 in a C-shaped arc, sometimes fewer at stipe apex.

Rachis Pale to light green, narrowly grooved; scaly, especially toward base of pinnae.

Blade Ovate lance-shaped; 2-pinnate; leathery, deep green to felty blue-green; underside scaly, upper side without scales.

Pinnae 12 to 16 pairs, lance-shaped, short-stalked, lowest pair usually a little shorter than middle pinnae; basal pinnules longer than adjacent pinnules, lower pinnule longer than adjacent upper pinnule; costae grooved above, continuous from rachis to costae; margins shallowly round-toothed to nearly entire; veins free, forked, mostly not reaching margin.

Sori Round, in 1 row near the margin on the top two-thirds of the blade; **indusium** kidney-shaped, silvery, attached at a sinus; **sporangia** lead gray, maturing to dark brown.

SIMILAR SPECIES
- In **male fern** (*Dryopteris filix-mas*), sori not marginal, usually more than 20 pairs of pinnae, with lowest pair much shorter than middle pinnae.

NORTH AMERICA

MIDWEST

NAME

Sent to Linnaeus from Canada and named by him *Polypodium marginale* in 1753; assigned to *Dryopteris* by Gray in 1848.

From Latin, *marginatus*, enclosed with a margin, referring to position of sori.

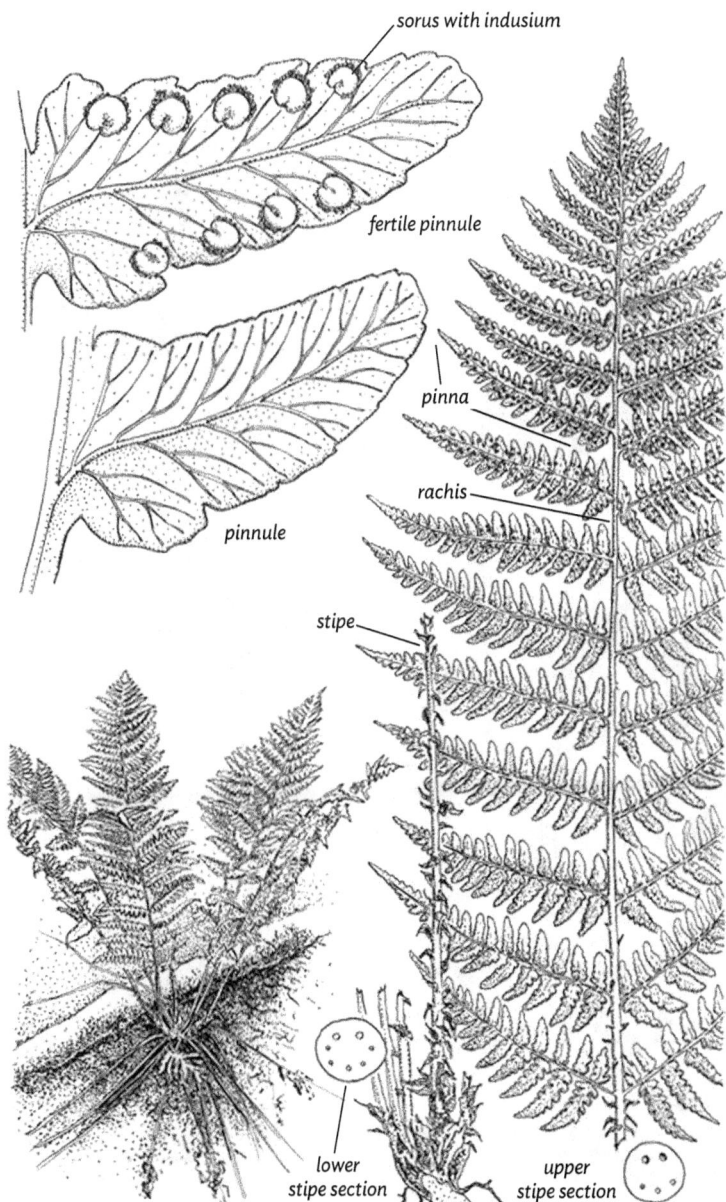

sorus with indusium

fertile pinnule

pinnule

pinna

rachis

stipe

lower stipe section

upper stipe section

Spinulose wood fern (*Dryopteris carthusiana*), underside of fertile pinnule.

Spinulose wood fern (*Dryopteris carthusiana*), stipe with scales.

Log fern (*Dryopteris celsa*).

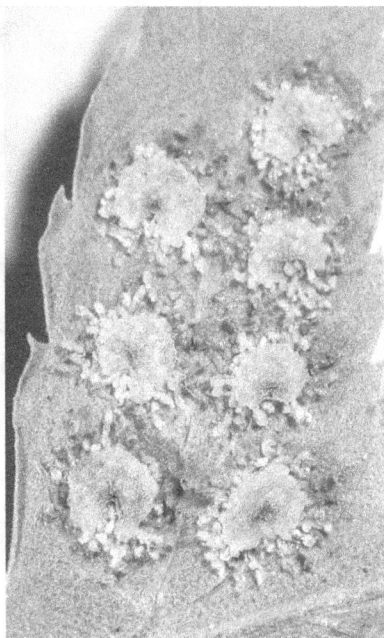

Clinton's wood fern (*Dryopteris clintoniana*).

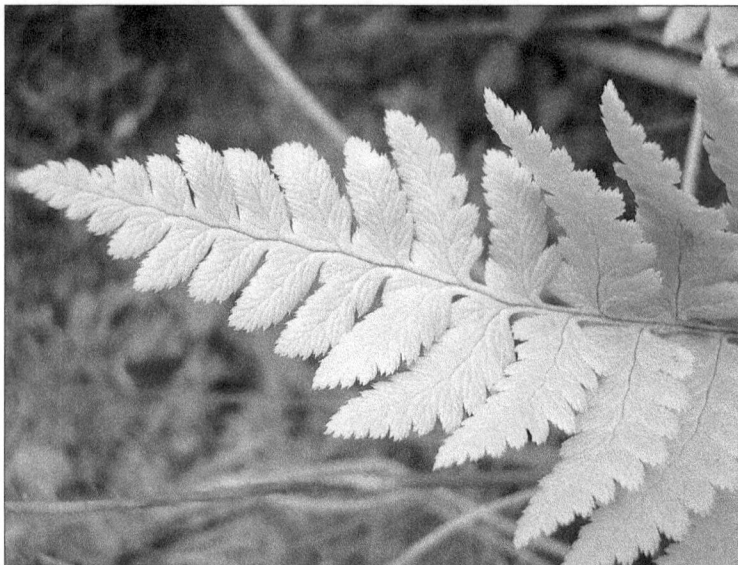

Crested wood fern (*Dryopteris cristata*).

Crested wood fern (*Dryopteris cristata*).

Spreading wood fern (*Dryopteris expansa*).

Male fern (*Dryopteris filix-mas*).

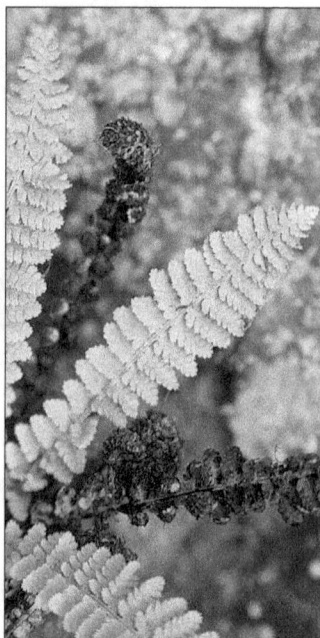

Fragrant wood fern (*Dryopteris fragrans*), note the persistent, curled, dead fronds.

Goldie's wood fern (*Dryopteris goldiana*).

Evergreen wood fern (*Dryopteris intermedia*).

Marginal wood fern (*Dryopteris marginalis*).

Polystichum

HOLLY FERN, SWORD FERN

Leathery, evergreen ferns that are generally monomorphic (except in *P. acrostichoides* with smaller fertile pinnae). Habitats vary from forests to rock cliffs and talus. *Polystichum* includes about 180 species occurring worldwide, 15 in North America, 3 species in our flora.

KEY CHARACTERS

- Similar to *Dryopteris,* but distinguished by the upward ear on the pinnae, commonly spiny margins, the peltate (round) indusia, and discontinuous grooving between rachis and costa. Also similar to *Cyrtomium,* but that genus has netted veins (and in the Midwest only known from s Indiana); veins in *Polystichum* are free and not net-like.

- **Fronds** evergreen, sterile and fertile fronds alike except in *P. acrostichoides;* **stipes** scaly; **leaf blades** 1-3 pinnate, lance-shaped, glossy; **pinnae** usually with an upper side lobe (auricle), costae grooved above, grooves discontinuous from rachis to costae.

- **Sori** round, in 1 (2) rows between midrib and margin, **indusium** peltate or rarely absent.

NAME

From Greek, *polys*, many, and *stichos*, row, referring to the several parallel rows of sori.

Northern holly fern (*Polystichum lonchitis*), crowded sori on pinna underside.

KEY TO POLYSTICHUM | HOLLY FERN, SWORD FERN

1   Pinnae of two types, terminal fertile pinnae distinctly narrower and smaller than sterile pinnae; leaf blade evergreen; Ill, Ind, Iowa, Mich, se Minn, Ohio, Wisc   **1. Polystichum acrostichoides**
CHRISTMAS FERN, page 170

1   Pinnae uniform; fertile and sterile pinnae similar in size and shape **2**

2   Leaf blade tardily deciduous, 2-pinnate, the pinnae spaced; Mich, ne Minn, n Wisc   **2. Polystichum braunii**
BRAUN'S HOLLY FERN, page 172

2   Leaf blade evergreen, 1-pinnate, the pinnae close-set; lowermost pair of pinnae triangular, much smaller than pinnae of middle leaf blade; n Mich, n Wisc   **3. Polystichum lonchitis**
NORTHERN HOLLY FERN, page 174

Polystichum acrostichoides
CHRISTMAS FERN

Polystichum braunii
BRAUN'S HOLLY FERN

Polystichum lonchitis
NORTHERN HOLLY FERN

Polystichum acrostichoides (Michx.) Schott
CHRISTMAS FERN

FIELD TIPS
- clumped evergreen fern
- fronds dark green, satiny
- fertile pinnae reduced in size, on upper portion of blade
- pinnae with upward-pointing lobe
- stipe scaly

MIDWEST RANGE
Ill, Ind, e Iowa, Mich, se Minn (threatened), Ohio, Wisc; common in southern Midwest region.

HABITAT
Shaded, moist to dry forests, rocky slopes and ravines; soils acidic to neutral.

DESCRIPTION
Rootstock Ascending or creeping, with old stipe bases and wilted fronds attached, pale brown, densely scaly.

Crozier Bent backwards, 1 cm wide, covered with silvery-white scales.

Frond Evergreen, clustered; sterile and fertile fronds somewhat different, those with fertile pinnae taller and more rigid than the completely sterile fronds, the fertile pinnae reduced in size and only in upper portion of blade; fronds to 80 cm long or more.

Stipe Grooved, green, densely scaly; scales light brown, smaller in size upwards; vascular bundles 5 in an arc.

Rachis Green, grooved, very scaly.

Blade Linear lance-shaped, base narrowed, 1-pinnate, glossy; underside with numerous small, fine fine scales.

Pinnae 20 to 30 pairs, lower pinnae bend forward and down; upper auricle well developed; margins entire to toothed and spiny; veins free, forked.

Sori Distinct, green at first, confluent later, completely covering lower surface of fertile pinnae (on the upper, reduced pinnae only); **indusium** round, attached at center, shriveling with age; **sporangia** golden brown.

SIMILAR SPECIES
- Fronds of **glade fern** (*Diplazium pycnocarpon*) also 1-pinnate, but that species deciduous and pinnae without upward-pointing lobes

NOTES
Used by early New England settlers as a Christmas decoration.

NAME
Michaux observed this fern during his explorations in "Pennsylvania, Carolina,

NORTH AMERICA

MIDWEST

and Tennessee," and named it *Nephrodium acrostichoides* in 1803. It was transferred to *Polystichum* by Schott in 1834.

Acrostichoides means resembling *Acrostichum*, a tropical genus, where the sori completely cover the pinna underside.

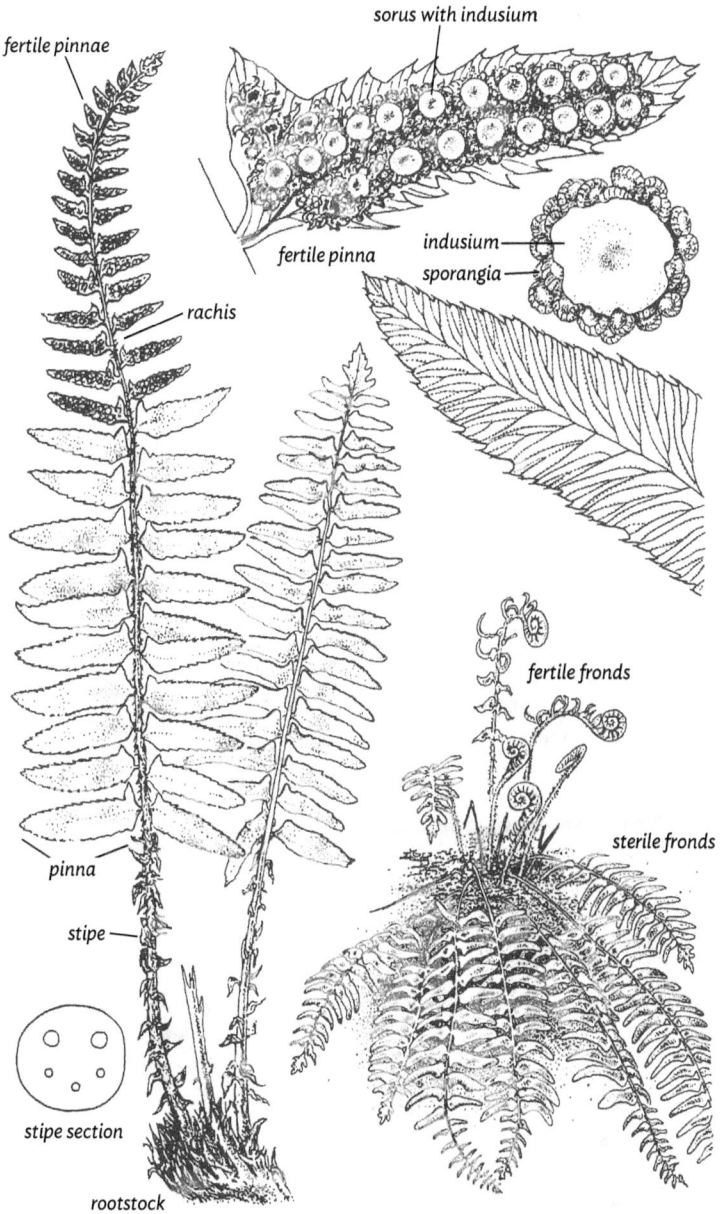

sorus with indusium

fertile pinnae

fertile pinna

indusium

sporangia

rachis

fertile fronds

sterile fronds

pinna

stipe

stipe section

rootstock

Polystichum braunii (Spenner) Fée
BRAUN'S HOLLY FERN

FIELD TIPS

• clumped, arched, semi-evergreen fern
• fronds dark green, shiny, tapered to tip and base, lower pinnae very small
• stipe and rachis densely scaly
• pinnule margins with bristle-tipped teeth
• in rocky woods in mostly northern portions of region

MIDWEST RANGE

Mich, ne Minn (endangered), n Wisc (threatened).

HABITAT

Cool, moist, shaded woods; especially where rocky; moist cliffs and boulders; soils circumneutral.

DESCRIPTION

Rootstock Erect, stout, with old stipe bases, very scaly; fronds arise in vase-like cluster.

Crozier For next year's fronds formed in late summer, these covered with silvery to light brown scales.

Frond Evergreen (or nearly so), dark green, clustered; sterile and fertile fronds alike, to 100 cm long.

Stipe Grooved; very scaly, the scales large and also hairlike, silvery at first, then light brown; vascular bundles 5-7 in an arc.

Rachis Light brown, grooved, densely scaly.

Blade Ovate, broadest at middle, tapered gradually to base, lowest pinnae only 2 cm long; upper pinnae lobed, lower pinnae divided to costa to form 6-18 pairs of pinnules; dark green, leathery; with whitish to brown hairlike scales on both sides.

Pinnae 20 to 40 pairs, oblong; pinnules short-stalked, eared near the rachis, the first upper pinnule the same size or only slightly larger than the next; margins toothed, bristle-tipped; veins free, forked.

Sori Round, distinct and separate, in 1 row between midrib and margin; **indusium** round, attached at center, light brown, shriveling with age; **sporangia** gray-brown to black.

NOTES

One of our most attractive ferns.

NAME

Discovered in Europe, and named *Aspidium braunii* by Spenner in 1825, then transferred to genus *Polystichum* by Fée around 1850.

Alexander Carl Heinrich Braun, 1805-1877, German professor of botany.

NORTH AMERICA

MIDWEST

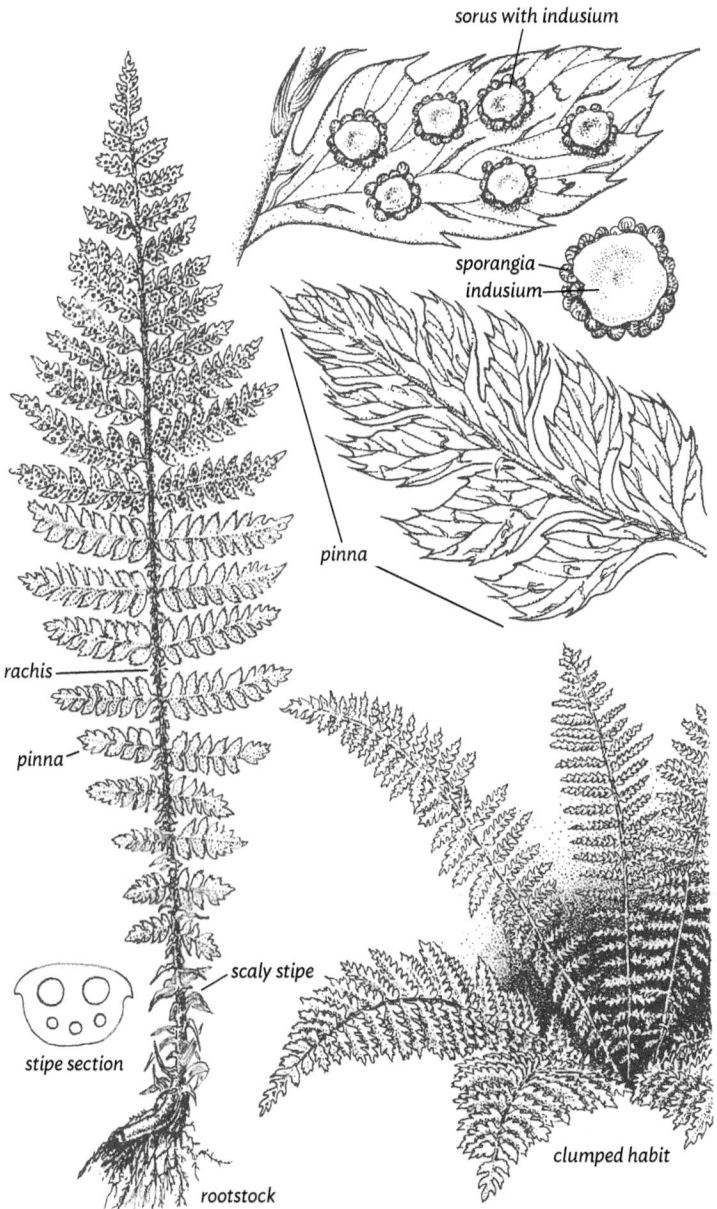

sorus with indusium

sporangia
indusium

pinna

rachis

pinna

scaly stipe

stipe section

rootstock

clumped habit

pinna

Polystichum lonchitis (L.) Roth
NORTHERN HOLLY FERN

FIELD TIPS
- clumped evergreen fern
- blade narrow, often widest above middle
- pinnae with upward-pointing lobe
- northern portions of region, often where calcareous

MIDWEST RANGE
n Mich, n Wisc.

HABITAT
Northern hardwood and mixed conifer-hardwood forests, calcareous rock outcrops and boulders.

DESCRIPTION

Rootstock Erect; scales pale brown, ovate to lance-shaped.

Frond Evergreen, persisting through the following summer, clustered; sterile and fertile fronds alike; to 40 cm long.

Stipe Grooved, green or straw colored with age, scales light brown, gradually diminishing in size upward; vascular bundles 5, in an arc.

Blade Linear, often widest above middle, base narrowed, drooping at tip; 1-pinnate; glossy; underside with many fine, small scales.

Pinnae 25 to 35 pairs, oblong to lance-shaped, bending towards the apex, lowest pinnae ± triangular, rarely overlapping; upper ear well developed; margins finely sharp-toothed, the teeth spreading; veins free, forked.

Sori Round, between costae and margins, only on upper half of blade; **indusium** peltate, grayish white; **sporangia** dark brown or black; **spores** spiny (magnification needed) and distinguish northern holly fern from dwarfed forms of other 1-pinnate species.

SYNONYMS
- *Polystichum mohrioides* (Bory) C. Presl var. *lemmonii* (Underw.) Fernald

SIMILAR SPECIES
- **Christmas fern** (*Polystichum acrostichoides*) similar but in *P. lonchitis* sterile and fertile fronds alike, and lowest pinnae small and triangular, not lance-shaped.

NAME
From Greek, *logch* or *loncha*, meaning spear, referring to spear-shaped fronds or pinnae.

NORTH AMERICA

MIDWEST

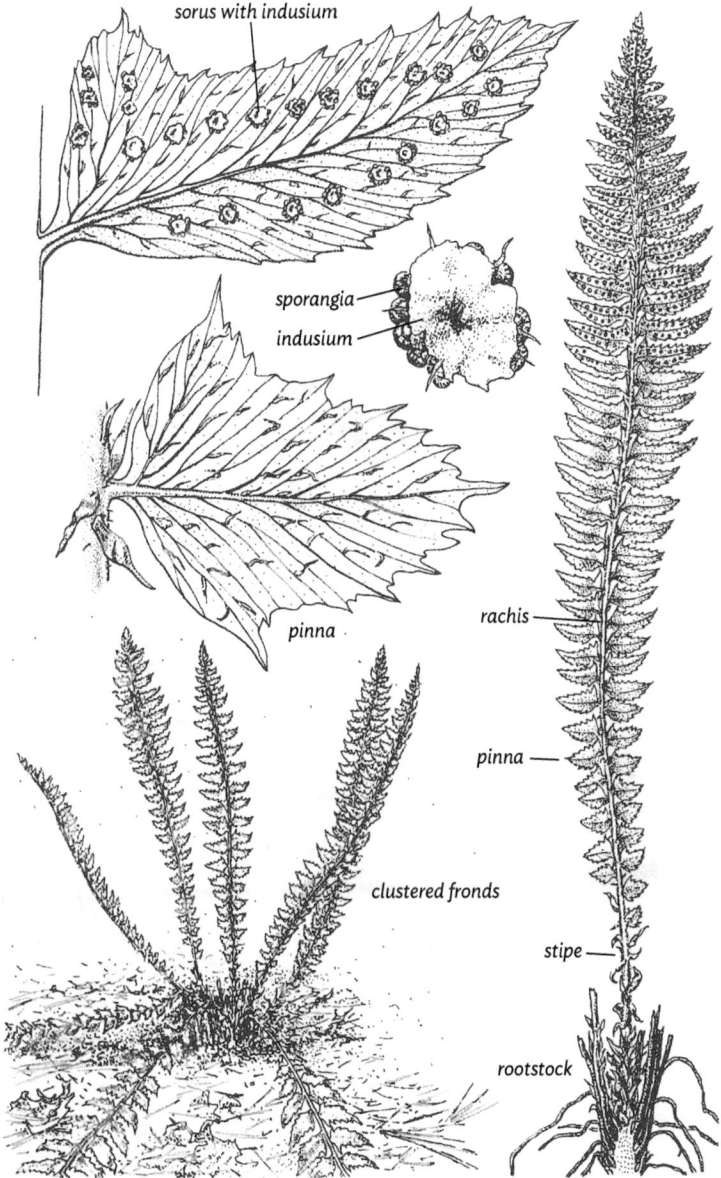

sorus with indusium

sporangia

indusium

pinna

rachis

pinna

clustered fronds

stipe

rootstock

Christmas fern (*Polystichum acrostichoides*), underside of fertile pinnae.

Braun's holly fern (*Polystichum braunii*), typical habit forming a circular clump.

Braun's holly fern (*Polystichum braunii*), underside of fertile pinnae.

Northern holly fern (*Polystichum lonchitis*), note eared pinnae and spine-tipped margins.

EQUISETACEAE | Horsetail Family
Equisetum
HORSETAIL, SCOURING-RUSH

Evergreen or deciduous perennials from creeping rhizomes. Modern classifications describe 1-3 genera within the family and 28 species of *Equisetum*, with 11 species in North America, and 9 species in our flora; hybrids also occur.

KEY CHARACTERS

- **Stems** rush-like, jointed, sometimes hollow; branched or unbranched; internodes of stems commonly ridged longitudinally, with stomata in rows or bands in the grooves, and the ridges with siliceous bumps or bands, giving stems of some species (the scouring rushes) a gritty feel.

- **Leaves** small, whorled, fused into a collarlike sheath at each node; teeth on the sheath represent tips of the leaves; when present, the branches are slender extensions of the sheaths.

- **Spores** green, spherical, wrapped with 4 elaters, and borne in sporangia on sporophylls in cones. The hexagonal plates on the cone separate when the spores ready to be shed; spores green and viable for only a few days.

- **Cones** terminal on vegetative stems, or in some species on specialized early season shoots that lack chlorophyll.

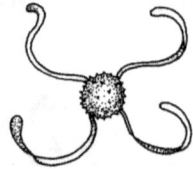

Elaters around spore of *Equisetum arvense*

NAME

The common name **scouring rush** was given to unbranched species of *Equisetum* by American pioneers who used the silica-roughened stems to scrub pots and pans. Before commercial scouring powders were developed for cleaning pots and pans, pulverized stems of these species were sometimes used. The common name **horsetail** was given to species with whorled branches, in reference to their bushy, bristly appearance resembling a horse's tail. The genus name from Greek, *equis*, horse, and *seta*, bristle.

EQUISETUM HYBRIDS

- *Equisetum* × *ferrissii* Clute: *Equisetum hyemale* and *E. laevigatum*, common in much of Midwest region.

- *Equisetum* × *nelsonii* (A.A. Eat) Schaffn.: *E. laevigatum* and *E. variegatum*, mostly near Lake Michigan in Wisc and in lower Mich.

- *Equisetum* × *litorale* Kuehl ex Rupr.: *Equisetum arvense* and *E. fluviatile*, known from sw Wisc, se Minn, n Ill and e Iowa.

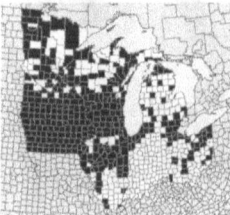

Equisetum xferrissii
E. hyemale x E. laevigatum

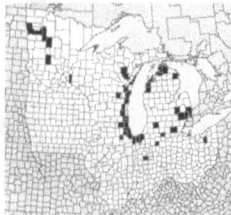

Equisetum xnelsonii
E. laevigatum x E. variegatum

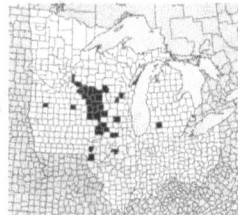

Equisetum xlitorale
E. arvense x E. fluviatile

KEY TO EQUISETUM | HORSETAIL, SCOURING-RUSH

1   Stems evergreen (annual in *E. laevigatum*); unbranched or with a few scattered branches, branches not in regular whorls (**scouring rushes**). **2**

1   Stems deciduous; usually with regular whorls of branches, sometimes unbranched (**horsetails**) **5**

2   Stems solid (central cavity absent); stems small, slender and sprawling; n Ill, ne Iowa, Mich, Minn, Wisc
1. Equisetum scirpoides
DWARF SCOURING-RUSH, page 198

2   Stems hollow (central cavity present); stems larger, usually upright **3**

3   Stems 1–3 dm tall, with 5–12 ridges, central cavity to 1/3 diameter of stem; Ill, n Ind, Mich, Minn, n Ohio, Wisc
2. Equisetum variegatum
VARIEGATED SCOURING-RUSH, page 202

3   Stems usually taller, with 16–50 ridges, central cavity more than half diameter of stem. **4**

4   Cones tipped with a distinct, small sharp point; stem sheaths with a black band at tip and base; regionwide   **3. Equisetum hyemale**
COMMON SCOURING-RUSH, page 190

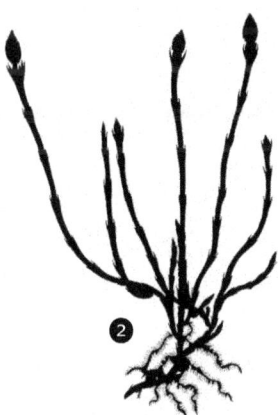

4   Cones blunt-tipped, sheaths with black band at tip only; regionwide   **4. Equisetum laevigatum**
SMOOTH SCOURING-RUSH, page 192

5   Stems unbranched **6**

5   Stems with regular whorls of branches **10**

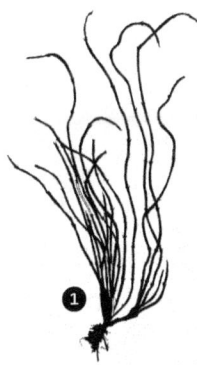

Equisetum scirpoides
DWARF SCOURING-
RUSH

Equisetum hyemale
COMMON SCOURING-RUSH

Equisetum variegatum
VARIEGATED SCOURING-RUSH

Equisetum laevigatum
SMOOTH SCOURING-RUSH

KEY TO EQUISETUM, CONTINUED

6 Stems green **7**

6 Stems brown or flesh-colored (fertile stems) **8**

7 Stems with 9-25 shallow ridges; central cavity more than half diameter of stem; sheath teeth entirely black or with narrow white margins; regionwide **5. Equisetum fluviatile**
WATER HORSETAIL, page 188

7 Stems with 5-10 strongly angled ridges; central cavity less than 1/3 diameter of stem; sheath teeth with white margins and dark centers; Ill, Mich, Minn, Wisc **6. Equisetum palustre**
MARSH HORSETAIL, page 194

8 Sheath teeth papery and red-brown, teeth joined and forming several broad lobes; n Ill, n Ind, Iowa, Mich, Minn, ne Ohio, Wisc **7. Equisetum sylvaticum**
WOODLAND-HORSETAIL, page 200

8 Sheath teeth black or brown, not papery, separate or joined in more than 4 small groups **9**

9 Fertile stems withering after spores mature, remaining unbranched; common regionwide **8. Equisetum arvense**
FIELD HORSETAIL, page 186

9 Fertile stems persistent, becoming branched and green; n Ill, Iowa, Mich, Minn, Wisc **9. Equisetum pratense**
MEADOW HORSETAIL, page 196

10 First internode of each branch shorter than the subtending sheath of the main stem. **11**

10 First internode of each branch equal or longer than the subtending sheath of the main stem **12**

11 Stems with 9–25 shallow ridges; central cavity more than half diameter of stem; sheath teeth more than 12, entirely black or with narrow white margins; regionwide **5. Equisetum fluviatile**
WATER HORSETAIL, page 188

11 Stems with 5–10 strongly angled ridges; central cavity about same size as outer cavities; sheath teeth 5–6, with white margins and dark centers; Ill, Mich, Minn, Wisc **6. Equisetum palustre**
MARSH HORSETAIL, page 194

12 Stem branches themselves branched; sheath teeth papery and red-brown, teeth joined to form several broad lobes; n Ill, n Ind, Iowa, Mich, Minn, ne Ohio, Wisc **7. Equisetum sylvaticum**
WOODLAND HORSETAIL, page 200

12 Stem branches unbranched; sheath teeth black or brown, not papery, separate or joined in more than 4 small groups **13**

13 Stem branches ascending; teeth of branch sheaths gradually tapering to a slender tip; common regionwide
8. Equisetum arvense
FIELD HORSETAIL, page 186

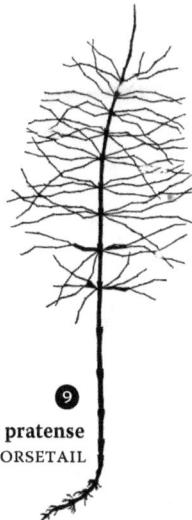

13 Stem branches spreading; teeth of branch sheaths broadly triangular; n Ill, Iowa, Mich, Minn, Wisc **9. Equisetum pratense**
MEADOW HORSETAIL, page 196

Equisetum palustre
MARSH HORSETAIL

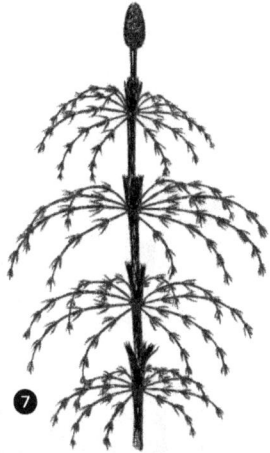

Equisetum sylvaticum
WOODLAND-HORSETAIL

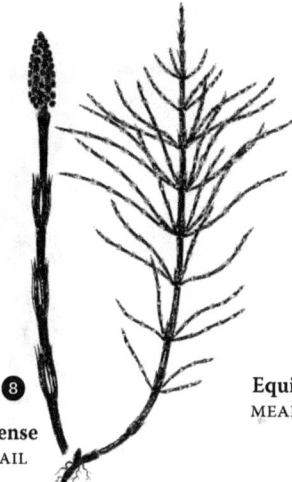

Equisetum fluviatile
WATER HORSETAIL

Equisetum arvense
FIELD HORSETAIL

Equisetum pratense
MEADOW HORSETAIL

Equisetum arvense
FIELD HORSETAIL

Equisetum fluviatile
WATER HORSETAIL

Equisetum hyemale
TALL SCOURING-RUSH

Equisetum laevigatum
SMOOTH SCOURING-RUSH

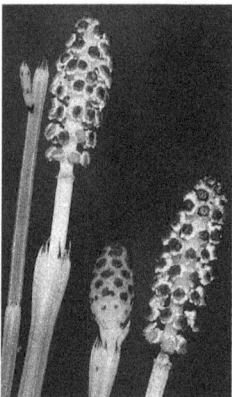

Equisetum palustre
MARSH HORSETAIL

Equisetum pratense
MEADOW HORSETAIL

Equisetum scirpoides
DWARF SCOURING-RUSH

Equisetum sylvaticum
WOODLAND HORSETAIL

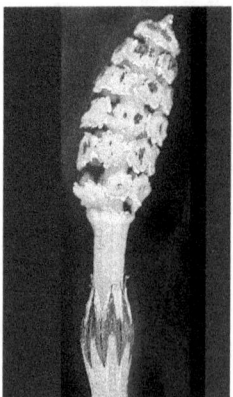

Equisetum variegatum
VARIEGATED
SCOURING-RUSH

Equisetum arvense
FIELD HORSETAIL

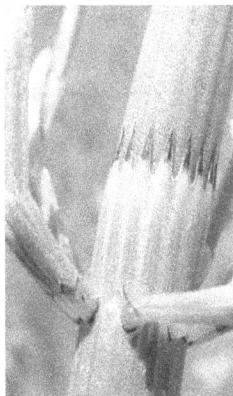

Equisetum fluviatile
WATER HORSETAIL

Equisetum hyemale
TALL SCOURING-RUSH

Equisetum laevigatum
SMOOTH SCOURING-RUSH

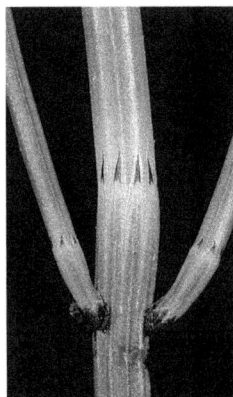

Equisetum palustre
MARSH HORSETAIL

Equisetum pratense
MEADOW HORSETAIL

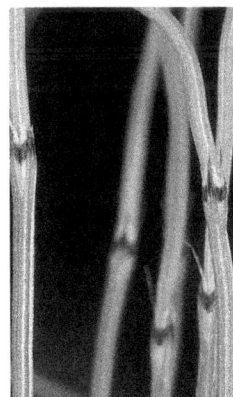

Equisetum scirpoides
DWARF SCOURING-RUSH

Equisetum sylvaticum
WOODLAND HORSETAIL

Equisetum variegatum
VARIEGATED
SCOURING-RUSH

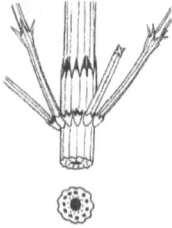

Equisetum arvense
FIELD HORSETAIL

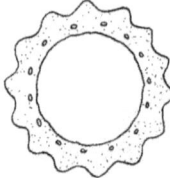

Equisetum fluviatile
WATER HORSETAIL

Equisetum hyemale
TALL SCOURING-RUSH

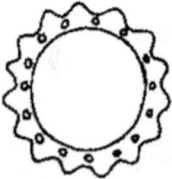

Equisetum laevigatum
SMOOTH SCOURING-RUSH

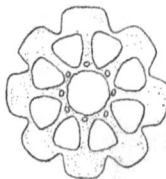

Equisetum palustre
MARSH HORSETAIL

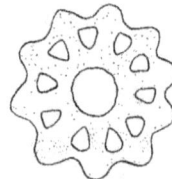

Equisetum pratense
MEADOW HORSETAIL

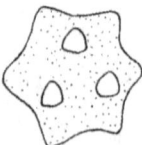

Equisetum scirpoides
DWARF SCOURING-RUSH

Equisetum sylvaticum
WOODLAND HORSETAIL

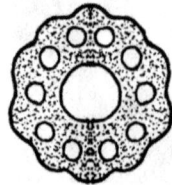

Equisetum variegatum
VARIEGATED
SCOURING-RUSH

Field horsetail (*Equisetum arvense*), fertile stems.

Equisetum arvense L.

FIELD HORSETAIL

FIELD TIPS

- very common, colony-forming
- stems deciduous
- sterile stems with dense whorls of branches from the nodes
- fertile stems smaller, nongreen, and appear in spring before sterile stems, then withering soon after shedding spores
- cones rounded at tip

MIDWEST RANGE

Regionwide; very common across North America.

HABITAT

Variety of sunny or shaded habitats (apart from where extremely wet or dry), including fields, woods, shores; often in somewhat disturbed places such as ditches, roadsides, railroad banks; sometimes weedy in yards.

DESCRIPTION

Rootstock Dark brown to black, hairy, sometimes bearing tubers; colony-forming.

Stems Deciduous, of two kinds, sterile and fertile, the sterile stems appearing as the fertile stems wither.

Sterile stems Upright to prostrate, 10-60 cm tall, 3-5 mm thick; central cavity usually less than 1/2 stem diameter (to sometimes slightly more than 1/2), with a ring of smaller cavities alternating with the 4-14 ridges; silica in dots on the ridges; sheaths with 4-14 short, narrow, dark, chaffy-margined teeth, these separate or occasionally joined; branches solid, in regular whorls, spreading or ascending, the branches themselves mostly unbranched, 3- or 4-angled in cross-section; the first internode of branch longer than the corresponding stem sheath.

Fertile stems Without chlorophyll, appearing in spring before sterile stems (and conspicuous because of their numbers along some roadsides), about 10-30 cm tall and generally shorter than the sterile stems, pale brown or pinkish, fleshy, unbranched, with 4-6 pale brown sheaths, these tipped with 6-12 darker teeth; dying after the spores are shed in spring

Cones Long-peduncled, rounded on top and not tipped with a sharp point.

SYNONYMS

- *Equisetum calderi* B. Boivin

NORTH AMERICA

MIDWEST

SIMILAR SPECIES

• Sterile stems of *Equisetum arvense* are perhaps most frequently confused with *E. pratense* and *E. palustre*:

• **Meadow horsetail** (*Equisetum pratense*) is more delicate in aspect, and the stems are whitish green; also, the teeth are traingular rather than narrowly lance-shaped as in *E. arvense*.

• In *E. arvense* the first internode of the branch is longer than the subtending teeth, whereas in **marsh horsetail** (*E. palustre*) the first internode is shorter than the subtending teeth.

NAME

Named by Linnaeus in 1753, with reference to occurrences in Europe and Virginia.

From Latin, *arvum*, field or cultivated land.

ABOVE Sterile stem with sheaths and whorls of branches.

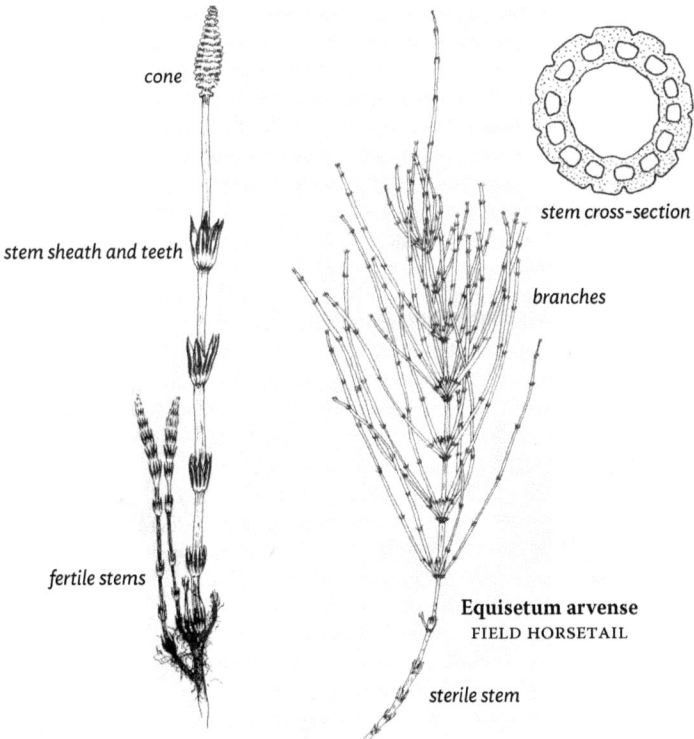

cone

stem sheath and teeth

stem cross-section

branches

fertile stems

Equisetum arvense
FIELD HORSETAIL

sterile stem

Equisetum fluviatile L.
WATER HORSETAIL

FIELD TIPS

- colony-forming, shallow water and muddy shores
- stems deciduous, unbranched or with small branches
- large central cavity, stems easily crushed when squeezed
- cones rounded at tip

MIDWEST RANGE

Regionwide.

HABITAT

In quiet, shallow water to about 60 cm deep of ponds, lakes and rivers, wet shores, ditches, bog margins, typically in full sun.

DESCRIPTION

Rootstock Smooth, shiny, light brown to reddish, branching; the stems single but often forming dense colonies from the spreading rhizomes.

Stems Deciduous, 30-150 cm tall, 3-8 mm thick; central cavity more than 4/5 diameter of stem; smaller outer cavities small or absent; 10-30 smooth ridges present; sheaths tightly appressed, with 15-20 teeth; teeth dark brown, narrow, acuminate, persistent. Stems unbranched, or with branches occurring sporadically, or sometimes whorled; branches to 15

cm long, hollow, with 4-6 ridges, the first internode shorter than the stem sheath.

Cones To 2.5 cm long, yellow to brown, rounded at top, stalked, deciduous, shedding spores late spring through the summer.

SYNONYMS

- *Equisetum limosum* L.

NOTES

A hybrid with *Equisetum arvense* (*E.* × *litorale* Kuehl ex Rupr.) is frequent in places where both species occur.

NAME

This plant in Europe, as in America, occurs in both a branchless form and one with conspicuous whorls of branches; not recognizing their identity, Linnaeus in 1753 applied the name *limosum* to the first and *fluviatile* to the second. The latter was validated by Ehrhart in 1792.

From Latin, *fluviatile*, pertaining to a river.

NORTH AMERICA

MIDWEST

ABOVE Fertile stems.

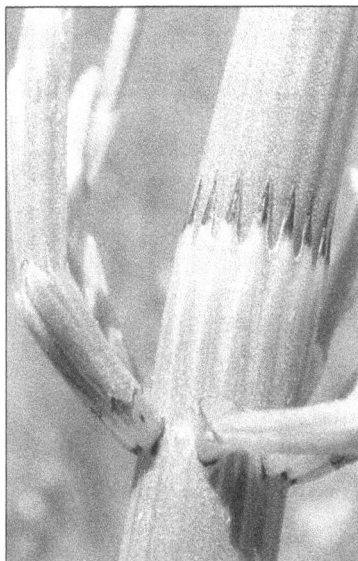

ABOVE Stem with small branches and toothed sheath.

ABOVE Stems in typical shallow water habitat.

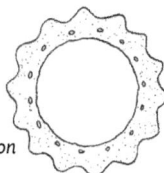

stem cross-section

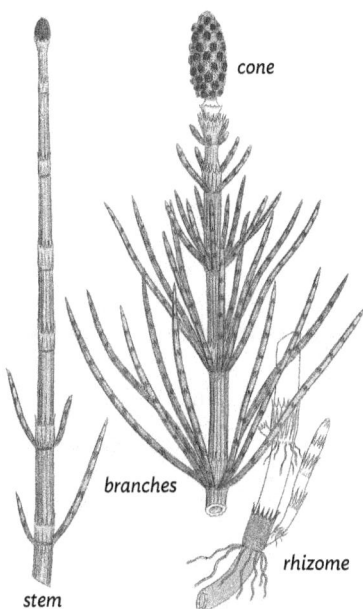

cone

branches

rhizome

stem

Equisetum fluviatile
WATER HORSETAIL

Equisetum hyemale L.
TALL SCOURING-RUSH

FIELD TIPS
- common, colony-forming, sandy places
- stems evergreen and rough, unbranched
- sheaths with wide, ash-gray band bordered by a lower black band and an upper black rim
- cones tipped by a small point

MIDWEST RANGE
Regionwide, common.

HABITAT
Sandy meadows and old fields, shores, dunes, roadsides, ditches, railroad embankments; sites wet to dry and in sun or shade.

DESCRIPTION
Rootstock Thick, blackish or dark brown, dull, rough, the stems arising singly or several together from the deep rhizome, and sometimes forming large colonies.

Stems Evergreen, 30-120 cm long, 4-6 mm thick, dark green, usually unbranched; central cavity more than 2/3 diameter of stem; small outer cavities alternating with 14-50 ridges; ridges broad, flat, or rounded, with prominent cross-bands; sheaths constricted at base, same color as the stem when young, but soon developing dark bands at base and top, with area between base and summit white or ashy gray; teeth lance-shaped, dark brown with broad chaffy margins, usually soon deciduous, leaving only darker traces.

Cones To 2 cm long when mature, yellow to black, short-stalked, tipped by a small point, shedding spores from summer to early fall, or persisting unopened until the following spring.

SYNONYMS
- *Equisetum affine* Engelm.
- *Equisetum praealtum* Raf.
- *Equisetum robustum* A. Braun
- *Hippochaete hyemalis* (L.) Bruhin

NOTES
The rough stems of tall scouring rush have been used to scour or clean pots, and as a sandpaper.

NAME
Initially considered distinct from the European scouring-rush (*Equisetum hyemale*) and named *E. prealtum* by Rafinesque in 1817, with reference to colonies along the Mississippi River in Louisiana.

From Latin, *hiemis*, winter; referring to the evergreen stems.

NORTH AMERICA

MIDWEST

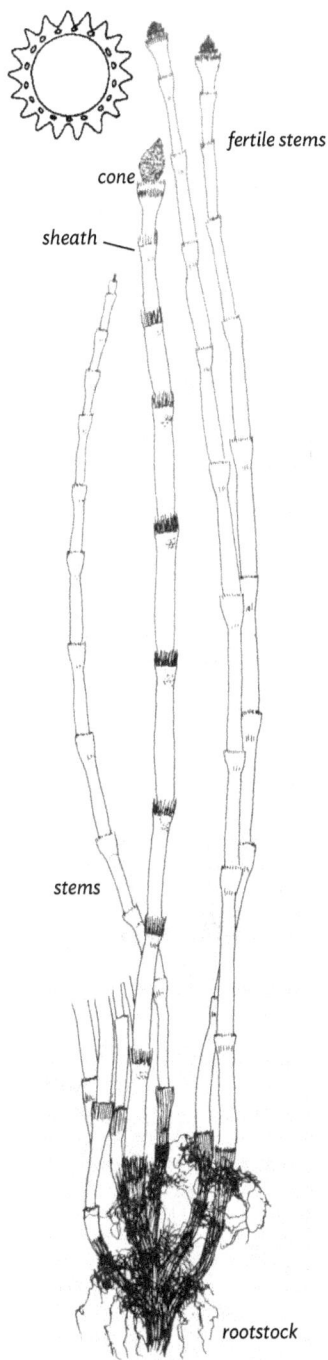

fertile stems

cone

sheath

stems

rootstock

ABOVE Stems tipped by sharp-tipped cones.

ABOVE Stem and sheath tipped by deciduous teeth.

Equisetum laevigatum A. Braun
SMOOTH SCOURING-RUSH

FIELD TIPS

- common, colony-forming; open, sandy places
- stems smooth, deciduous, unbranched
- sheaths green with a black rim; sheath teeth early deciduous
- cones blunt-tipped

MIDWEST RANGE

Regionwide, especially common westward.

HABITAT

Moist to dry sandy river terraces, shores and meadows, open swamps, moist dunes, prairies, ditches, roadsides, along railways; usually where open or partially shaded.

DESCRIPTION

Rootstock Thick, dull, rough, dark brown; stems single or several together from the creeping rhizome.

Stems Deciduous, light-green, 40-80 cm tall, 2-7 mm thick, usually unbranched; central cavity 2/3 to 3/4 diameter of stem; smaller outer cavities alternating with 14-26 ridges; ridges rounded, smooth or with cross-bands of silica; sheaths constricted at the base, flaring towards the top, same color as the stem except for narrow band at top, or in old stems the lower sheath becoming girdled with brown; teeth narrowly lance-shaped, dark with chaffy margins, soon deciduous and leaving only small depressions at the sheath tips.

Cones To 2 cm in length when mature, yellow to brown, short-stalked, the tip with a small sharp point or rounded; shedding spores from late spring into summer.

SYNONYMS

- *Equisetum funstonii* A.A. Eaton
- *Equisetum kansanum* Schaffn.
- *Hippochaete laevigata* (A. Braun) Farw.

SIMILAR SPECIES

- **Tall scouring-rush** (*Equisetum hyemale*) is similar, but has two dark bands on each leaf sheath, and often found in somewhat wetter habitats.

NOTES

A hybrid with *Equisetum hyemale* is common: *E.* × *ferrissii* Clute; also a hybrid with *E. variegatum* occurs, mostly near Lake Michigan in Wisconsin and in lower Michigan: *E.* × *fnelsonii* (A.A. Eat) Schaffn.

NORTH AMERICA

MIDWEST

NAME

Named in 1844 by A. Braun from specimens collected along banks of Mississippi River south of St. Louis, Missouri.

From Latin, *laevigatum*, smooth, referring to the relatively smooth stems.

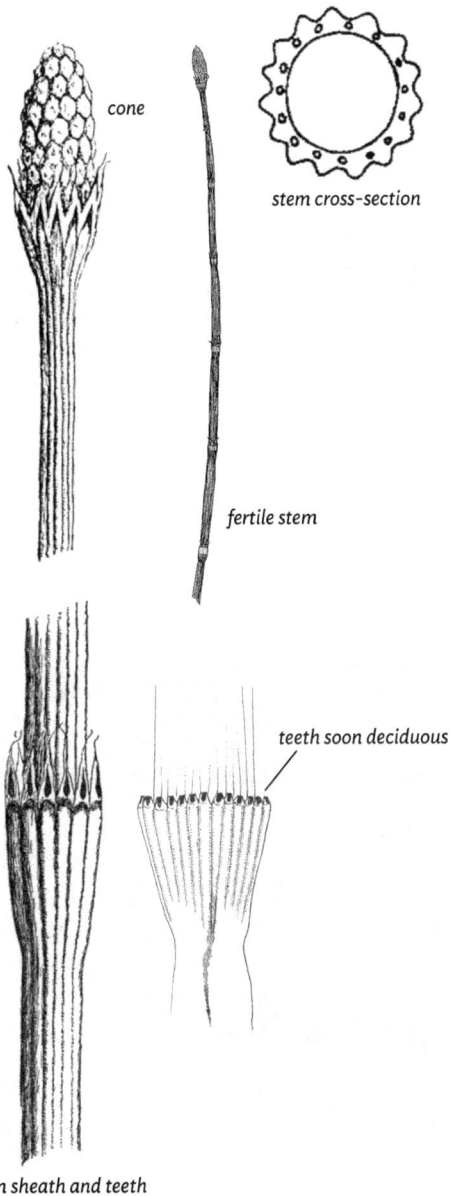

cone

stem cross-section

fertile stem

teeth soon deciduous

stem sheath and teeth

Equisetum palustre L.
MARSH HORSETAIL

FIELD TIPS
- in shallow water or wet places
- stems deciduous, branched
- small central cavity
- cones blunt-tipped

MIDWEST

RANGE
n Ill, Mich, Minn, Wisc.

HABITAT
Wet woods and meadows, swamps, ditches, shores, cold water seeps, soils wet or inundated up to about 20 cm deep.

DESCRIPTION
Rootstock Black to brown, shiny, occasionally tuber-bearing; stems single or clustered from the rhizomes.

Stems Deciduous, mostly 20-60 cm long, 1-3 mm thick, erect, solitary or clustered; central cavity less than 1/2 the stem diameter; outer cavities about same size as central cavity and alternating with 5-10 prominently angled smooth or rough ridges; sheaths green with long, narrow, black and chaffy-margined teeth. Branches spreading in regular whorls from the middle nodes (sometimes few to none), with the first branch internode shorter than the subtending stem sheath.

Cones 1-3.5 cm long, rounded at tip, deciduous, on a stalk at end of main stems; spores shed in summer.

NOTES
Marsh horsetail usually found in wetter situations than the similar **field horsetail** *(Equisetum arvense)*.

NAME
In 1753 Linnaeus gave the name *Equisetum palustre* to a European plant; it was later extended to include similar American specimens.

From Latin, *palustris*, marshy.

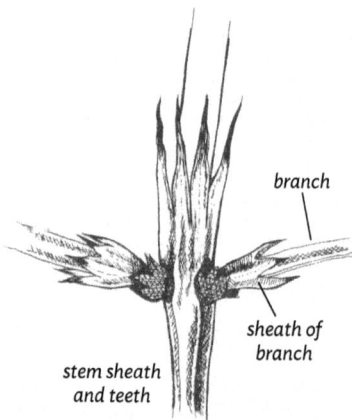

branch

sheath of branch

stem sheath and teeth

NORTH AMERICA

MIDWEST

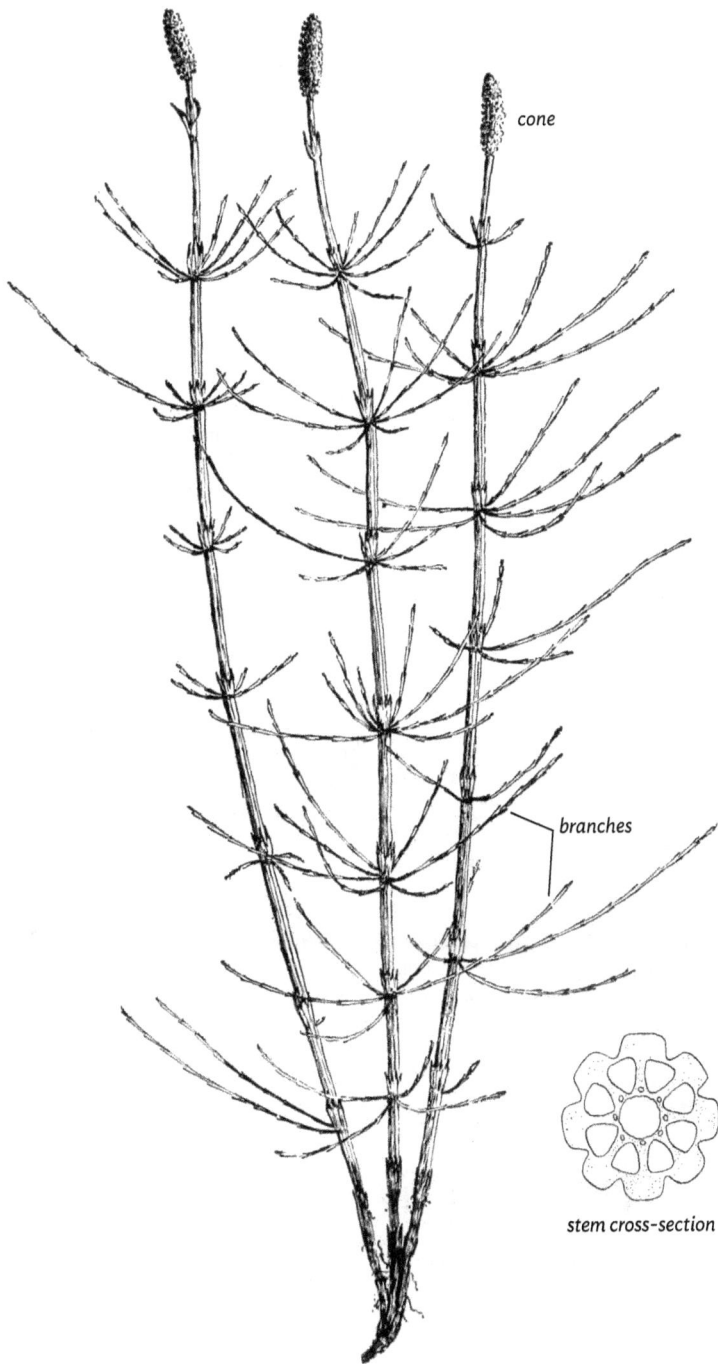

cone

branches

stem cross-section

Equisetum pratense Ehrh.
MEADOW HORSETAIL

FIELD TIPS
- colony-forming, wet places
- stems deciduous, branched, either sterile or fertile, the fertile stems seldom produced
- sheath teeth white-margined, persistent
- cones rounded at tip

MIDWEST RANGE
n Ill, Iowa, Mich, Minn, Wisc.

HABITAT
Conifer swamps, moist to wet deciduous or mixed conifer-hardwood forests, often near springs or seeps; in sun or partial shade.

DESCRIPTION
Rootstock Dull, black; the stems mostly growing singly from the rhizome.

Stems Deciduous, of two kinds, sterile and fertile; the fertile stems rarely seen.

Sterile stems Whitish green, to 60 cm tall and 1-3 mm thick, regularly and abundantly branched; central cavity less than 1/2 diameter of stem, with small outer cavities alternating with the 8-18 ridges; silica spicules in 3 rows on the ridges of the middle and upper internodes; sheaths pale, persistent, with 10-20 narrow teeth, the teeth dark with white margins;

branches in regular whorls, horizontal to drooping, triangular in cross-section, the first internodes of each branch shorter than the corresponding sheaths; the branch teeth slightly incurved, with thin white margins.

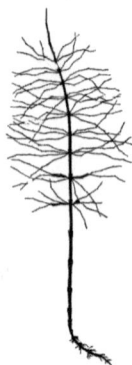

Fertile stems Apparently not common, at first unbranched and lacking chlorophyll, fleshy, becoming green and branched after the spores are shed; sheaths and teeth about twice as long as those of the sterile stems.

Cones Peduncled, to 3 cm long, rounded at the tip, deciduous, shedding spores from late spring into summer.

SIMILAR SPECIES
- Separated from **field horsetail** (*Equisetum arvense*) by the fertile stems being persistent, turning green, and developing branches.

NAME
Well known in northern Europe, this horsetail was named by Ehrhart in 1784.

From Latin, *pratum*, meadow, referring to its habitat.

NORTH AMERICA

MIDWEST

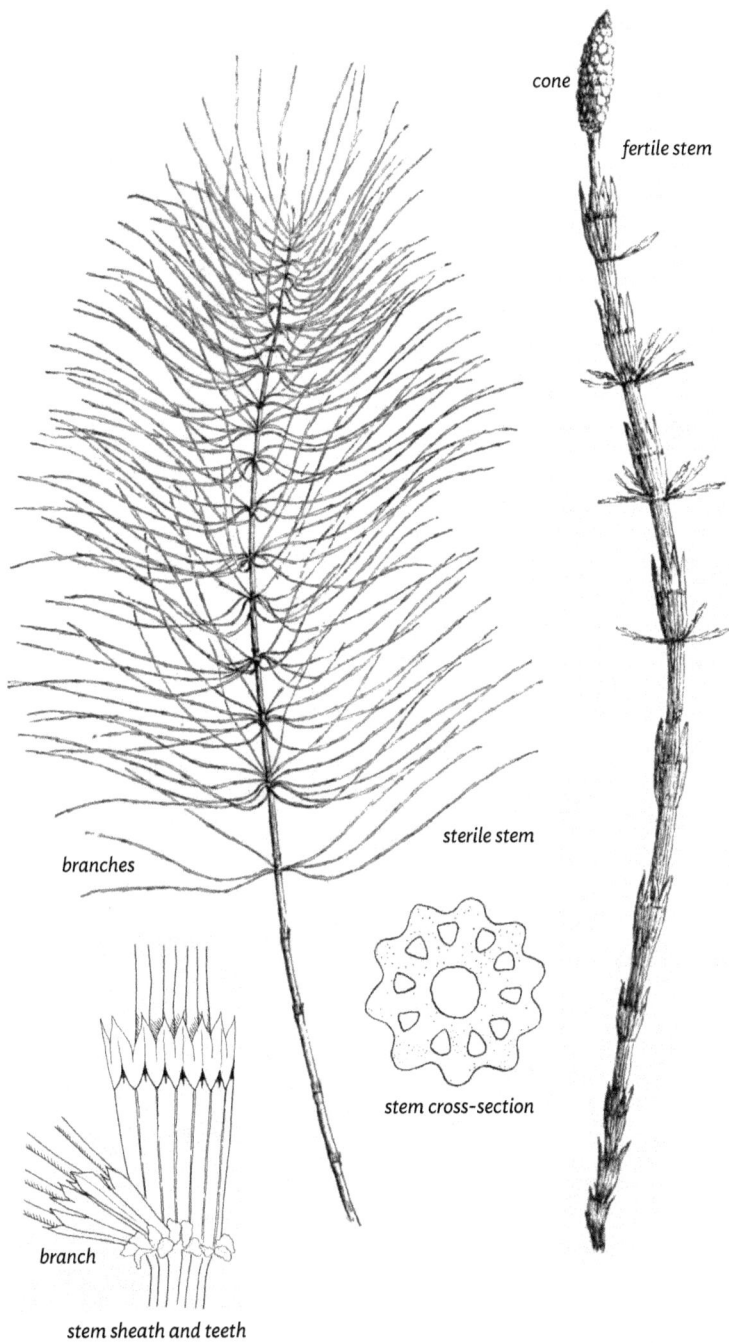

cone

fertile stem

branches

sterile stem

branch

stem sheath and teeth

stem cross-section

Equisetum scirpoides Michx.
DWARF SCOURING-RUSH

FIELD TIPS
- plants small, evergreen, clumped
- stems kinked at the nodes; each stem node with three teeth
- stems solid, not hollow, unique among our *Equisetum*
- cones small, tipped with a small point

MIDWEST RANGE
n Ill (endangered), e Iowa, Mich, Minn, Wisc.

HABITAT
Cedar swamps, moist mixed forests, usually in shade; sometimes on open north-facing bluffs and ledges; may form crowded mats, the stems partly buried in humus.

DESCRIPTION
Rootstock Shallow, branching, the stems in tufts and forming colonies via the spreading rhizomes.

Stems Evergreen, mostly 10-20 cm long, slender, 0.5-1.0 mm thick, dark green, usually unbranched, ascending or prostrate, arched and recurving; center of stem solid, the central cavity absent, replaced by 3 (rarely 4) cavities alternating with the deeply grooved ridges; silica rosettes in lines on the crests of the ridges; sheaths green below, black above, loose, with 3 (rarely 4) triangular teeth; teeth dark with white margins, persistent, but their awl-like tips usually soon breaking off.

Cones Small, 2-3 mm long, black, tipped by a small point, shedding spores in summer, or persisting unopened until the following spring.

NAME
From Latin, *scirpus*, rush or bulrush, Greek *eidos*, like; the small tufted stems similar to a rush.

NORTH AMERICA

MIDWEST

ABOVE Small, slender stems typical of *Equisetum scirpoides.*

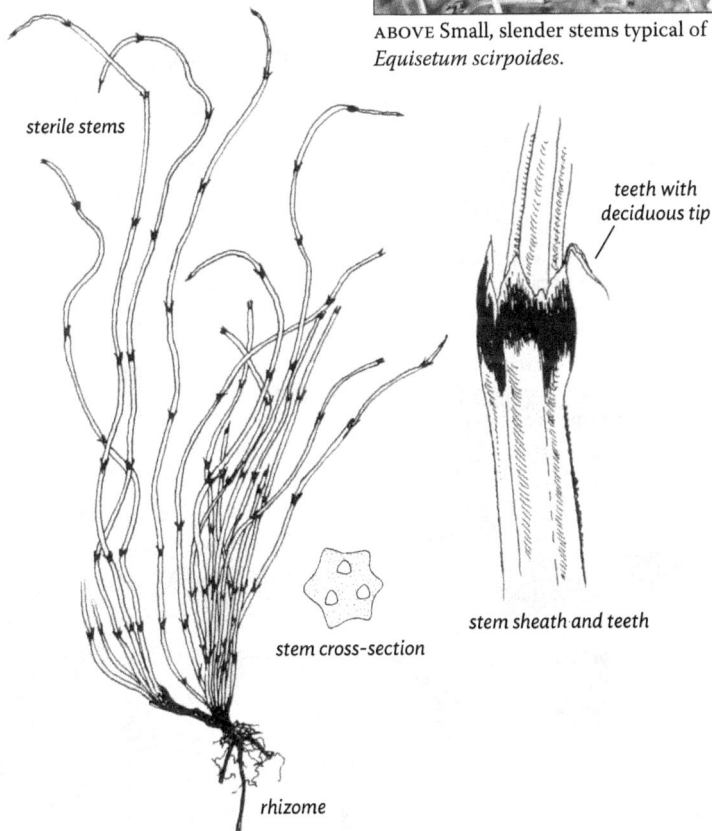

sterile stems

teeth with deciduous tip

stem cross-section

stem sheath and teeth

rhizome

Equisetum sylvaticum L.
WOODLAND HORSETAIL

FIELD TIPS

- colony-forming, moist to wet woods
- stems deciduous, either sterile or fertile; sheaths loose and reddish brown.
- secondary branching of most primary branches
- cones rounded at tip

MIDWEST RANGE

n Ill (endangered), n Ind, Iowa (threatened), Mich, Minn, ne Ohio, Wisc.

HABITAT

Swamps, moist forests and forest margins, wet thickets, moist open meadows.

DESCRIPTION

Rootstock Light brown, shiny, occasionally tuber-bearing; forming colonies, the stems mostly solitary from the rhizome.

Stems Annual, of two kinds, sterile and fertile.

Sterile stems To 70 cm long and 1.5-3 mm thick; central cavity 1/2 to 2/3 diameter of stem, with prominent outer cavities alternating with the 10-18 ridges; silica tubercules in 2 rows on the ridges; sheaths loosely inflated, the reddish brown papery teeth persistent, usually joined to form 3 or 4 groups; branches whorled, arched, appearing lacy due to the secondary branches; branches with 3 or 4 (rarely 5) ridges, and with narrow, pointed and spreading teeth.

Fertile stems At first unbranched and lacking chlorophyll, fleshy, becoming green and branched after the spores released; sheaths and teeth usually larger than those of sterile stems.

Cones Stalked, to 3 cm in length, rounded at tip, shedding spores in spring then falling.

NAME

Named by Linnaeus from European specimens in 1753.

From Latin, *sylva*, woods or forest, referring to its habitat.

NORTH AMERICA

MIDWEST

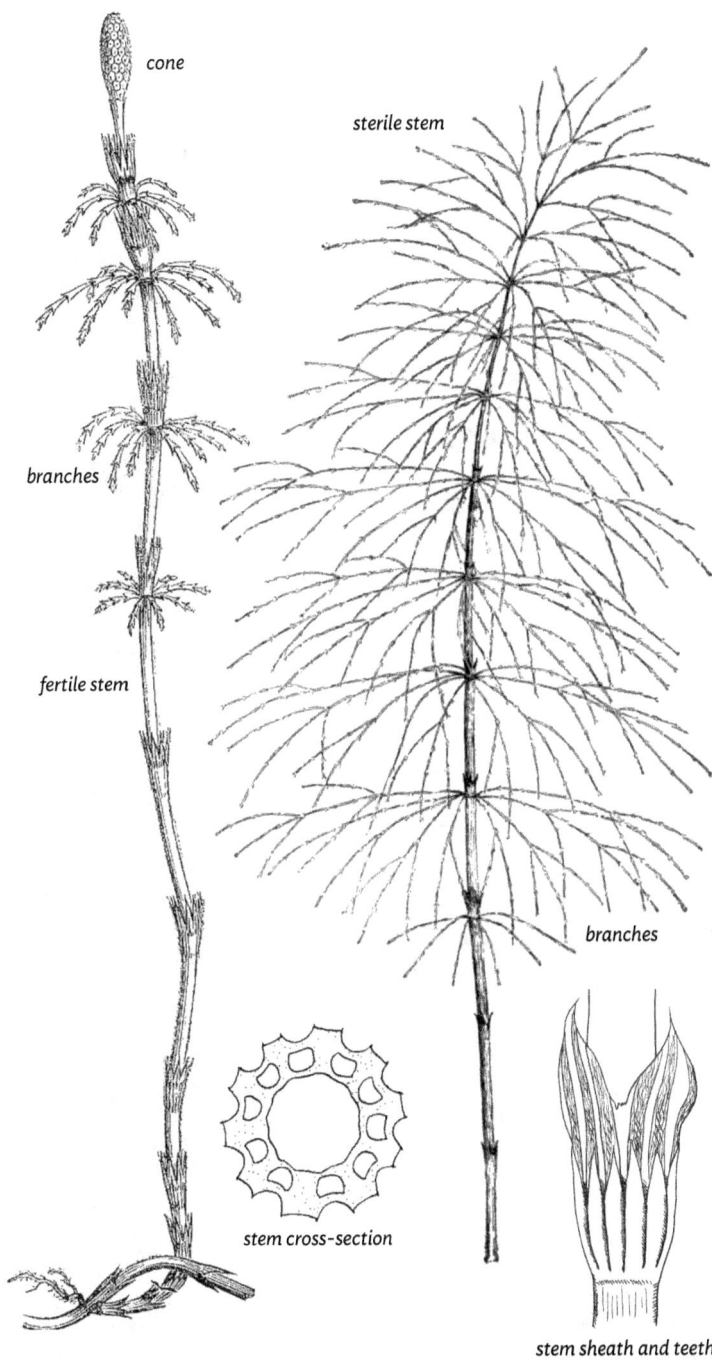

cone

sterile stem

branches

fertile stem

branches

stem cross-section

stem sheath and teeth

Equisetum variegatum
Schleich. ex F. Weber & D.M.H. Mohr
VARIEGATED SCOURING-RUSH

FIELD TIPS
- colony-forming in moist, sandy, open places
- stems evergreen, slender, unbranched
- minutely grooved sheaths have 3–12 teeth with conspicuous white margins
- cones tipped by a small point

MIDWEST RANGE
Ill, n Ind (endangered), Mich, n Minn, n Ohio (endangered), Wisc.

HABITAT
Moist sand, shores, beaches, interdunal flats, roadside ditches, borrow pits; often forming dense colonies.

DESCRIPTION
Rootstock Smooth, blackish, shiny, branching; the stems in clusters from the rhizomes.

Stems Evergreen, 10-50 cm long, 0.5-3 mm thick, usually unbranched, ascending; central cavity 1/3 to 2/3 diameter of stem; large outer cavities alternating with 3-12 ridges; silica tubercules in two lines on ridges; sheaths green at base, black above, slightly spreading; teeth lance-shaped, persistent, with or without thread-like tips, and with a brown central portion and wide white margins.

Cones Small, 5-10 mm long, tipped by a small sharp point, shedding spores in late summer, or more often persisting unopened until the following spring.

SYNONYMS
- *Hippochaete variegata* (Schleich. ex F. Weber & D. Mohr) Bruhin

NAME
This plant was discovered in Europe and named by Schleicher in 1807; American plants found soon afterward proved so similar that they are regarded as representing the same species.

From Latin, *variego*, to be different colors.

NORTH AMERICA

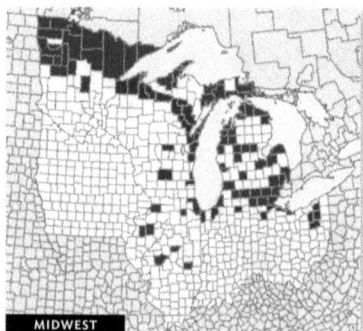

MIDWEST

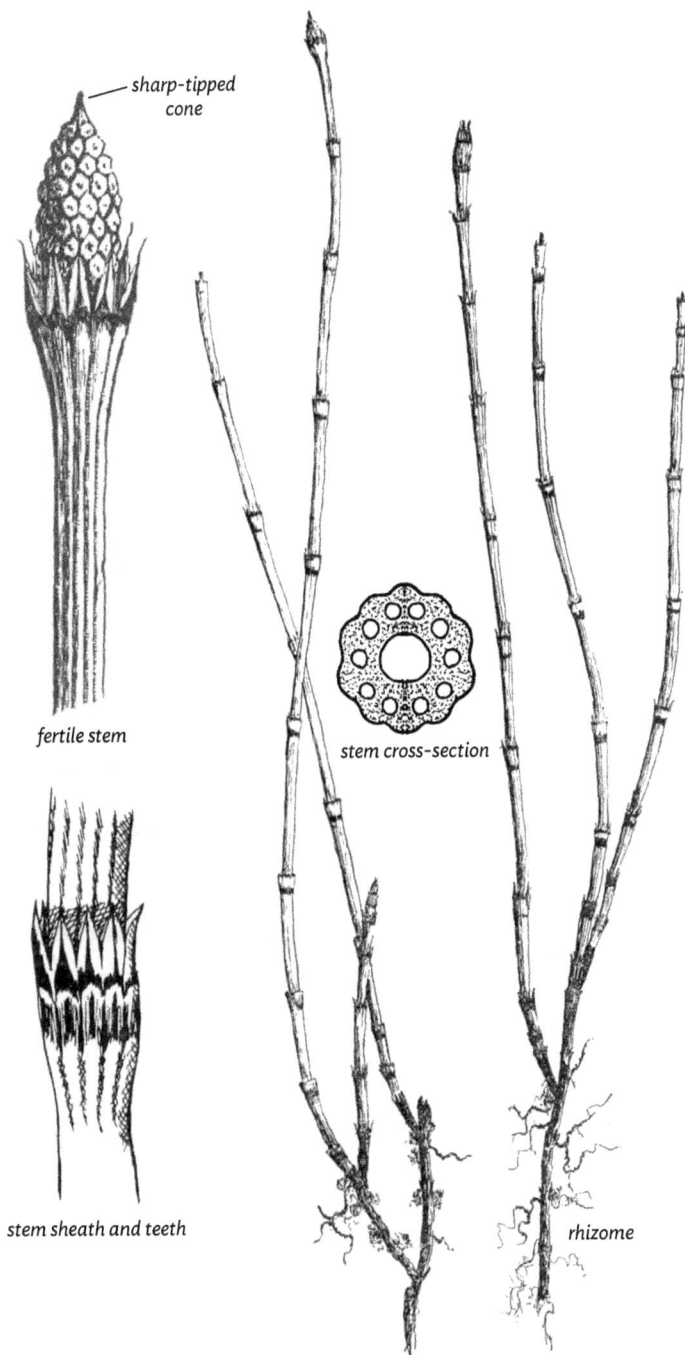

sharp-tipped cone

fertile stem

stem sheath and teeth

stem cross-section

rhizome

HYMENOPHYLLACEAE
Filmy Fern Family
Trichomanes
BRISTLE FERN

Worldwide, the Filmy Fern Family contains 31 genera, mostly of tropical regions; *Trichomanes* includes ca. 118 species, with 8 in North America, and a single species found in southern portions of our region. Bristle ferns are very small, delicate ferns of shaded crevices and hollows in noncalcareous rock; plants are colony-forming and usually growing on the ceiling of the hollow.

KEY CHARACTERS
- Delicate evergreen fern of dark, moist rock crevices, caves, and hollows.
- **Fronds** translucent, 1-celled thick, adapted to low-light conditions.
- Forming small, dense colonies.
- **Rhizome** slender, creeping and branching, covered by black hairs and roots, producing a row of small, often drooping fronds.
- **Sporangia** in a cluster on a marginal bristlelike extension of a vein, surrounded at base by a cone-shaped sheath.
- Uncommon in southern portion of region (s Illinois, s Indiana, s Ohio).

NOTES
One of our smallest and most delicate ferns.

NAME
An ancient Greek term for some now unknown small hairy fern was described as a genus by Linnaeus in 1753.

From Greek, *thrix*, hair, and *manes*, cup, referring to bristlelike receptacle extending from a cuplike indusium.

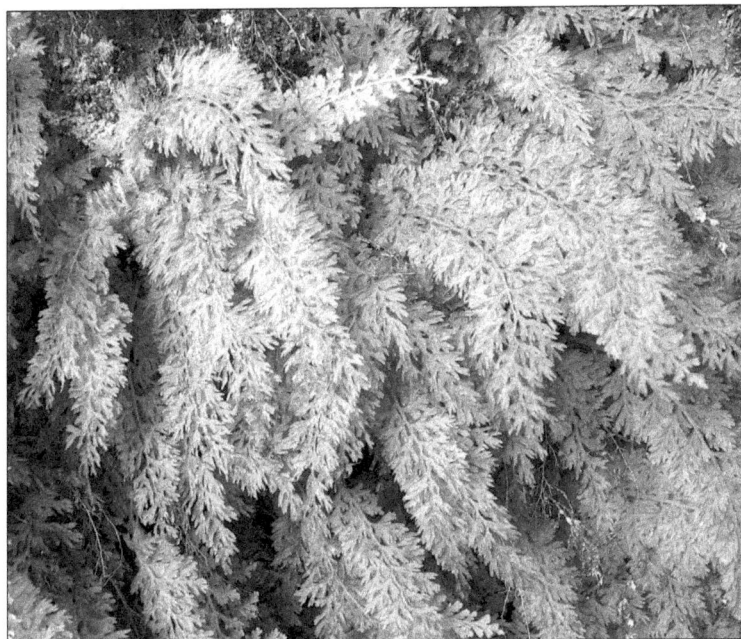

Appalachian bristle fern (*Trichomanes boschianum*).

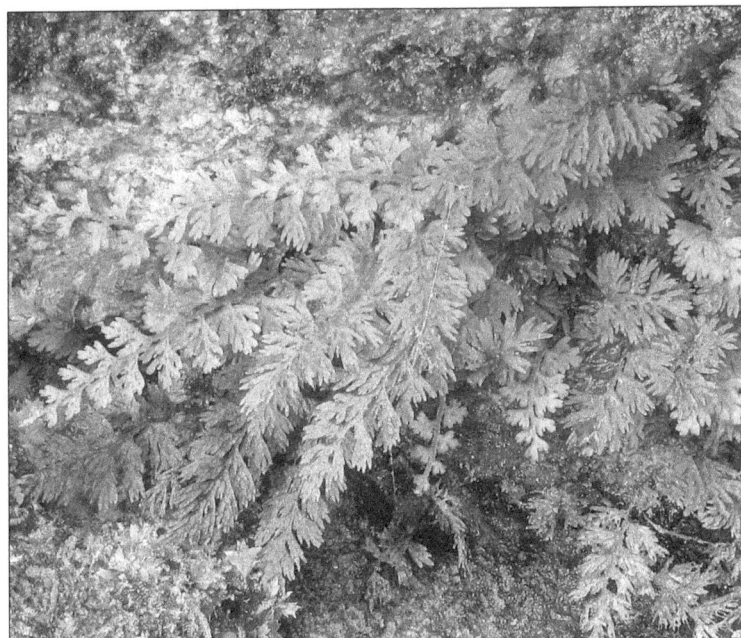

Appalachian bristle fern (*Trichomanes boschianum*).

Trichomanes boschianum Sturm
APPALACHIAN BRISTLE FERN

FIELD TIPS

- delicate evergreen fern of dark, moist rock crevices and caves, forming colonies
- fronds translucent, 1-celled thick
- stipe and rachis winged
- sporangia clustered in a bristlelike extension of a vein, surrounded at base by a cone-shaped sheath
- uncommon in southern portion of region

MIDWEST RANGE

s Ill (endangered), s Ind (endangered), s Ohio (endangered).

HABITAT

Shaded, moist, humid crevices and cave-like hollows in acidic sandstone and gneiss rock; plants often hanging from rock ceiling.

DESCRIPTION

Rootstock Long-creeping, slender, black, hairy; roots few.

Fronds Widely spaced along rhizome; lance-shaped, 4-20 cm long and 1-4 cm wide.

Blade 1- to 2-pinnate-pinnatifid, only 1-cell thick between veins, veins with scattered hairs.

Stipe Shorter than blade, green, flattened and winged; covered with dark hairs of 2 kinds: unbranched gland-tipped hairs, and nonglandular branched or unbranched hairs.

Rachis Green, winged.

Sori Terminal on projecting bristlelike veins at base of pinna lobes, surrounded at base by cone-shaped indusial sheath, but not flared at mouth. **Sporangia** tiny, green, and extending about halfway up the bristle

SYNONYMS

Previously considered synonymous with the tropical *Trichomanes radicans* Swartz, but now considered a distinct species endemic to e USA.

NAME

Named in honor of R. B. van den Bosch, a 19th-century botanist.

NORTH AMERICA

MIDWEST

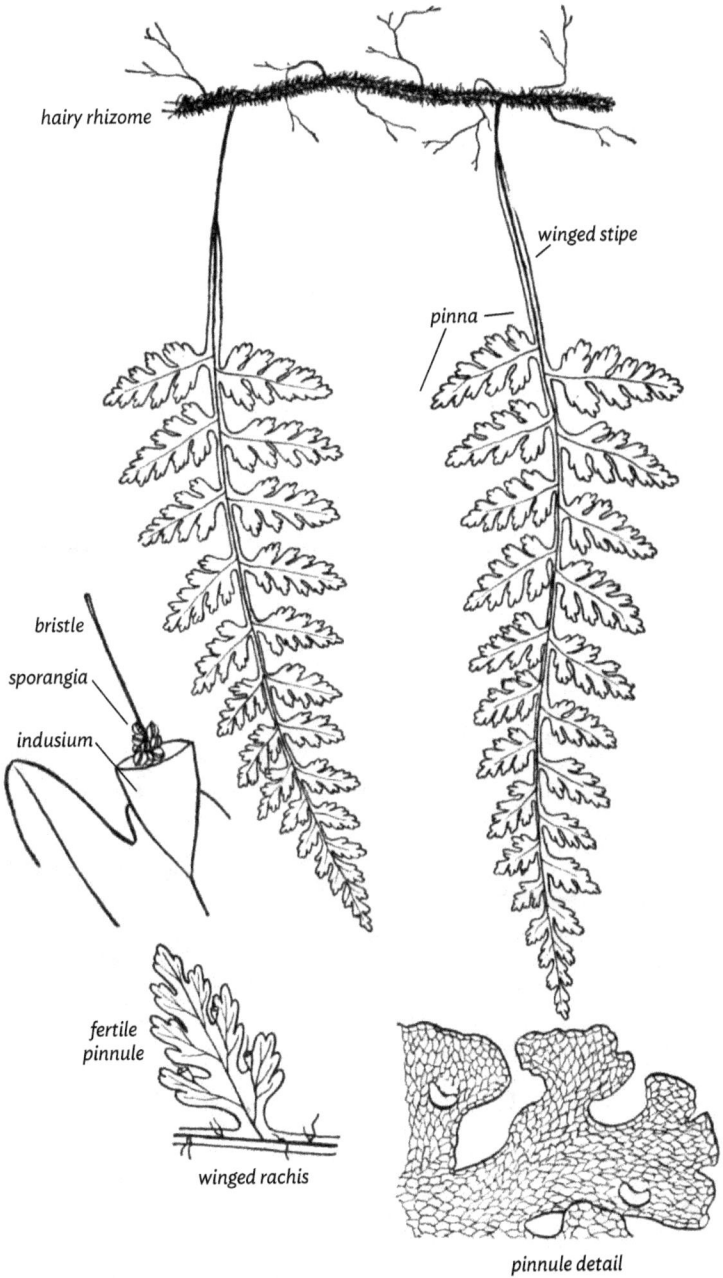

hairy rhizome

winged stipe

pinna

bristle

sporangia

indusium

fertile
pinnule

winged rachis

pinnule detail

LYGODIACEAE | Climbing Fern Family
Lygodium
CLIMBING FERN

Fronds vinelike, of indeterminate growth, **pinnae** reduced to short stalks, each bearing a pair of opposite pinnules, **sporangia** in 2 rows, 1 on each side of midvein of contracted, oblong, marginal lobes of ultimate segments, covered by hoodlike flap of tissue serving as the indusium.

Lygodium is the sole member of this family; the genus includes about 20 species, found mostly in tropical regions. North America is home to a single native species—*Lygodium palmatum*—and 2 species which have escaped from cultivation (below). The *Lygodium* species are a true anomaly within the fern world; they are climbing, vine-like ferns exhibiting indeterminate growth in which the rachis can reach 3-10 meters in length.

Sometimes placed in Schizeaceae, the Curly-Grass Fern Family.

KEY CHARACTERS
• The region's only twining and climbing, vine-like fern.

NAME
From Greek, *lygodes*, flexible, referring to the rachis bending alternately from side to side.

ADDITIONAL MEMBERS OF GENUS
The two species below are not known from our region, but **Japanese climbing fern** is reported from western Pennsylvania. In the USA, *Lygodium microphyllum* is known only from Florida.

• **Japanese climbing fern** [*Lygodium japonicum* (Thunb. ex Murr.) Sw.], originally introduced to the USA as an ornamental plant, is established in moist places in the se USA, ranging from e Texas to North Carolina, with isolated occurrences in w Pennsylvania.

• **Old World climbing fern** [*Lygodium microphyllum* (Cav.) R. Br.], native to Australia, southern Asia and Africa, was first found in Florida in the 1960s and is now established on an estimated 25,000 hectares (about 100,000 acres) in central and southern Florida. An aggresive, kudzu-like weed, it infests cypress swamps, engulfing tree islands with fronds up to 30 meters long. Due to the warm climate, plants do not die back in winter, allowing for luxuriant growth. Efforts are ongoing to control the spread of this plant using biological agents including several moths, a beetle, and a mite.

rachis

pinna

Japanese climbing fern
LYGODIUM JAPONICUM

American climbing fern
(*Lygodium palmatum*)

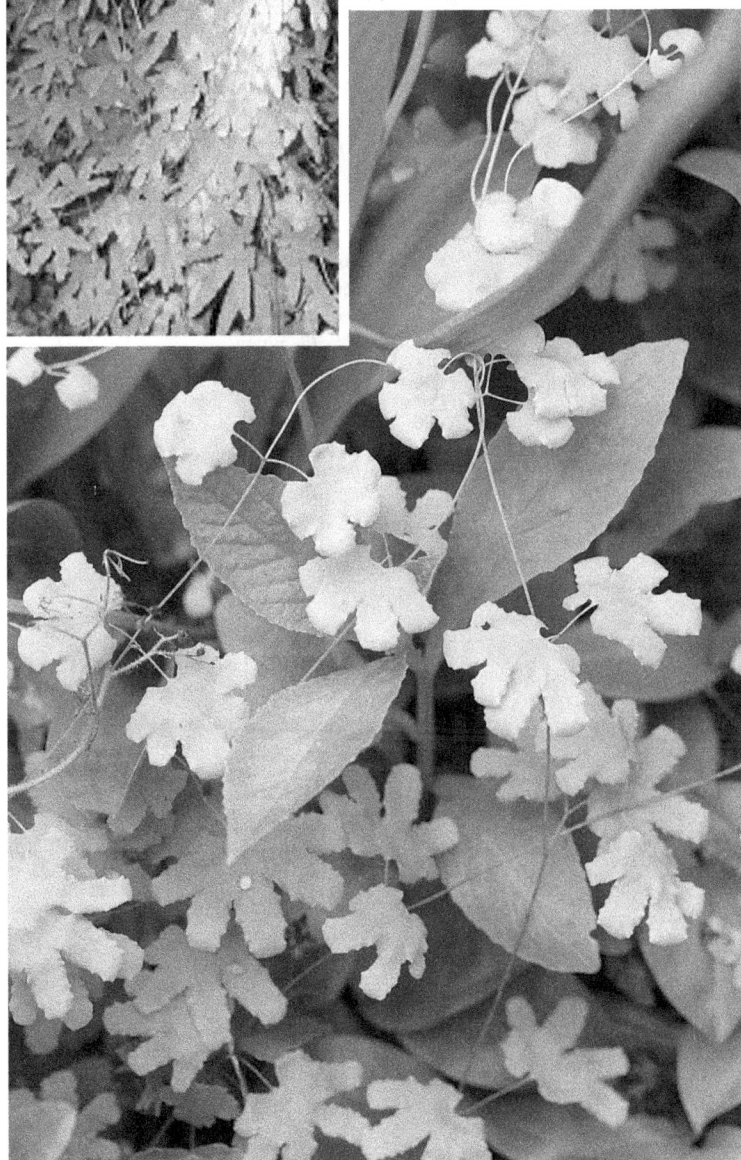

Lygodium palmatum (Bernh.) Sw.
AMERICAN CLIMBING FERN

FIELD TIPS

- region's only climbing fern; rare in southern Midwest
- rachis vinelike, to 100 cm or more long
- leaf blades sterile or fertile
- sterile blades 4-7-lobed
- fertile blades small, in terminal panicle

MIDWEST RANGE

s Ind (endangered), sw Mich (endangered), Ohio; more common in se USA.

HABITAT

Woodland margins, wet thickets, in sun to light shade; soils moist and acidic.

DESCRIPTION

Rootstock Creeping, branched, slender, black, with sparse short, blackish hairs; scales absent.

Frond Divided into sterile and fertile portions, usually 40-100 cm long; climbing, forming apparently a "climbing stem" with palmately lobed "leaves" but actually a twining rachis bearing the pinnae.

Rachis Twining, 2-forked, each fork with a palmately 4-7-lobed blade 2-7 cm wide; round and dark at base, upwards green or brownish and somewhat flattened and winged.

Blade 2-pinnate, fertile blades uppermost, contracted and several times forked, forming a terminal panicle.

Sterile Pinnae Evergreen until following year's growth, alternate, each divided into two stalked pinnules.

Fertile Pinnae Deciduous, above the sterile pinnae at tips of vine, several times branched.

Sori In a double row on fertile segments, under an indusiumlike flap of tissue at lobe edges of the smaller fertile pinnae.

NAME

First found by William Bartram in Georgia, but he failed to give it a valid name. Plants from e USA collected by Muhlenberg were named *Hydroglossum* by Willdenow, and *Gisopteris palmata* by Bernhardi in 1801. Swartz transferred the species to his genus *Lygodium* 5 years later.

Also known as **Hartford fern**, referring to its having been the subject of the first conservation law applied to a fern, passed by Connecticut legislature in 1869.

From Latin, *palmatum*, referring to the palmate shape of the leaf blade.

NORTH AMERICA

MIDWEST

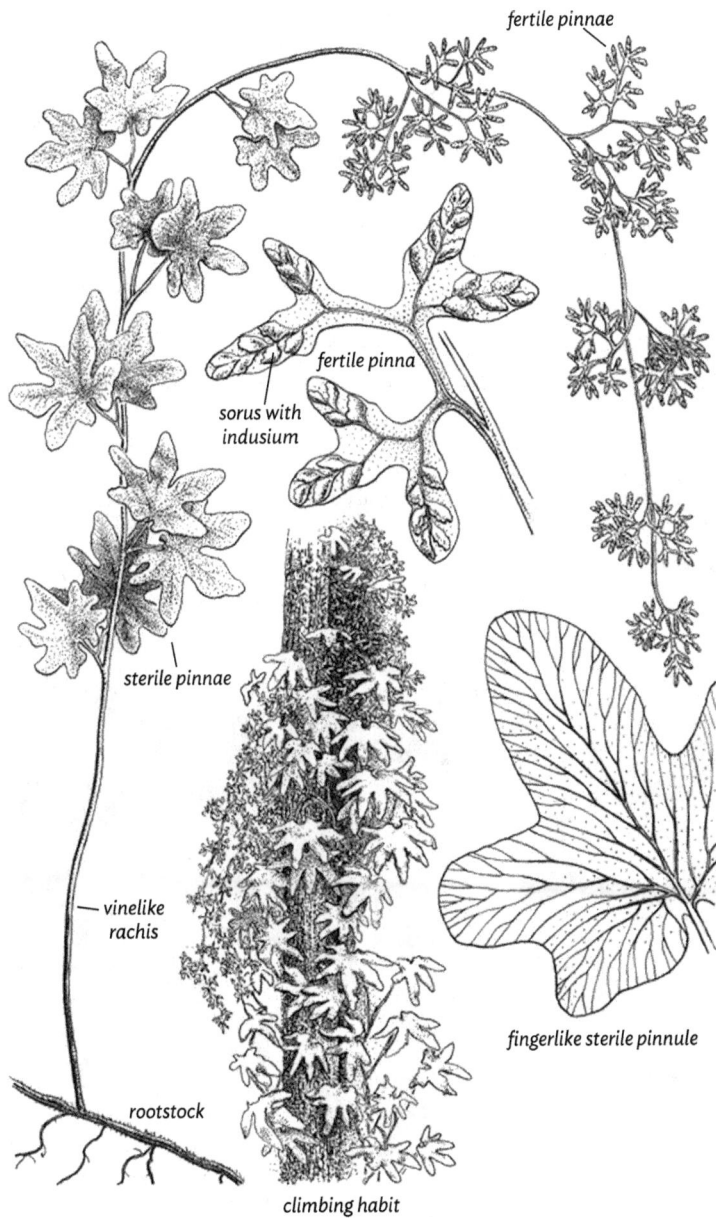

fertile pinnae

fertile pinna

sorus with
indusium

sterile pinnae

vinelike
rachis

rootstock

climbing habit

fingerlike sterile pinnule

MARSILEACEAE | Water-Clover Family
Marsilea
WATER-CLOVER

Perennial aquatic ferns with **leaf blade** divided into 4-segments. Family includes 5 genera and 17 species of *Marsilea*, mostly of tropical regions; 2 species in our flora, one of which is introduced from Europe and Asia.

KEY CHARACTERS

- **Aquatic ferns** of quiet, fresh water or sometimes exposed on shores; roots and often rhizomes anchored in bottom muck.
- **Leaf blade** 4-parted like a 4-leaf clover, atop a slender stipe of variable length, (length depending on depth of water).
- **Pinnae** floating on water surface or raised slightly above it, sometimes submerged.
- **Spores** of 2 kinds, the female, egg-bearing megaspores, and the male, sperm-bearing microspores; in 1-several hard, bean- or pea-like structures (sporocarps) near base of stipes.
- **Sporocarps** divided into several compartments, each surrounded by a gelatinous ring which serves to hold the megaspore until fertilization occurs by the microspore.

NAME

After Count Luigi Fernando Marsigli (1658-1730), an Italian naturalist.

European water-clover
(*Marsilea quadrifolia*)

Hairy water-clover (*Marsilea vestita*), plants stranded on exposed shore.

KEY TO MARSILEA | WATER-CLOVER

1 Pinnae nearly smooth; sporocarps often in pairs borne on a long stalk arising from stipe above its base; introduced and local in Ill, Ind, Iowa, Mich, Ohio **1. Marsilea quadrifolia**
EUROPEAN WATER-CLOVER, page 214

1 Pinnae usually shairy; sporocarps borne singly on short stalks from the rhizome or base of stipe; native and uncommon in Minn and nw Iowa **2. Marsilea vestita**
HAIRY WATER-CLOVER, page 216

Marsilea vestita
HAIRY WATER-CLOVER

Marsilea quadrifolia
EUROPEAN WATER-CLOVER

Marsilea quadrifolia L.

EUROPEAN WATER-CLOVER

FIELD TIPS

- deciduous aquatic fern
- pinnae 4-parted, emergent to slightly below water surface
- sporocarps 1-2 on short stalks from near base of stipe

MIDWEST RANGE

Ill, s Ind, s Iowa, Mich LP, Ohio; introduced from Europe.

HABITAT

In mud or shallow water of quiet lakes, ponds and rivers.

DESCRIPTION

Rhizome Creeping and rooted in bottom of lakes and ponds, less than 1 mm in diameter, smooth or with a few hairs; scales absent; roots forming both at nodes and sparsely (1-3) along internodes.

Frond Deciduous, sterile and fertile fronds alike,

Stipe Green, slender, to about 15 cm long, depending on water-depth.

Blade Floating, submerged or emergent, horizontal, divided into 4 clover-like pinnae; smooth.

Pinnae 1.5-2.5 cm wide, smooth or with a few hairs; margins rounded; veins forked.

Sporocarp Usually 2 on stalks from lower portion of stipe, shaped like a small bean, 2-5 mm long, with yellow hairs when young, hairless when mature.

NOTES

Introduced to the USA from Europe in 1862, now naturalized and sporadic in eastern and midwestern states.

SIMILAR SPECIES

- Differs from **hairy water-clover** (*Marsilea vestita*) by larger pinnae, the pinnae and stipe nearly hairless, sporocarps usually 2 (sometimes 3) on a long, hairless stalk from near base of stipe (and not from axil of stipe and rhizome as in *M. vestita*); hairy water-clover known in Midwest only from nw Iowa and w Minnesota.

NORTH AMERICA

MIDWEST

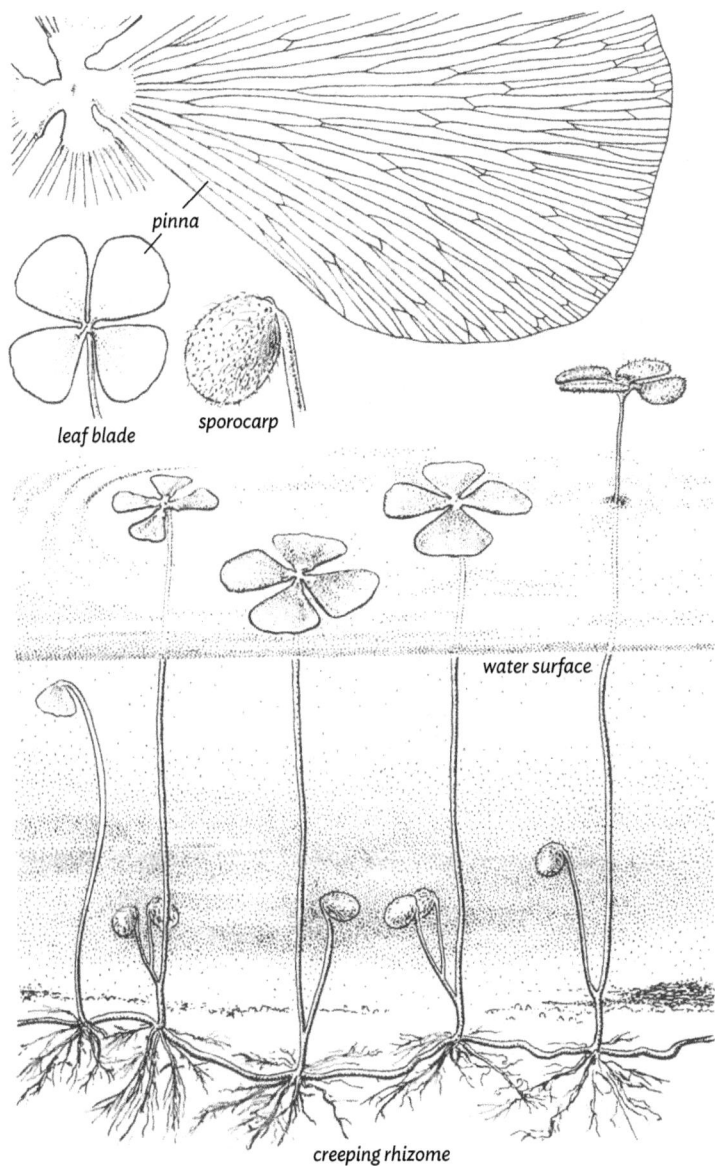

pinna

leaf blade

sporocarp

water surface

creeping rhizome

Marsilea vestita Hook. & Grev.
HAIRY WATER-CLOVER

FIELD TIPS
- uncommon fern of shallow water and shores in w Midwest
- pinnae 4-parted
- sporocarps single on short stalks from rhizome or base of stipe

MIDWEST RANGE
nw Iowa (threatened), w Minn (endangered); more common westward.

HABITAT
In our region, found in moist soil margins of shallow prairie pools; also in temporary rainwater pools on rock outcrops.

DESCRIPTION
Rhizome Creeping, slightly branched, less than 1 mm in diameter, smooth or sparsely hairy; scales absent; roots forming only at nodes.

Frond Deciduous; of two types, the sterile floating, submerged or elevated slightly above water surface; the fertile in or on bottom muck.

Stipe Green to straw-colored, 2 mm wide, to about 20 cm long depending on water depth, smooth, occasionally waxy or with a few scales at base, vascular bundles one V-shaped bundle.

Blade Floating, submerged or slightly emergent, with 4 pinnae, round in outline or sometimes folded; sparsely appressed hairy.

Pinnae 0.3 to 2.5 cm long and to 2 cm wide when mature.

Sporocarp Borne singly on a short, sometimes drooping stalk arising from rhizome or from base of stipe; soft and green when young, becoming hard and brown at maturity, hairy, with 2 teeth near base; with 7-10 sori in each of the 2 cells, and about 10-20 megasporangia in each sorus.

SYNONYMS
- *Marsilea fournieri* C. Chr.
- *Marsilea minuta* Fourn.
- *Marsilea mucronata* A. Braun
- *Marsilea tenuifolia* Engelm. ex A. Braun
- *Marsilea uncinata* A. Braun

NAME
From Latin, *vestitus*, clothed, referring to the hairy leaf blades.

NORTH AMERICA

MIDWEST

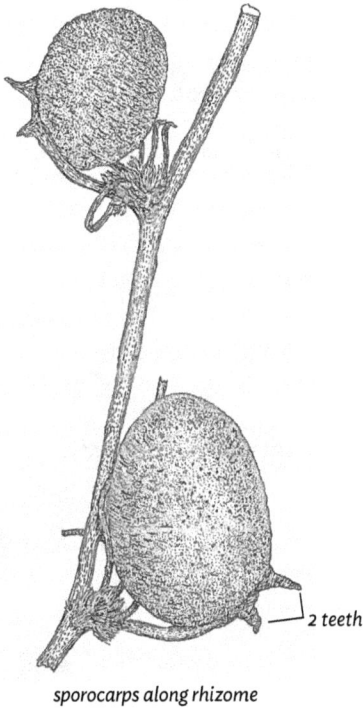

sporocarps along rhizome

2 teeth

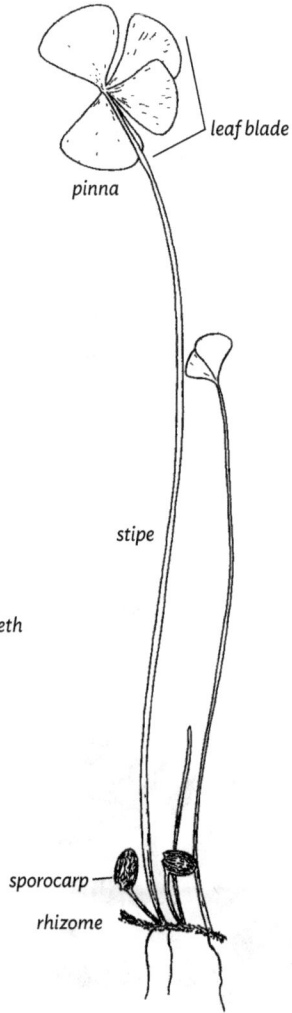

leaf blade

pinna

stipe

sporocarp

rhizome

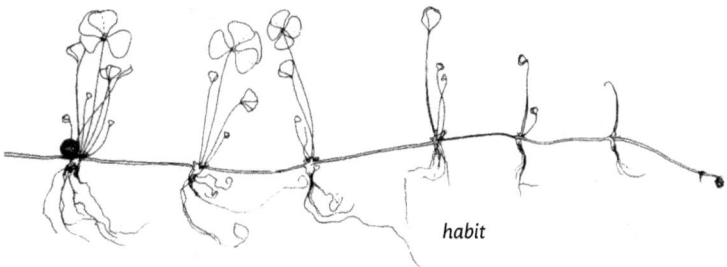

habit

ONOCLEACEAE
Sensitive Fern Family

Large coarse ferns with creeping hairy rhizomes (*Onoclea*) or with stolons on ground surface (*Matteuccia*); sterile and fertile fronds strongly different, the **sterile fronds** deciduous, pinnatifid to 1-pinnate-pinnatifid; **fertile fronds** persistent. **Sori** enclosed under recurved margin of pinna segment (outer false indusium) and a tiny true inner **indusium** (membranous or of hairs). Small family of 4 genera (or sometimes treated as having a single genus *Onoclea*); we have 2 genera, each with a single species.

KEY CHARACTERS
- Colony-forming ferns of moist to wet places.
- Sterile and fertile **fronds** strongly dissimilar
- **Stipes** with 2 vascular bundles, uniting upwards to form U-shape.
- **Sori** in beadlike clusters (*Onoclea*), or under rolled margins of fertile pinnae (*Matteuccia*).

KEY TO ONOCLEACEAE | SENSITIVE FERN FAMILY

1 Sterile blades solitary from creeping rhizomes, deeply divided into lobes (or the lowermost divisions pinnae); common, regionwide
1. Onoclea sensibilis
SENSITIVE FERN, page 222

1 Sterile blades in a circle from a thick crown; pinnate with lobed pinnules; n Ill, Ind, Iowa, Mich, Minn, Ohio, Wisc
2. Matteuccia struthiopteris
OSTRICH FERN, page 220

Onoclea sensibilis
SENSITIVE FERN

Matteuccia struthiopteris
OSTRICH FERN

Matteuccia

OSTRICH FERN

One of the region's largest ferns, forming round tufts of bright green fronds. One (-3) circumboreal species in genus.

KEY CHARACTERS

• Large fern forming vase-shaped clumps, may form colonies via creeping stolons.
• Sterile and fertile **fronds** strongly different.
• **Sterile frond** 1-pinnate-pinnatifid, shaped like ostrich feather in outline, long-tapering to base.
• **Fertile fronds** shorter than sterile, turning dark brown, persistent into the following year; sori covered by inrolled margin of pinnae.

NAME

Named in honor of Carlo Matteucci (1811-1868), an Italian physicist.

Onoclea

SENSITIVE FERN

Genus includes 1 species of eastern North America and eastern Asia.

KEY CHARACTERS

• Coarse fern of moist to wet places, forming colonies from creeping rhizome.
• Sterile and fertile **fronds** strongly different.
• **Sterile fronds** arising singly from rhizome, mostly pinnatifid.
• **Fertile fronds** turning dark brown; sori enclosed by the inrolled margins to form beadlike covering.

NAME

From Greek *onos*, vessel, and *kleio*, to close, referring to closely rolled fertile fronds.

ABOVE Early spring harvest of edible croziers (fiddleheads) of **ostrich fern** (*Matteuccia struthiopteris*).

Matteuccia struthiopteris (L.) Todaro
OSTRICH FERN

FIELD TIPS
- large, colony-forming, vase-like fern of wet places
- sterile fronds arched, widest in upper portion, lowest pinnae much smaller than upper
- fertile fronds upright from center of cluster, persistent

MIDWEST RANGE
n Ill, Ind, Iowa, Mich, Minn, Ohio, Wisc.

HABITAT
Streambanks, wet ravines, seeps and springs, wet woodlands, and swamps, soils typically alluvial, mucky or sandy; may form large colonies via creeping rhizomes.

DESCRIPTION
Rootstock Erect, stout, densely scaly, forming a vase-like cluster; also with spreading runners which produce several new plants each year.

Crozier Green with deciduous tan scales, 2-4 cm wide; edible.

Frond Several in a cluster; sterile and fertile fronds very different; sterile fronds deciduous, 50-100 cm long, pinnate-pinnatifid. Fertile fronds 15-30 cm long, pinnate, dark green to blackish, drying to brown, often persist to next year.

Stipe Green, with white hairs; vascular bundles 2.

Rachis Green, deeply grooved, with sparse whitish hairs, scales, or both.

Sterile Blade Elliptic, widest in upper part, 1-pinnate-pinnatifid.

Fertile Blade 1-pinnate, dark green to blackish, drying to brown.

Pinnae Lowest pinnae (several pairs) greatly reduced in size; sessile; costae shallowly grooved above, grooves not continuous from rachis to costae; veins free, not forked, extending to margin.

Sori On margins, covered by inrolled edge of fertile pinnae; **indusium** vestigial; **sporangia** green, persist through winter and released in spring before sterile leaves expand.

SYNONYMS
- *Matteuccia pensylvanica* (Willd.) Raymond
- *Onoclea struthiopteris* (L.) Hoffm. p.p.
- *Pteretis nodulosa* (Michx.) Nieuwl.
- *Pteretis pensylvanica* (Willd.) Fernald

SIMILAR SPECIES
- Fronds of **cinnamon fern** (*Osmunda cinnamomea* and **interrupted fern** (*O. claytoniana*) similar but do not have very small pinnae at base of frond.

NORTH AMERICA

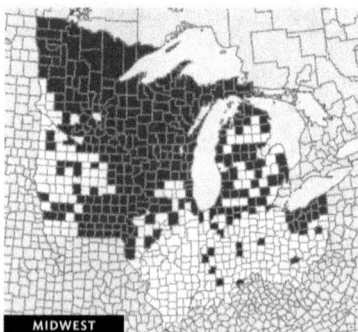

MIDWEST

NOTES

Croziers (fiddleheads) edible and prepared like asparagus; sometimes harvested commercially (especially in New England).

NAME

In 1803 Michaux described *Onoclea nodulosa*, mistakenly ascribing it to Carolina. The identity of this fern remained a mystery for some time, but his type specimen proved to be ostrich fern, and to have come from Montreal.

From Greek, *struthos*, ostrich, and *pteris*, fern.

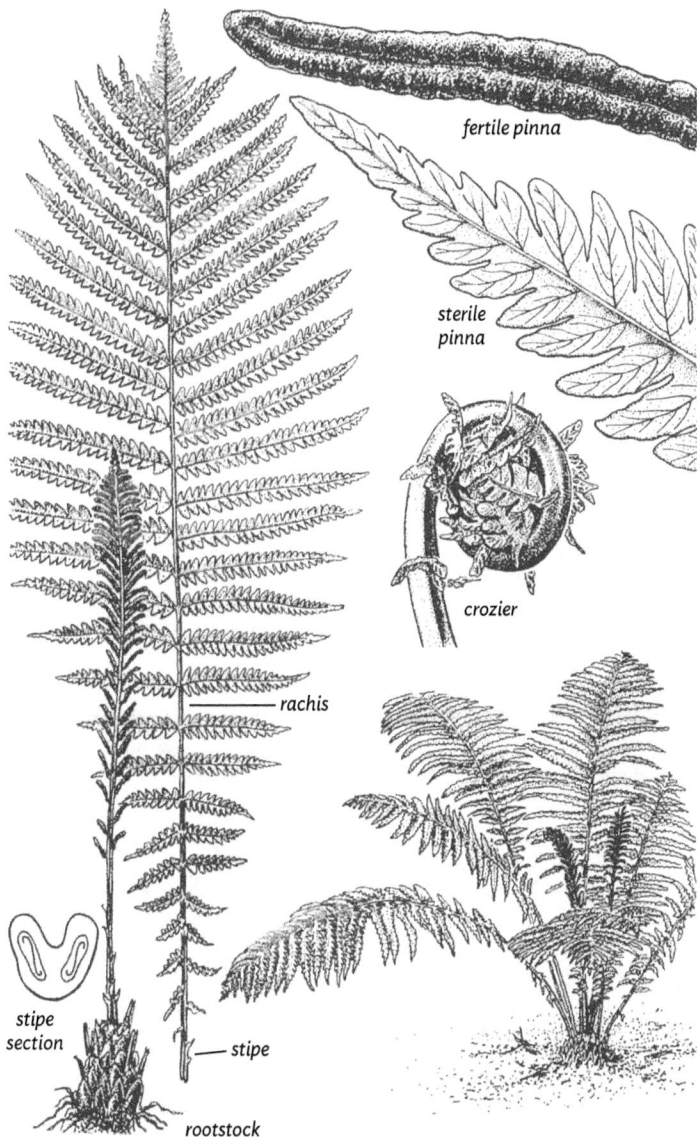

fertile pinna

sterile pinna

crozier

rachis

stipe section

stipe

rootstock

Onoclea sensibilis L.

SENSITIVE FERN

FIELD TIPS

- coarse fern, may form large patches
- sterile fronds deciduous, triangular, not cut to rachis
- fertile fronds brown, with bead-like segments enclosing sori, persistent

MIDWEST RANGE

Ill, Ind, Iowa, Mich, Minn, Ohio, Wisc; common.

HABITAT

Swamps, streambanks, thickets, marshes, moist woods; in sun or shade, often forming dense colonies.

DESCRIPTION

Rootstock Long-creeping, branched, producing a fibrous mat near the soil surface, smooth or with a few scales.

Crozier Pale red.

Frond Sterile and fertile fronds different, arising singly from points along the rhizome; sterile fronds deciduous, often tilted up and back, turning brown in late-sumer to early fall, even before frost, to about 60 cm long. Fertile fronds shorter than sterile fronds, not evergreen but persistent and erect overwinter and sometimes for several years.

Stipe Usually longer than blade, swollen and dark brown at base, with a few light-brown scales, these deciduous; vascular bundles 2.

Rachis Winged (sterile fronds), especially upward; smooth.

Sterile blades Broadly triangular, to 40 cm long and about as wide; 1-pinnate-pinnatifid at base, pinnatifid above.

Fertile blades 2-pinnate, dark brown with buff rachises

Pinnae 5 to 11 pairs, lowest pinnae largest or nearly so, stemless or attached laterally to rachis; upper pinnae connected to rachis by winglike tissue; fertile pinnules rolled into bead-like segments, enclosing the sporangia; margins entire to wavy; veins of sterile pinnae netted.

Sori Round, covered by rolled-up margins; **indusium** vestigial, not seen; **sporangia** green, the enveloping fertile pinnules becoming black in maturity; spores green in maturity, persist through winter and released in spring before sterile leaves expand.

NAME

John Clayton found this fern in Virginia in the early 1700s, and named by Linnaeus in 1753.

From Latin, *sensibilis*, sensitive, in reference to the sterile fronds quickly turning brown in late-summer or fall.

NORTH AMERICA

MIDWEST

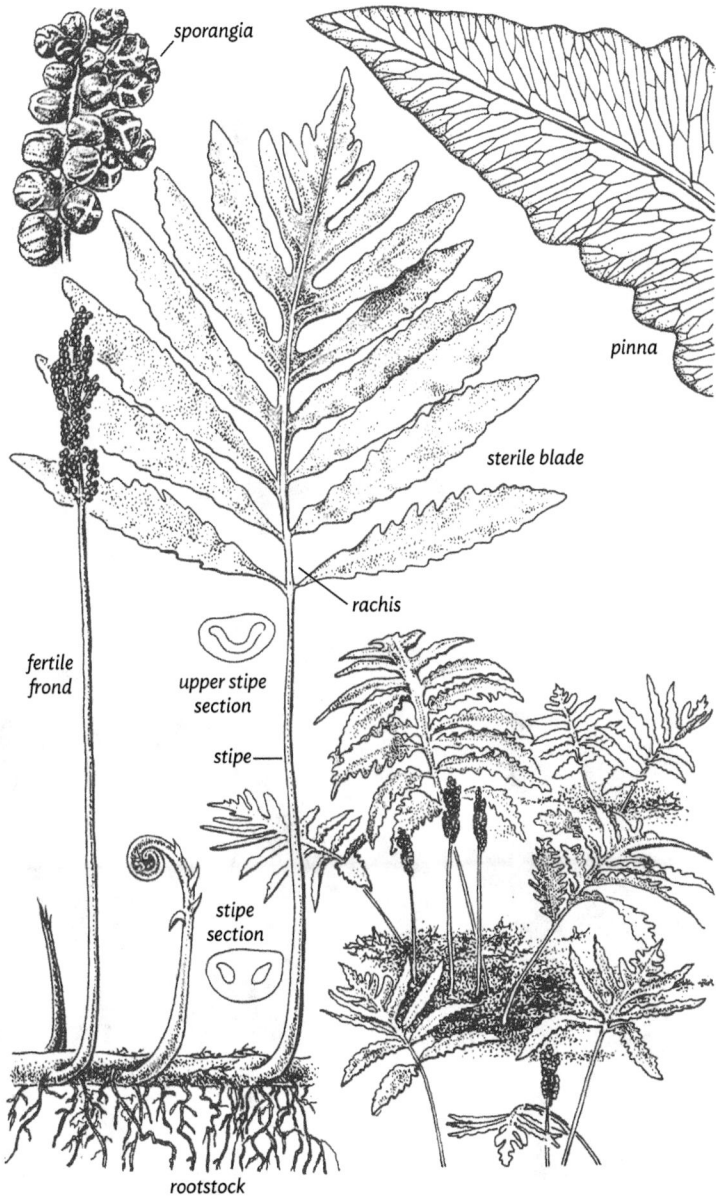

sporangia

pinna

sterile blade

rachis

fertile frond

upper stipe section

stipe

stipe section

rootstock

Ostrich fern (*Matteuccia struthiopteris*), new fronds unfolding to form vaselike cluster.

Ostrich fern (*Matteuccia struthiopteris*), fertile fronds (l), detail of sterile frond (r).

Sensitive fern (*Onoclea sensibilis*), sterile frond.

Sensitive fern (*Onoclea sensibilis*), fertile frond (l), netlike veins of sterile frond (r).

OPHIOGLOSSACEAE
Adder's-Tongue Family

Small to medium, perennial, somewhat succulent ferns. Roots few, fleshy, bearing a **common stalk** divided into a sterile blade portion (**trophophore**) and a fertile sporangia-bearing portion (**sporophore**); blade simple, divided, or compound; fertile portion without blade-like tissue, typically composed of a long stalk with a terminal, branched or unbranched, sporangia-bearing area; **sporangia** sphere-shaped, large (in comparison with those of most other ferns), about 1 mm in diameter, thick-walled, in 2 rows on the branches or on the unbranched sporangia-bearing stalk. The sphere-shaped sporangia are not covered by an indusium, lack an annulus or ring, and when mature, open via a transverse slit to release spores.

Unlike most ferns, new fronds do not emerge in the familiar fiddlehead form (circinate vernation), instead, the new shoot emerges uncoiled and straight in *Botrychium*, and somewhat bent in rattlesnake fern (*Botrypus*) and grape fern (*Sceptridium*). Gametophyte located underground, usually without chlorophyll and associated with mycorrhizal fungi.

In contrast to most true ferns (the "leptosporangiate" ferns), members of this family are part of the "eusporangiate" group which produce large, thick-walled sporangia, and perhaps only distantly related to other true ferns. Overall, a cosmopolitan family of 5-6 genera and ca. 70-80 species, sometimes subdivided into 2 subfamilies: **Botrychioideae** (including genera *Botrychium, Botrypus, Sceptridium*) and **Ophioglossoideae** (genus *Ophioglossum*); these also sometimes recognized as distinct families.

KEY CHARACTERS
- Mostly small ferns, rhizomes and stalks fleshy.
- New shoots not coiled like a fiddlehead, instead emerging straight from the ground.
- Fronds divided into a single **sterile blade** (trophophore) and a spike-like or panicle-like **fertile stalk** (sporophore).
- **Sporangia** large, thick-walled.
- Habitats diverse, from open to shaded, moist to dry, sometimes where somewhat disturbed so that the litter layer is not very deep.

NOTES
Taxonomically, **rattlesnake fern** (*Botrypus*) and **grape ferns** (*Sceptridium*) often included in *Botrychium*; however, given the large number of moonworts and readily apparent differences amongst species, some classifications separate the family into 3 subgenera or 3 distinct genera (as incorporated here).

NAME
Family name from Greek, *ophis*, snake, *glossa*, tongue; the narrow fertile segment suggesting the name adder's-tongue.

KEY TO OPHIOGLOSSACEAE | ADDER'S-TONGUE FAMILY

1   Sterile blades simple, entire; veins netlike; sporangia imbedded in rachis of spike **1. Ophioglossum**
ADDER'S-TONGUE, page 260

1   Sterile blades pinnately lobed or dissected; veins forked; sporangia exposed, often on a branched structure. **2**

2   Sterile blades somewhat leathery, persisting over winter; blades with distinct stalk, usually 5-25 cm long in fertile plants; fertile portion of frond joining sterile portion at or near ground level
2. Sceptridium
GRAPE FERN, page 269

2   Plants deciduous, withering in fall; blades usually ± unstalked, large (more than 5 cm long) or much smaller (in some species of *Botrychium*), herbaceous or sometimes fleshy; fertile portion of frond joining sterile portion well above ground level **3**

3   Sterile blades triangular, 3–4x pinnate, stalkless and mostly 5–25 cm wide; fertile portion erect, appearing to be a continuation of stipe **3. Botrypus virginianus**
RATTLESNAKE FERN, page 258

3   Sterile blades short-triangular, oblong, or linear; lobed (simple) to 3-pinnate (usually 1-pinnate to 2-pinnate-pinnatifid), mostly 1-5 cm wide; fertile portion of blade upright or spreading
4. Botrychium
MOONWORT, page 228

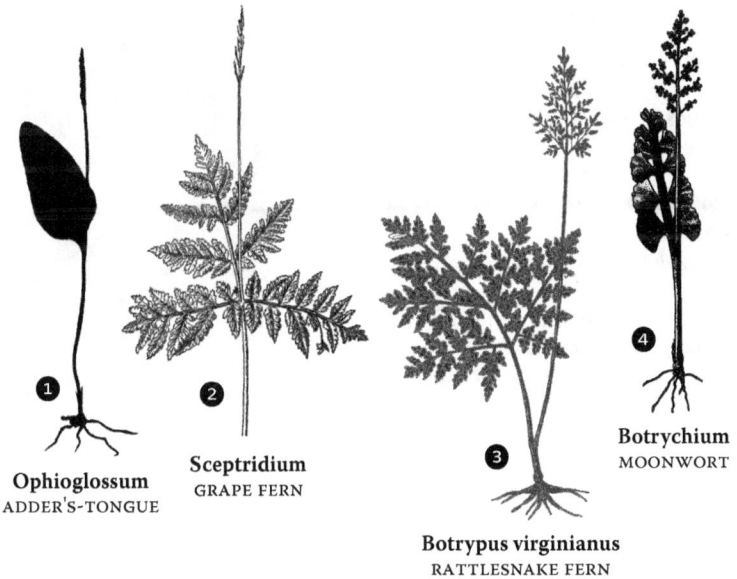

Ophioglossum
ADDER'S-TONGUE

Sceptridium
GRAPE FERN

Botrypus virginianus
RATTLESNAKE FERN

Botrychium
MOONWORT

Botrychium
MOONWORT

Small, smooth ferns with fleshy stems, a single **sterile blade** (trophophore) and a **fertile blade** (sporophore). Plants typically produce only one leaf each year from an underground rhizome. Sterile blade divided or lobed, ovate to triangular in outline; fertile portion of leaf consisting of an elongate stalk terminated by a sporangia-bearing region.

Moonworts have a close association with mycorrhizal fungi on which they depend for their water, mineral, and carbohydrate needs. *Botrychium* distribution is probably highly related to the occurrence of these underground fungi.

A nearly cosmopolitan genus of approx. 50-60 species, with greatest diversity at high latitudes and high elevations; about 30 species in North America; 14 in our flora, some very rare.

KEY CHARACTERS

• Small, somewhat fleshy, deciduous ferns.

• **Fronds** divided into a lobed or divided sterile blade and a fertile portion bearing sporangia on its surface.

• **Sporangia** relatively large, often yellowish when mature.

NOTES

Most species quite variable in their form, and more than one species may occur in suitable habitats, resulting in considerable taxonomic confusion; use the key, species descriptions, habitat information, and distribution maps to confirm your identification. If collecting plants, cut rather than pinch the frond at ground level as underground parts are not necessary and are important to leave in the soil for future growth.

NAME

From Greek, *botrys,* a cluster, referring to the clusters of sporangia. The common name is in reference to the crescent-moon shape of the pinnae of the type species, *Botrychium lunaria,* and the Old English *wyrt,* herb, for alleged medicinal properties.

Botrychium lunaria was first described in 1542 by Fuchs as *Lunaria minor.* Linnaeus recognized two species of *Botrychium* in his 1753 Species Plantarum, *B. lunaria* and *B. virginiana.* He placed both in the genus *Osmunda.* Presl (1845) was the first to use the name *Botrychium,* recognizing 17 species in his treatment of the genus. The first modern comprehensive treatment of the family was that of Clausen in his 1938 Monograph of the Ophioglossaceae.

Botrychium lunaria
COMMON MOONWORT

MOONWORT FORMS

The shape of the trophophore (sterile portion) is the best diagnostic character for *Botrychium*; our species have one of three basic forms:

Ⓐ Blade once-dissected (1-pinnate), pinnae fan-shaped;

Ⓑ Blade triangular, twice-dissected (2-pinnate);

Ⓒ Blade intermediate (pinnate-pinnatifid), form derived from ancestral hybridization between *B. lanceolatum* and species of the fan-leaflet group.

Forms B and C sometimes referred to as the **midribbed species** because their pinnae have strong central veins, whereas those of form A (pinnae fan-shaped) have multiple parallel veins of equal size. Presence of a midrib in the basal pinnae is a good way to identify plants in forms B and C when they are too small to have developed pinna lobing.

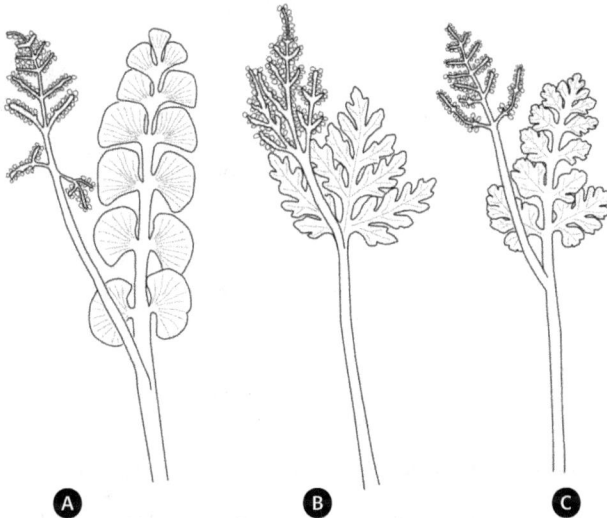

A	**B**	**C**
BLADE 1-PINNATE, PINNAE FAN-SHAPED	BLADE TRIANGULAR, 2-PINNATE	BLADE 1-PINNATE-PINNATIFID

• *Botrychium campestre*
  • *B. lunaria*
  • *B. minganense*
  • *B. mormo*
  • *B. pallidum*
  • *B. simplex*

• *Botrychium hesperium*
  • *B. lanceolatum*

• *Botrychium acuminatum*
  • *B. matricariifolium*

KEY TO BOTRYCHIUM | MOONWORT

1 Sterile blade (trophophore) simple to lobed, lobes rounded to square and angular, stalks usually 1/2 to 2/3 length of sterile blade; rare plant of shaded woods *(Botrychium mormo)*, or more common and often in open grassy fields *(Botrychium simplex)* **2**

1 Sterile blade pinnately lobed (either if actual pinnae or simply lobed), lobes of varying shapes, stalk usually less than 1/4 length of sterile blade; plants often in open sunny places; dunes, streambanks, roadsides, on trails and openings in forests, etc. **3**

2 Segments of sterile leaves rounded, margins mostly entire; plants herbaceous in texture, green; habitats various, in forests or open ground; n Ill, n Ind, Iowa, Mich, Minn, Ohio, Wisc

1. Botrychium simplex
LEAST MOONWORT, page 252

2 Segments of sterile leaves angular, outer margins often coarsely dentate; plants ± succulent; shiny yellow-green; rare in shady forest understories, often concealed by leaf litter; n Mich, Minn, Wisc

2. Botrychium mormo
LITTLE GOBLIN MOONWORT, page 248

3 Basal pinnae or segments of sterile blade with venation like the ribs of a fan; midrib absent **4**

3 Basal pinnae or segments of sterile blade with pinnate venation; midrib present
. . . . **8**

4 Basal pinnae broadly fan-shaped (almost perfect half moons) with narrow stalks; Mich, n Minn, n Wisc **3. Botrychium lunaria**

COMMON MOONWORT, page 242

4 Basal pinnae narrowly fan- or wedge-shaped to nearly linear **5**

5 Sterile blade at least partially folded longitudinally when alive (conduplicate), usually not more than 4 cm long by 1 cm wide; pinnae up to 5 pairs; basal pinnae usually 2-parted **6**

5 Sterile blade flat or folded only at base when alive, usually up to 10 long by 2.5 cm wide; pinnae up to 10 pairs; basal pinnae unlobed, or if lobed, not usually 2-parted. **7**

6 Sterile blade very fleshy; fertile portion of blade usually less than 1.5 times length of sterile blade; pinnae mostly linear; basal pinna lobes ± equal; plants appearing in late spring; n Ill, Iowa, Mich, Minn, Wisc **4. Botrychium campestre**

IOWA MOONWORT, page 236

Botrychium simplex
LEAST MOONWORT

Botrychium mormo
LITTLE GOBLIN MOONWORT

Botrychium lunaria
COMMON MOONWORT

Botrychium campestre
IOWA MOONWORT

KEY TO BOTRYCHIUM | MOONWORT, CONTINUED

6 Sterile blade herbaceous; fertile portion of blades usually 1.5–4 times the length of vegetative blades; pinnae asymmetrically fan-shaped; basal pinna lobes unequal; plants appearing in summer; Mich, n Minn, nw Wisc   **5. Botrychium pallidum**
PALE MOONWORT, page 250

7 Sterile blade narrowly oblong (sterile blade widest above base), firm to herbaceous; pinnae fan-shaped, margins shallowly crenate; proximal fertile portion of blade branches 1-pinnate; Mich, n Minn, n Wisc   **6. Botrychium minganense**
MINGAN MOONWORT, page 246

7 Sterile blade narrowly deltate (sterile blade widest at lowest pinna pair); pinnae spatulate to linear spatulate, margins entire to very coarsely and irregularly dentate; most proximal fertile portion of blade branches usually 2-pinnate; n Mich, Minn, Wisc (Door County)   **7. Botrychium spathulatum**
SPOON-LEAF MOONWORT, page 254

8 Fertile portion of blade 3-parted, with 3 major branches from near base of stalk at sterile blade **9**

8 Fertile portion of blade unbranched or with loosely pinnate branches smaller than the single main stem **10**

9 Sterile blade triangular in outline, margins toothed throughout; mostly in forests; occasional in other habitats; Mich, Minn, Ohio, Wisc   **8. Botrychium lanceolatum**
LANCE-LEAF MOONWORT, page 240

9 Sterile blade ovate or oblong to linear with (usually) enlarged basal pinnae; remaining pinnae ± entire-margined; dunes and inland areas of sand, occasional in forests; n Mich
          9. Botrychium hesperium
WESTERN MOONWORT, page 238

10 Pinnae of small to medium sterile blades at least wavy-margined; larger blades with long narrow "teeth" especially on basal pinnae; sand dunes, occasional inland; Mich, ne Minn
          10. Botrychium acuminatum
TAILED MOONWORT, page 244

10 Pinnae of small sterile blades mostly smooth margined but appearing unevenly branched; large blades regularly multi-pinnate; n Ill, n Ind, e Iowa, Mich, Minn, Ohio, Wisc
          11. Botrychium matricariifolium
DAISY-LEAF MOONWORT, page 244

Botrychium spathulatum
SPOON-LEAF MOONWORT

Botrychium minganense
MINGAN MOONWORT

Botrychium pallidum
PALE MOONWORT

Botrychium lanceolatum
LANCE-LEAF MOONWORT

Botrychium hesperium
WESTERN MOONWORT

Botrychium matricariifolium
DAISY-LEAF MOONWORT

ADDITIONAL MIDWEST SPECIES

Four uncommon species of moonwort are known from northern Minnesota:

❶ **Triangle-lobe moonwort** (*Botrychium ascendens* W.H. Wagner), St. Louis and Crow Wing Counties, northern Minnesota. Pinnae of sterile blade wedge-shaped, angled upward, outer margins toothed, lower pinnae sometimes divided into 2 lobes and also bearing sporangia; found in open, grassy fields.

❷ **Frenchman's Bluff moonwort** (*Botrychium gallicomontanum* Farrar & Johnson-Groh), known from northwestern Minnesota on Frenchman's Bluff in Norman County and several locations in Kittson County(state endangered), and from Glacier National Park in Montana. Habitats are dry, sandy to gravelly, mid to tall grass prairie. Found growing with *B. campestre* and *B. simplex* and intermediate in form and possibly a hybrid between these 2 species. Distinguishing features include the large space between the lowest pinnae pair and the next pinnae pair, the strongly ascending pinnae, and the distinctly stalked sterile blade.

❸ **Narrow-leaf moonwort** (*Botrychium lineare* W.H. Wagner), St. Louis County, northern Minnesota. Plants small, leathery, pale green; sterile blade with 4-6 widely spaced pinnae pairs, the pinnae very narrowly spoon-shaped, sometimes divided into 2 spreading lobes; somewhat similar to *B. campestre* but pinnae in that species wider and less deeply lobed. Reports from other regions suggest an affinity for gravelly or limestone habitats.

❹ **False daisy-leaf moonwort** (*Botrychium pseudopinnatum* W.H. Wagner), known from St. Louis County, Minnesota, plus reported from Douglas County in northwest Wisconsin. May be more widespread in northern parts of our region as often confused with the more common **daisy-leaf moonwort** (*B. matricariifolium*). The sporophore of *B. pseudopinnatum* tends to be more branched, and plants are a lustrous deep green vs. the dull paler green of *B. matricariifolium*.

Botrychium ascendens
Triangle-lobe moonwort

Botrychium gallicomontanum
Frenchman's Bluff moonwort

Botrychium lineare
Narrow-leaf moonwort

Botrychium pseudopinnatum
False daisy-leaf moonwort

❶ Botrychium ascendens
TRIANGLE-LOBE MOONWORT

sporophore

Botrychium gallicomontanum
FRENCHMAN'S BLUFF MOONWORT

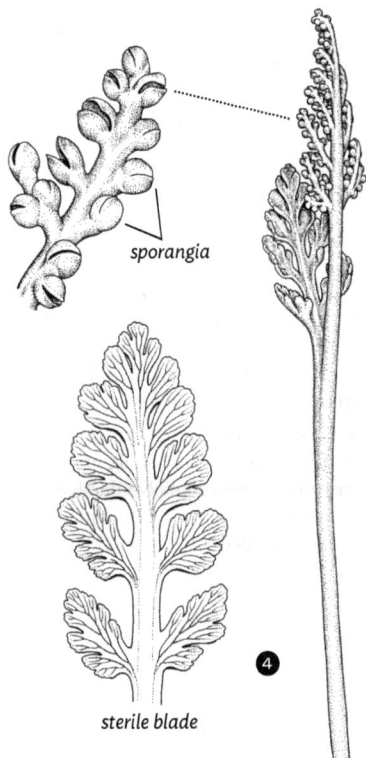

sporangia

sterile blade

Botrychium pseudopinnatum
FALSE DAISY-LEAF MOONWORT

Botrychium lineare
NARROW-LEAF MOONWORT

Botrychium campestre

W.H. Wagner & Farrar

IOWA MOONWORT

FIELD TIPS

- very small and inconspicuous fern of prairies, sand dunes and other sunny, calcareous sites
- stems and leaves succulent or fleshy

MIDWEST RANGE

n Ill (endangered), Iowa, Mich (threatened), Minn, Wisc (endangered).

HABITAT

Prairies, dunes, grasslands, and other open, calcareous sites. In Michigan, known from sand dunes of Sleeping Bear National Lakeshore; also from sandy, open places, usually near Great Lakes. In Minnesota, found in dry prairies having coarse, shallow soil; associated species are little bluestem (*Schizachyrium scoparium*) and big bluestem (*Andropogon gerardii*).

DESCRIPTION

Frond Small, less than 5 cm tall and hard to find if surrounded by taller vegetation. Plants succulent, developing early in growing season, with aboveground portion drying by mid-summer.

Sterile blade Fleshy, once pinnate, sessile to the common stalk, oblong in outline, longitudinally folded; to about 4 cmlong and 1.3 cm wide. Usually divided into 5 pairs of linear or linear-spatulate segments; margins crenate or dentate and also usually notched into 2 or several smaller segments.

Common stalk Short, less than 3 cm long.

Sporophore Often large relative to size of sterile blade.

NOTES

Tiny plantlets (gemmae) may be produced on underground stems.

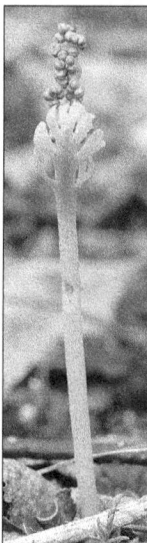

NORTH AMERICA

MIDWEST

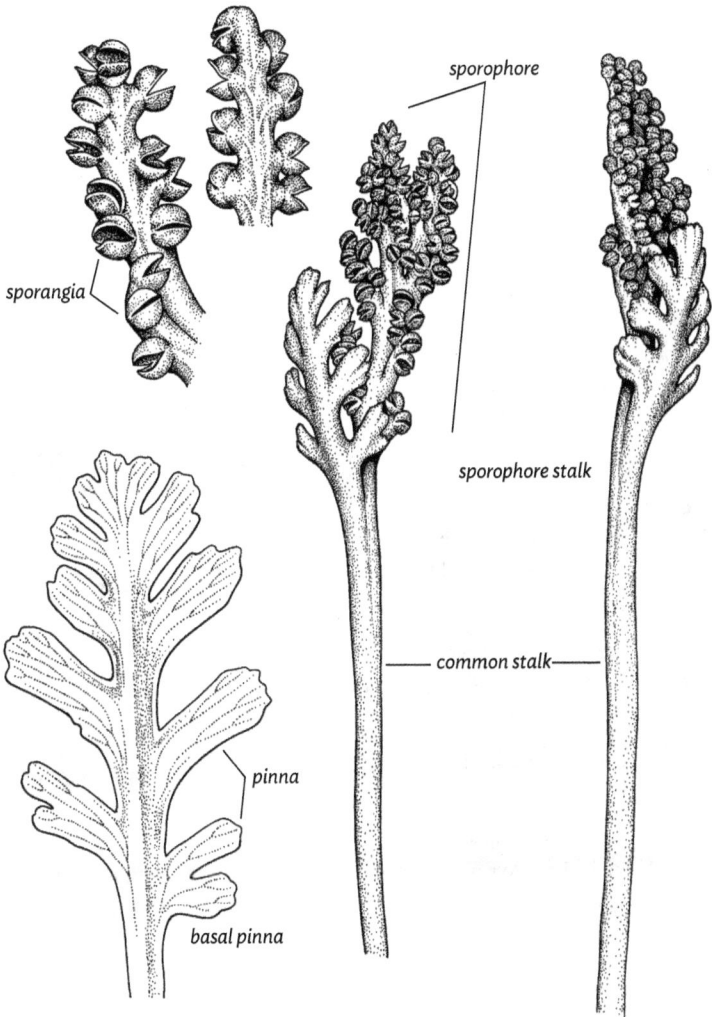

sporophore

sporangia

sporophore stalk

common stalk

pinna

basal pinna

Botrychium hesperium

(Maxon & Clausen) W.H. Wagner & Lellinger

WESTERN MOONWORT

FIELD TIPS

• basal pinnae exaggerated in size in contrast to rest of blade, which is often narrow and merely lobed
• plants succulent, dull gray-green
• rare in sand dunes and other open sandy habitats

MIDWEST RANGE

n Mich (threatened); disjunct from main range of Rocky Mountains.

HABITAT

Often on dunes, but rather local; also sandy open fields, sandy open woodlands; often with *Botrychium campestre* and *B. matricariifolium*.

DESCRIPTION

Frond Average 12 (5-20) cm in height.

Sterile blade Dull gray-green; oblong-linear to triangular in outline, 1- or 2-pinnate, 2.5 (1-5) cm long. Pinnae up to 6 pairs, usually crowded or overlapping, rachis relatively wide; basal pair of pinnae typically much larger and more divided than the others and with irregularly lobed margins; often slightly more distance between 1st and 2nd pinna pairs than between 2nd and 3rd pinna pairs; stalk absent or short, to 3 mm long.

Sporophore Relatively tall, to 2x length of sterile blade.

SYNONYMS

• *Botrychium lunaria* (L.) Sw. subsp. *occidentale* Á. Löve & D. Löve & Kapoor
• *Botrychium matricariifolium* (A. Braun ex Dowell) A. Braun ex W.D.J. Koch subsp. *hesperium* Maxon & R.T. Clausen

NOTES

Ours plants sometimes considered separate species **Michigan moonwort** (*Botrychium michiganense* Gilman & F.S. Wagner), with basal pinnae tending to be larger and narrower than those of the more western species *B. hesperium*.

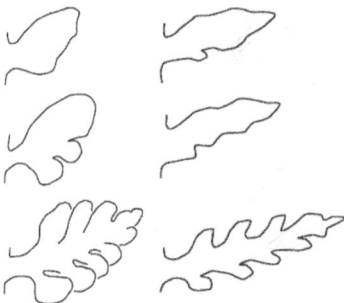

Upper, middle and basal pinnae of *B. hesperium* (l) and *B. michiganense* (r).

NORTH AMERICA

MIDWEST

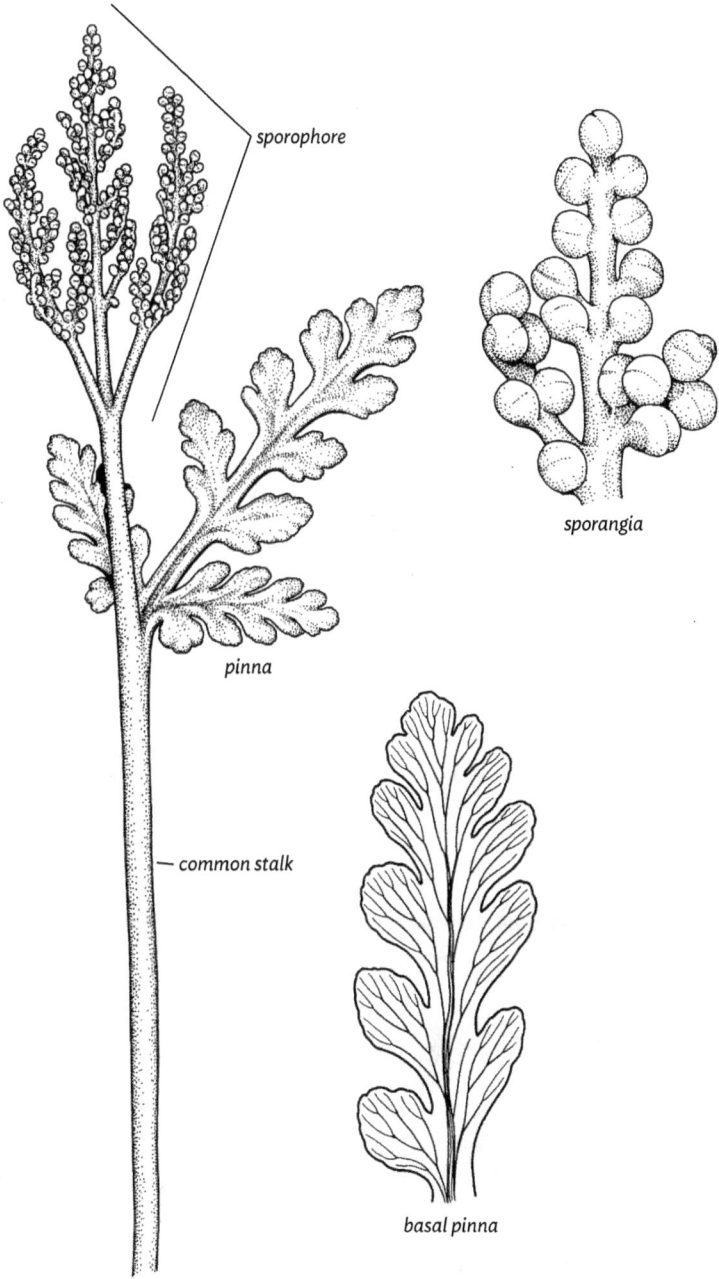

sporophore

sporangia

pinna

common stalk

basal pinna

Botrychium lanceolatum
(Gmel.) Angstr.

LANCE-LEAF MOONWORT

FIELD TIPS
- blade sessile, triangular, dark green and shiny
- pinnae deeply lobed

MIDWEST RANGE
Mich, Minn (threatened), Ohio (threatened), Wisc.

HABITAT
Most commonly in humus-rich leaf mold soils of moist deciduous and mixed conifer-hardwood forests, often with sugar maple (*Acer saccharum*).

DESCRIPTION
Frond 2-20 cm long or longer, fleshy.

Sterile blade Sessile, broadly triangular in outline, 2-3 cm long, dark green and shiny, from upper portion of common stalk; pinnae mostly 4-5 pairs, deeply lobed.

Common stalk Relatively long, brownish green.

Sporophore Divided into several more or less equal, slightly spreading branches. Spores mature in July and August.

SYNONYMS
Only variety *angustisegmentum* is present in our area; var. *lanceolatum* occurs in western North America.

- *Botrychium angustisegmentum* Pease & A.H. Moore) Fernald
- *Botrychium lanceolatum* (S.G. Gmel.) Angstr. subsp. *angustisegmentum* (Pease & A.H. Moore) R.T. Clausen

SIMILAR SPECIES
- **Daisy-leaf moonwort** (*Botrychium matricariifolium*) similar but in *B. lanceolatum* sterile blade stalkless, basal pinnae much larger than other pinnae, and blade triangular in outline rather than trowel-shaped.

NOTES
This is another of the species of grapeferns discovered by Gmelin in Russia, and assigned by him in 1768 to the genus *Osmunda*; transferred to *Botrychium* by Angström in 1854.

NAME
From Latin, *lanceola*, a small lance, referring to the lance-shaped pinnae.

NORTH AMERICA

MIDWEST

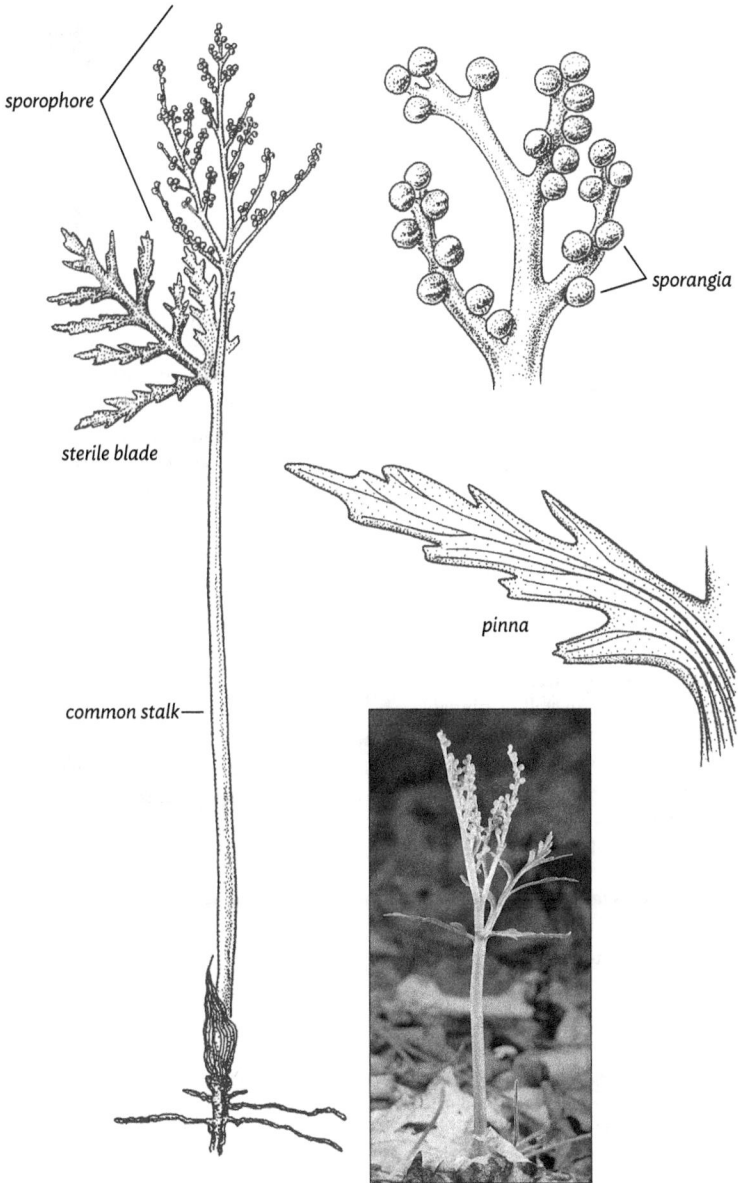

sporophore

sporangia

sterile blade

pinna

common stalk—

Botrychium lunaria (L.) Sw.
COMMON MOONWORT

FIELD TIPS
• small moonwort of open habitats, often where calcium-rich
• sterile blade stalkless, shiny, fleshy, deep green
• pinnae segments fan-shaped

MIDWEST RANGE
Mich, n Minn (threatened), n Wisc (endangered).

HABITAT
Open grassy fields and forests, usually on sandy or gravelly soil, sometimes over limestone and under trees of northern white cedar (*Thuja occidentalis*); also where somewhat disturbed.

DESCRIPTION
Fronds Mostly less than 15 cm long, somewhat leathery.

Sterile blade Deep green, fleshy, shiny, oblong, to 10 cm long, stalkless or nearly so, inserted near or below middle of common stalk, 1-pinnate. Pinnae 6- 9 pairs, spreading, mostly overlapping except in shaded forest forms, basal pinna pair about equal in size and cutting to adjacent pair, broadly fan-shaped, undivided to tip, margins entire or undulate, rarely dentate, veins fanlike, midrib absent.

Common stalk 3-7 cm from base to juncture with sterile blade.

Sporophore 1-2 pinnately branched, sometimes tip bending slightly downward; spores mature June-August.

SYNONYMS
• *Botrychium onondagense* Underw.

SIMILAR SPECIES
• **Mingan moonwort** (*Botrychium minganense*) similar but sterile blade stalked and pinnae pairs separated and not overlapping.

Most easily distinguished from other moonworts by spread of its pinnae; typically basal pinnae span an arc of nearly 180 degrees, 3rd pinna pair spans ca. 90 degrees.

NOTES
The subject of many tales in northern European folklore, including opening locked doors without a key; spores were said to make one invisible.

NAME
From Latin, *luna*, moon, referring to the crescent-shaped pinna segments.

NORTH AMERICA

MIDWEST

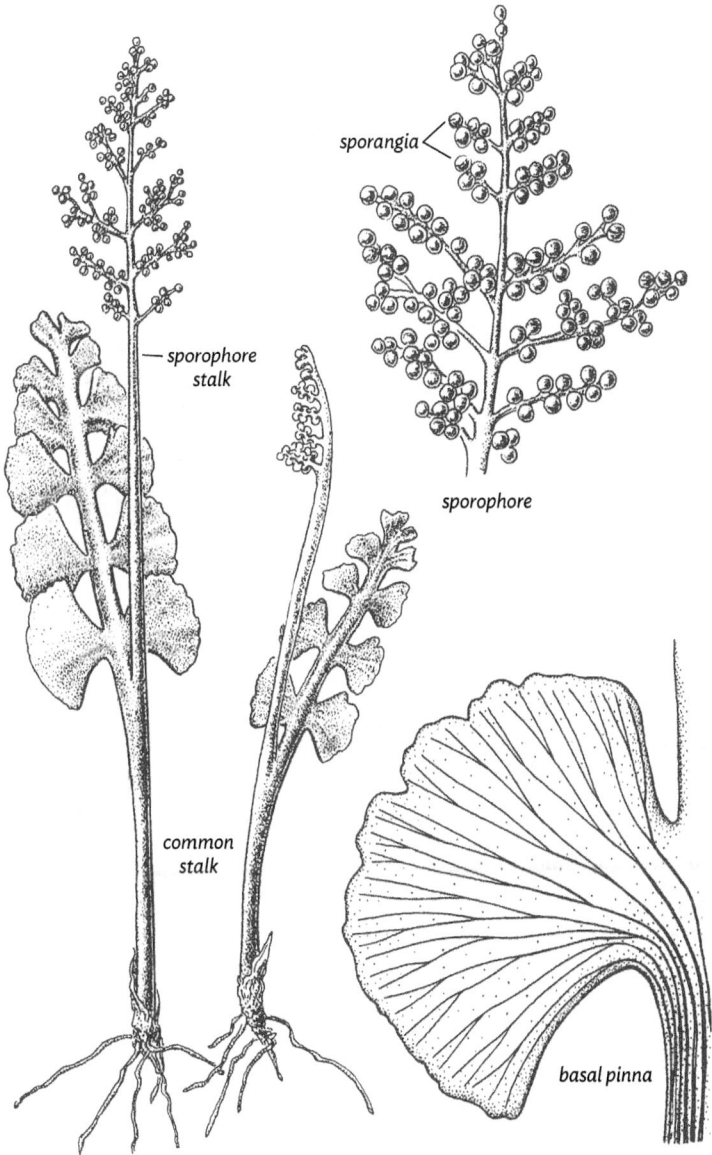

sporangia

sporophore stalk

sporophore

common stalk

basal pinna

Botrychium matricariifolium
(A. Braun ex Dowell) A. Braun ex Koch
DAISY-LEAF MOONWORT

FIELD TIPS
• plants to 20 cm tall
• sterile blade stalked, dull pale green, trowel-shaped in outline
• sporangia yellow

MIDWEST RANGE
n Ill (endangered), n Ind (threatened), e Iowa (endangered), Mich, Minn, Ohio, Wisc.

HABITAT
Most common on sandy, acidic soils such as old fields, clearings, and dry rocky woods; also reported from moist cedar woods or swamps.

DESCRIPTION
Fronds Variable in shape and form, 5-20 cm long, somewhat fleshy.

Sterile blade Dull, pale green, trowel-shaped, longer than wide; varies from 1-3x pinnate; short-stalked, inserted above middle of common stalk; pinnae 2-7 pairs, the segments often deeply lobed.

Common stalk Slender, fleshy, pale green, often with a pinkish stripe.

Sporophore Erect, usually with 3 main branches, these again divided; sporangia large, yellow.

SIMILAR SPECIES
• Lance-leaf moonwort (*Botrychium*

lanceolatum), but in that species sterile blade nearly stalkless, and triangular rather than trowel-shaped in outline.

NAME
Species name first suggested by A. Braun in 1843, and refers to similarity of blade to the dissected leaves of chamomile (*Matricaria*).

ADDITIONAL MIDWEST SPECIES
• **Tailed moonwort** (*Botrychium acuminatum* W.H. Wagner) similar to *B. matricariifolium*, and previously considered a form of that species, but the pinnae much more separated, and have much narrower lobes (or none). Known from Michigan (endangered), n Minnesota and Lake Superior region of Ontario; typical habitats are vegetated Great Lakes shoreline dunes and sandy places inland from dunes. In Michigan, associated species include dune grass (*Calamovilfa longifolia*), marram grass (*Ammophila breviligulata*), ground juniper (*Juniperus communis*), white spruce (*Picea glauca*), sand cherry (*Prunus pumila*), and jack pine (*Pinus banksiana*).

NORTH AMERICA

MIDWEST

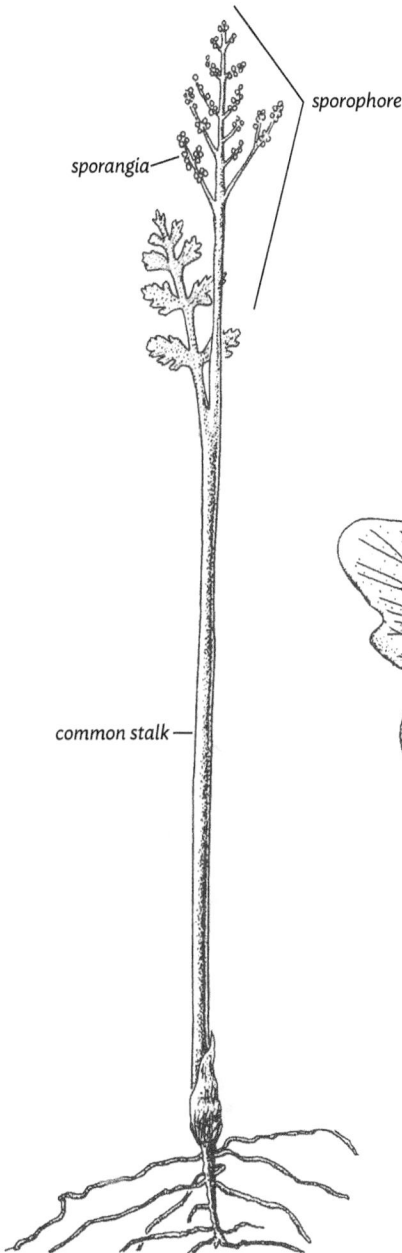

sporophore

sporangia

common stalk

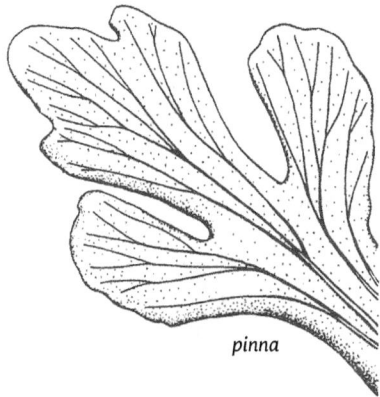

pinna

Botrychium minganense Victorin

MINGAN MOONWORT

FIELD TIPS
- plants to 20 cm tall but usually smaller
- sterile blade dull yellow-green, fleshy
- sporophore longer than sterile blade

MIDWEST RANGE
Mich, n Minn, n Wisc.

HABITAT
Moist hardwood forests, aspen-balsam fir woods, and old clearings; soils mostly circumneutral.

DESCRIPTION
Fronds Yellow-green, somewhat membranous, to about 20 cm tall, but usually much smaller.

Sterile blade Fleshy, dull yellowish green, stalked, emerges below or near middle of common stalk. Pinnae usually 6-10 pairs, ascending, fan-shaped or shallowly notched; sometimes with a few sporangia on margins of lower pinnae.

Sporophore 1-pinnate, longer than sterile blade.

SYNONYMS
- *Botrychium lunaria* (L.) Sw. subsp. *minganense* (Victorin) Calder & Roy L. Taylor

SIMILAR SPECIES
- Distinguished from **moonwort** (*Botrychium lunaria*) by its yellow-green color, stalked rather than sessile

sterile blade, and narrower basal pinnae segments, which are not overlapping or as broadly fan-shaped.

- **Pale moonwort** (*Botrychium pallidum*) differs in being distinctly smaller and the pinnae whitish green.

- Somewhat similar to **spoon-leaf moonwort** (*B. spathulatum*) but the pinnae in that species narrower and spoon-shaped.

NAME
Refers to Mingan Island in Gulf of St. Lawrence, where species first found as a new species (collected previously but confused with other species, especially *Botrychium simplex*).

Earlier studies considered *Botrychium minganense* and *B. spathulatum* to be forms of the more common *B. lunaria*.

dried specimens

NORTH AMERICA

MIDWEST

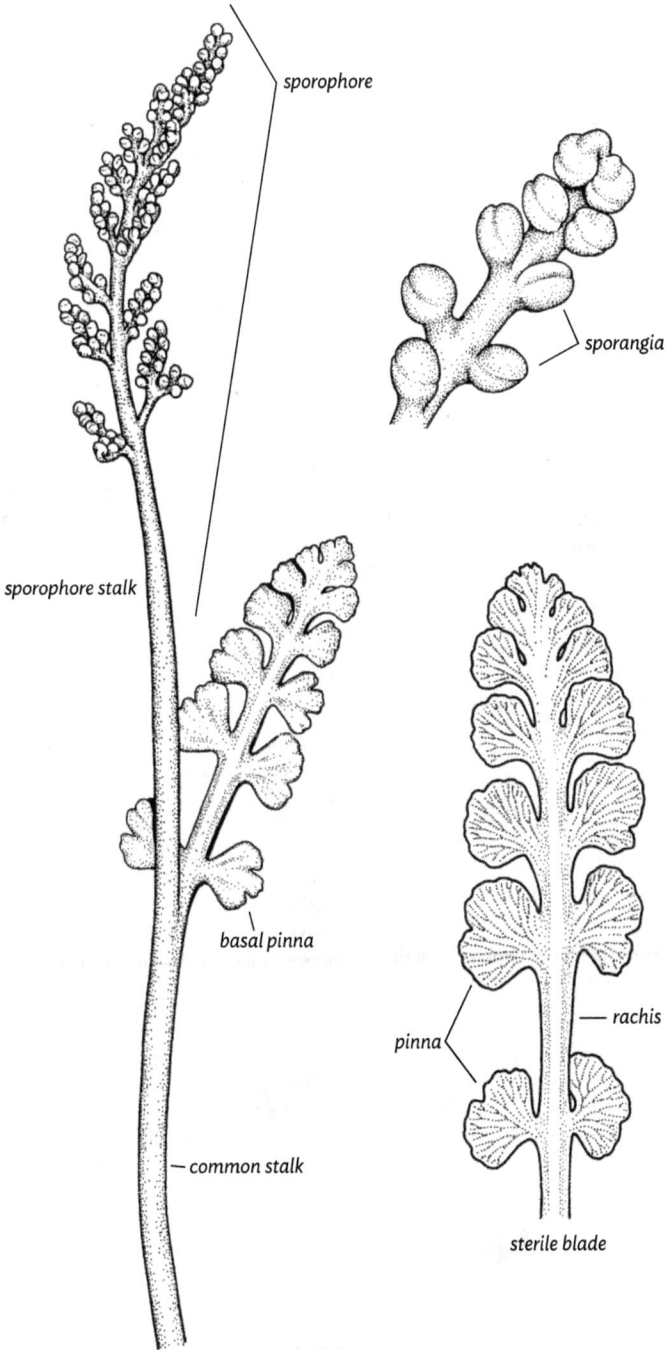

sporophore

sporangia

sporophore stalk

basal pinna

pinna

rachis

common stalk

sterile blade

Botrychium mormo W.H. Wagner
LITTLE GOBLIN MOONWORT

FIELD TIPS
- tiny moonwort of rich deciduous forests
- plants succulent, yellow-green
- pinnae short, blunt
- sporangia sunken into fleshy stalk of sporophore

MIDWEST RANGE
n Mich (threatened), Minn, Wisc (endangered).

HABITAT
Mature deciduous forests, typically dominated by sugar maple or basswood, and sometimes with eastern hemlock or northern white cedar. Sites shaded and moist; soils are loams, with a rich litter layer.

DESCRIPTION
Frond Our tiniest moonwort, uncommon and sometimes not appearing above the leaf litter; plants to about 8-10 cm high but often smaller; yellow-green, somewhat shiny. Gametophyte sometimes persisting at base of plant.

Sterile blade Variable; blade of well-developed plants with 2-3 pairs of small blunt lobes; blade in smaller plants may be nearly absent.

Sporophore To about 3 cm long, with several sporangia embedded in the fleshy stalk.

SIMILAR SPECIES
- **Least moonwort** (*Botrychium simplex*) similar but occurs in different habitats (dry fields, bogs, swamps, roadsides, ditches) than *B. mormo*. The most similar variety is *B. simplex* var. *tenebrosum* (A.A. Eaton) R. T. Clausen, distinguished by its generally non-fleshy appearance, duller, greener color, and high attachment of leaf blade to common stalk.

NOTES
While always uncommon and difficult to find, little goblin moonwort tends to disappear during droughty years, remaining dormant within the ground litter or producing very small plants without sterile blades. Plants growing in the litter layer appear whitish and lack chlorophyll.

One serious threat to *B. mormo* populations are **earthworms** which can completely consume the organic surface layer of forest soils upon which little goblin fern is dependent (it is interesting to note that native earthworms are absent from the northwoods but have been inadvertently introduced into forests by gardeners and fishermen).

NORTH AMERICA

MIDWEST

Declines of this *Botrychium* have been documented on national forests of the region.

NAME

Mormo refers to an evil, goblin-like spirit of Greek mythology, perhaps in reference to the ghostly white color of plants not growing out of the leaf litter.

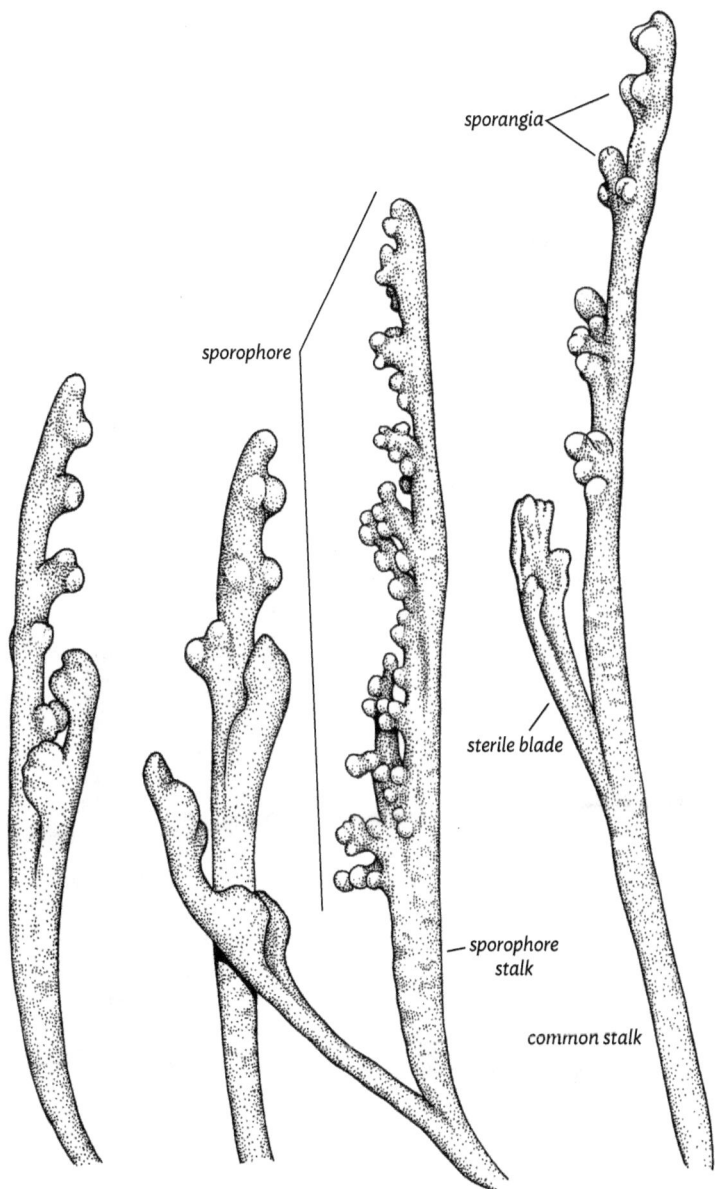

Variation in plant size and development

Botrychium pallidum W.H. Wagner
PALE MOONWORT

FIELD TIPS
- plants very small, waxy pale green to whitish
- pinnae usually folded lengthwise
- lower pinnae often 2-lobed, sometimes with sporangia

MIDWEST RANGE
Mich, n Minn (endangered), nw Wisc.

HABITAT
Open fields, dry sand and gravel ridges, roadsides, wet depressions, marshy lakeshores, tailings basins, second-growth forests; soils sandy.

DESCRIPTION
Frond Small, 2.5-7 cm long, waxy pale green to whitish.

Sterile blade To 4 cm long by 1 cm wide, 1-pinnate, with up to 5 pairs of fan-shaped pinnae, each pair of pinnae often folded towards each other; basal pinnae usually divided into 2 unequal lobes, the upper lobe larger; margins entire to irregularly toothed. Sporangia sometimes present on lobes of lower pinnae.

Sporophore Longer than sterile blade, the sporangia on short branches from main stalk.

SIMILAR SPECIES
- **Mingan moonwort** (*Botrychium minganense*) similar but larger and yellow-green rather than pale green as in *B. pallidum*; pinnae not folded as in *B. pallidum*.

NOTES
Often with dense clusters of tiny, spherical **gemmae** (plantlets) on underground roots.

NORTH AMERICA

MIDWEST

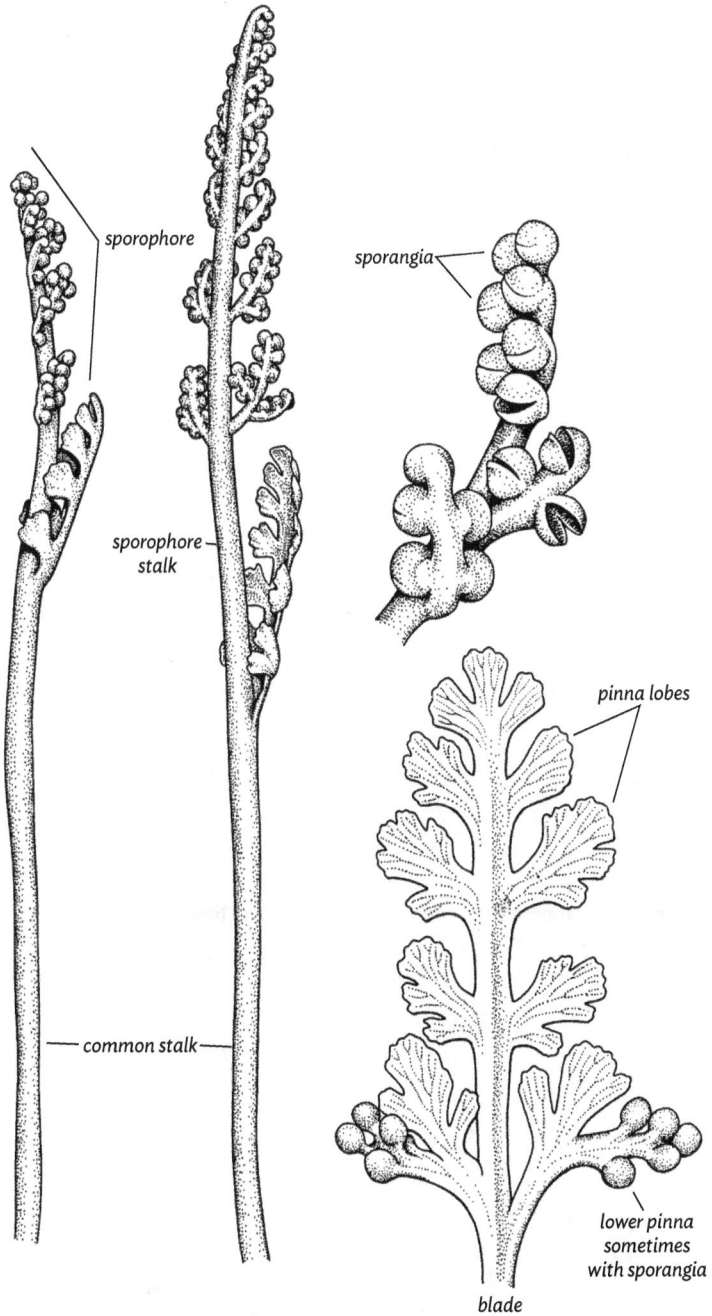

sporophore

sporangia

sporophore
stalk

pinna lobes

common stalk

lower pinna
sometimes
with sporangia

blade

Botrychium simplex E. Hitchc.

LEAST MOONWORT

FIELD TIPS

- small plants, dull to bright green to whitish green
- sterile blade simple to 1-pinnate; lowest pinnae usually largest, divided
- sporophore often unbranched

MIDWEST RANGE

n Ill (endangered), n Ind (endangered), Iowa (threatened), Mich, Minn, Ohio (endangered), Wisc.

HABITAT

Variety of open to shaded habitats: prairies, clearings, bracken fern grasslands, sandy woods, swamps and lakeshores. *Botrychium simplex* var. *tenebrosum* more common in forests, especially low places in moist hardwood forests or on mossy hummocks in conifer swamps.

DESCRIPTION

Frond Variable in size, to 15 cm long, often much smaller; rather fleshy; dull to bright green to whitish green.

Sterile blade Stalked; simple, lobed or pinnately divided with up to 7 pairs of well-developed, often overlapping lobes; small plants may have only 3 segments; basal pair of lobes usually much larger and more divided than upper pairs; margins nearly entire; sometimes with a few sporangia on margins of lower pinnae.

Common stalk Blades attached at base or towards middle.

Sporophore Simple or 1-pinnate, 1-8x length of sterile blade. Spores mature in late May and June.

SYNONYMS

- *Botrychium tenebrosum* A.A. Eaton

SIMILAR SPECIES

Very small plants may appear similar to **little goblin moonwort** (*Botrychium mormo*) which has fleshier leaves and stalk.

NOTES

Sterile portion variable, which has led to the naming of several taxonomic varieties, separated as follows:

- **var. *simplex*** (Lasch) R.T. Clausen Junction of sporophore and sterile blade near ground level, basal pinnae largest; stalk of sterile blade almost as long as blade.

- **var. *tenebrosum*** A.A. Eaton Junction of sporophore and sterile blade higher, well above ground level, basal pinnae more or less same size as upper pinnae; stalk of sterile blade much shorter than blade.

NORTH AMERICA

MIDWEST

NAME

Most circumboreal species have been named in Europe first, but the present one was overlooked there by the early workers, and described from Massachusetts by Hitchcock in 1823.

Simplex, simple, may refer to the shallowly lobed sterile blade or to the sometimes unbranched sporophore.

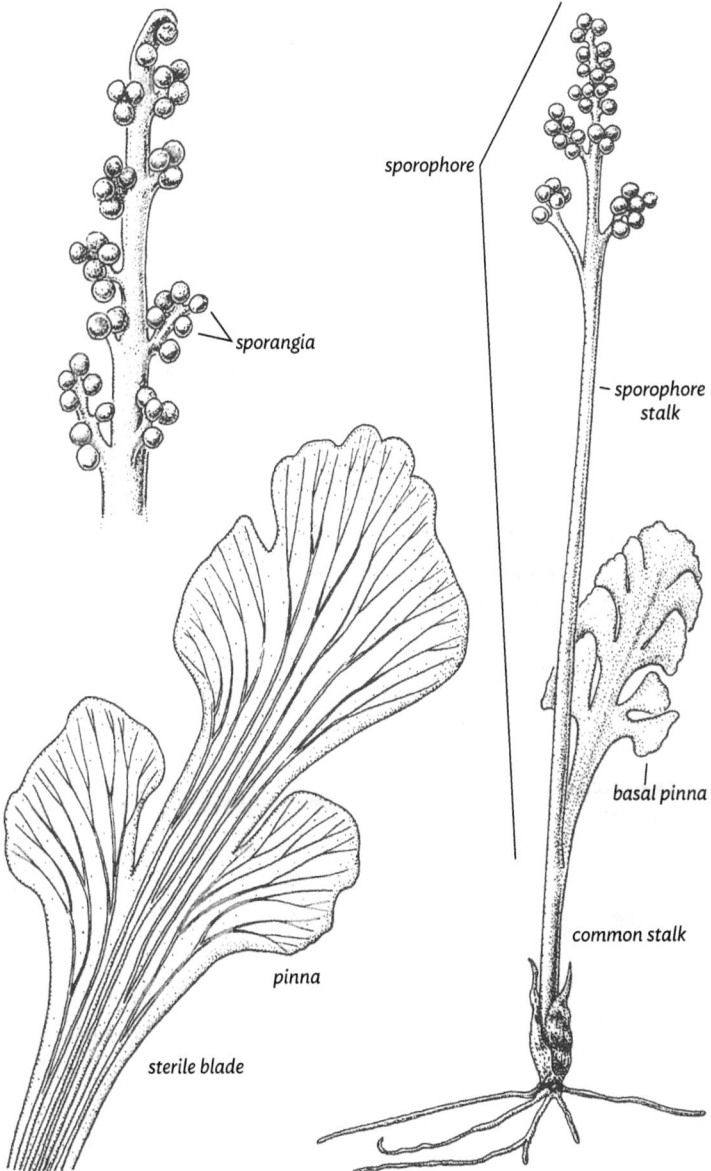

sporophore

sporangia

sporophore stalk

basal pinna

common stalk

pinna

sterile blade

Botrychium spathulatum
W.H. Wagner
SPOON-LEAF MOONWORT

FIELD TIPS
- small leathery plant; shiny yellowish green
- pinnae spoon-shaped, not overlapping
- lowest pinnae largest

MIDWEST RANGE
Northern Mich (threatened), Minn, Wisc (Door County).

HABITAT
Sandy, sunny, grassy fields and open woodlands, often where underlain by limestone.

DESCRIPTION
Frond Single, erect, to 12 cm long; shiny yellowish-green, leathery.

Sterile blade Sessile or short-stalked (less than 1 mm long); pinna pairs mostly 4-5 (7), spoon- or fan-shaped, widest at tip; lowest pinnae largest, commonly folded over rachis; pinnae mostly widely spaced and not overlapping, outer pinna margins entire or lobed.

Sporophore 1-2x length of the trophophore; 1-2 times pinnately divided into segments bearing the sporangia.

SIMILAR SPECIES
- **Moonwort** (*Botrychium lunaria*) has pinnae closely adjacent and more broadly fan-shaped than the widely separated, more narrowly fan-shaped pinnae of *B. spathulatum*.

- **Mingan moonwort** (*B. minganense*) has oval to fan-shaped pinnae in contrast to spoon-shaped pinnae of *B. spathulatum*; sporophore of *B. minganense* tends to be less divided than that of *B. spathulatum*.

NAME
Botrychium spathulatum, along with *B. pallidum*, was first described in 1990 (Wagner and Wagner 1990).

dried specimens

NORTH AMERICA

MIDWEST

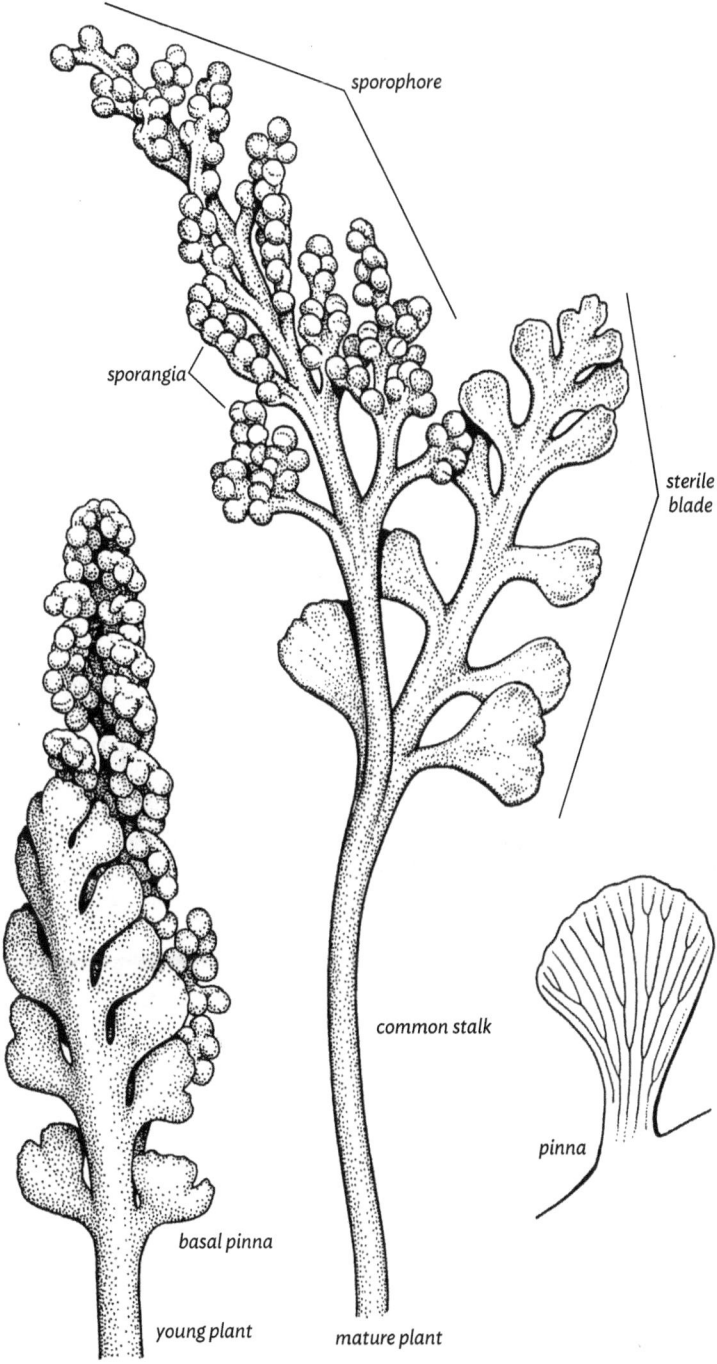

sporophore

sporangia

sterile blade

common stalk

pinna

basal pinna

young plant

mature plant

Botrypus
RATTLESNAKE FERN

A large and usually easily identified member of the family. Plants appear early in spring, the sterile blade often persisting into fall, the fertile portion (sporophore) withering in early summer. **Sterile blade** broadly triangular and usually more than 10 cm wide, and sometimes as much as 30 cm wide. **Fertile stalk** attached directly below base of sterile blade, and both the blade and fertile branch are held well above the ground. Sterile plants can usually be recognized by the large size and the elevated blade. Worldwide, the genus includes 7 species, 1 in North America and in our flora.

KEY CHARACTERS
• Our largest and most common member of Adder's-tongue Family.
• **Leaf blade** nearly horizontal, broadly triangular in outline, highly dissected, bright green and shiny, deciduous.
• **Sporophore** long-stalked, overtopping sterile blade, branched, not persistent.
• **Sporangia** bright yellow.
• Usually in rich, shaded woods, but also found in conifer swamps and drier upland deciduous forests.

NOTES
Botrypus and the **grape ferns** *(Sceptridium)* are distinguished from **moonworts** *(Botrychium)* by their generally larger size, and separated from one another by different points of attachment of their sporophores: in **rattlesnake fern** *(Botrypus)*, trophophore (sterile portion) and sporophore joined well above ground level, trophophore not stalked; in *Sceptridium*, trophophore and sporophore joined near ground level, trophophore stalked.

NAME
From Greek, *botry*, bunch of grapes, from resemblance of the sporangia to clusters of grapes. Common name from reported use of the mashed roots by Native Americans to treat bites of poisonous snakes. The clusters of unopened sporangia are also somewhat similar to the rattles on tail of a rattlesnake.

Rattlesnake fern (*Botrypus virginianus*), trophophore (lower), sporophore (upper).

Botrypus virginianus (L.) Holub
RATTLESNAKE FERN

FIELD TIPS
- our most common member of family
- plants large, bright green, smooth
- sterile blade broadly triangular
- sporophore above sterile blade, not persisting through summer
- sporangia bright yellow

MIDWEST RANGE
Pesent across the region and found in nearly all of North America (except for arid places).

HABITAT
Most common in moist deciduous forests, also in drier woods and less commonly on hummocks in cedar swamps.

DESCRIPTION
Frond Deciduous, erect, smooth or nearly so, to about 50 cm long.

Sterile blade Broadly triangular, usually more than 10 cm wide, blades of larger plants to 30 cm wide; membranous or slightly fleshy, attached above middle of the stipe, 2-pinnate to 3-pinnate; margins toothed; veins few.

Common stalk Erect, smooth, fleshy, pinkish at base; fertile branch attached directly below base of sterile blade and both are held well above ground level.

Sporophore Pinnately compound, produced in spring but withering in early summer; sporangia bright yellow.

SYNONYMS
- *Botrychium virginianum* (L.) Sw.

SIMILARSPECIES
- Small plants somewhat resemble **lance-leaf moonwort** (*Botrychium lanceolatum*), which also has glossy, triangular sterile blades. However, the sporophore of rattlesnake fern is long-stalked compared to the very short stalk of the sporophore of *B. lanceolatum*.

The large size, highly divided, nearly horizontal blade, and characteristic fertile segment make identification straight-forward.

NOTES
The most common and widespread member of the family, found in every state except Utah.

NAME
Although also found in Europe and Asia, species first named based on specimens from Virginia by Linnaeus in

NORTH AMERICA

MIDWEST

1753. His name for it was *Osmunda virginiana*, but transferred to genus *Botrychium* by Swartz in 1801. Now often placed in own genus *Botrypus*.

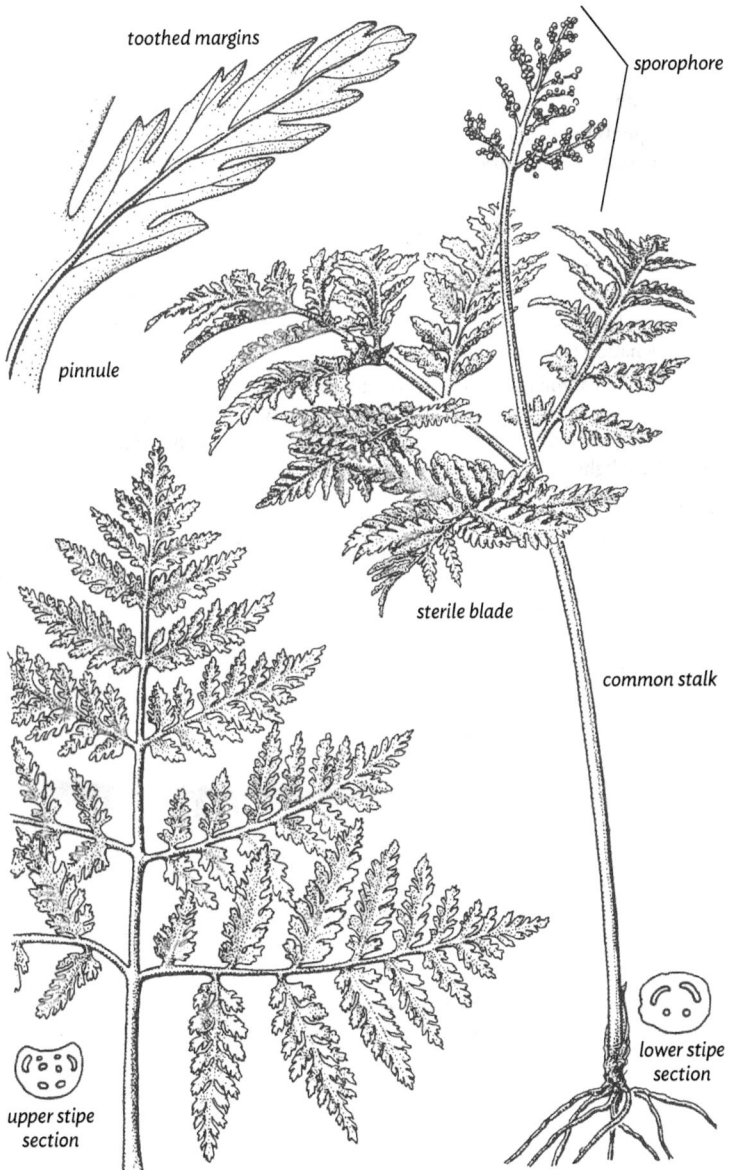

toothed margins

sporophore

pinnule

sterile blade

common stalk

upper stipe section

lower stipe section

Ophioglossum
ADDER'S-TONGUE

Small, deciduous fern with a single simple leaf and a stalked fertile portion. Worldwide, the genus contains ca. 20 species, mainly of tropical and subtropical regions; 8 species in North America, 3 in our flora.

KEY CHARACTERS

• Small plants of moist to dry, usually open, grassy habitats.

• **Leaf blade** fleshy, simple, midrib absent, the veins netlike.

• **Fertile portion** an elongate stalk tipped by an unbranched, spikelike, sporangia-bearing region; the **sporangia** deeply embedded and in 2 rows.

NOTES

Ophioglossum has the highest number of chromosomes known for vascular plants, with as many as 1,200 chromosomes in each cell reported.

Plants small and easily overlooked, and also resemble sterile lilies or plantains (*Plantago*). However, *Ophioglossum* species can be distinguished by the leaf blade's netlike venation (not parallel) and lack of a midrib.

Leaf venation can be useful for identifying some *Ophioglossum* species. However, as this may be hard to see in dried specimens, Wagner et al. (1984) suggested wetting the leaf with several drops of 95% ethanol and then observing using transmitted light.

NAME

From Greek, *ophis*, snake, *glossa*, tongue; the narrow fertile segment suggesting the name adder's-tongue.

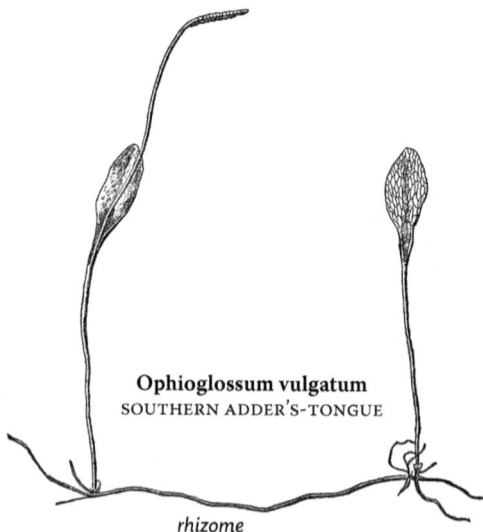

Ophioglossum vulgatum
SOUTHERN ADDER'S-TONGUE

rhizome

KEY TO OPHIOGLOSSUM | ADDER'S-TONGUE

1 Leaf blade abruptly narrowed a tip to a small sharp point; limestone regions of s Ill, s Ind, s Ohio
1. Ophioglossum engelmannii
LIMESTONE ADDER'S-TONGUE, page 262

1 Tip of leaf blade blunt; more widely distributed in our region in a variety of habitats **2**

2 Leaves gradually tapered to the base, elliptic in outline (widest near the middle); n Ill, n Ind, Iowa, Mich, Minn, n Ohio, Wisc
2. Ophioglossum pusillum
NORTHERN ADDER'S-TONGUE, page 264

2 Leaves abruptly tapered to the base, ovate in outline (clearly widest below the middle); Ill, Ind, se Iowa, s Mich, Ohio
3. Ophioglossum vulgatum
SOUTHERN ADDER'S-TONGUE, page 266

— *blade sharp-tipped*

Ophioglossum engelmannii
LIMESTONE ADDER'S-TONGUE

Ophioglossum vulgatum
SOUTHERN
ADDER'S-TONGUE

Ophioglossum pusillum
NORTHERN ADDER'S-TONGUE

Ophioglossum engelmannii Prantl
LIMESTONE ADDER'S-TONGUE

FIELD TIPS

- restricted to limestone in southern portions of region
- plants pale green, with 1 (rarely 2) simple leaves
- leaf blade tipped with small sharp point

MIDWEST RANGE

s Ill, s Ind, s Ohio (endangered).

HABITAT

Dry limestone and dolomite barrens, glades and open woods.

DESCRIPTION

Frond Pale green, somewhat fleshy, 10-20 cm tall, sometimes producing 2 leaves; typically forming colonies.

Sterile blade Simple, entire, elliptical, sharp-pointed at tip, smooth; 2.5 to 3.5 cm long and 1 to 4.5 cm wide.

Sporophore 5-10 cm long, the portion bearing sporangia to 5 cm; with 20-40 sporangia pairs, tapered at tip to a narrow point.

SYNONYMS

- *Ophioglossum vulgatum* var. *engelmanii* (Prantl) Clute

SIMILAR SPECIES

- Similar to **southern adder's-tongue** (*Ophioglossum vulgatum*), with overlapping range and sometimes sharing same habitat. However, plants of *O. engelmannii* have small, sharp points on the tip of blade, the upper half of the blades curve downward, and the main veins form large primary areoles that enclose several secondary areoles (as opposed to *O. vulgatum* which does not have secondary areoles). Fronds of limestone adder's-tongue are thick and almost succulent, and paler green than those of *O. vulgatum*.

NAME

Sent from Missouri to Prantl by botanist George Engelmann, and named for the latter in 1883.

NORTH AMERICA

MIDWEST

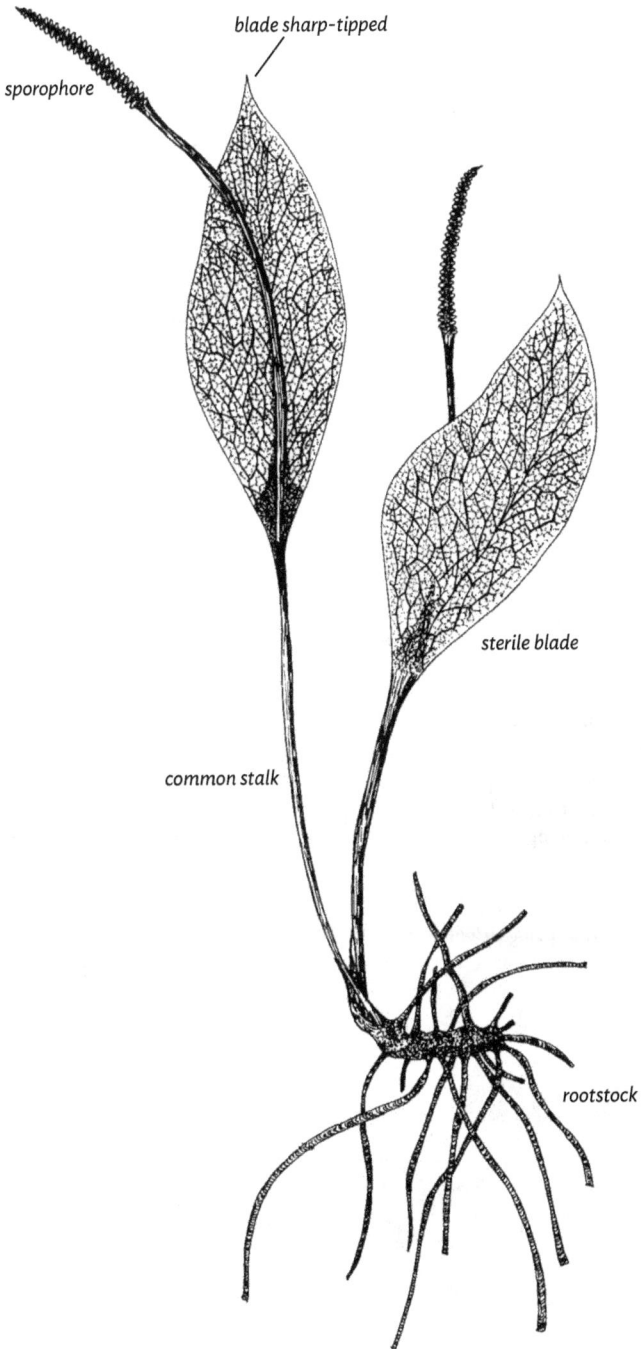

sporophore

blade sharp-tipped

sterile blade

common stalk

rootstock

Ophioglossum pusillum Raf.
NORTHERN ADDER'S-TONGUE

FIELD TIPS
- small plant of grassy, wet to moist places
- blade pale green, simple, widest near middle
- sporangia in 2 rows on fertile branch

MIDWEST RANGE
n Ill, n Ind, e Iowa, Mich, Minn, n Ohio (endangered), Wisc.

HABITAT
Moist sandy fields, wet meadows, ditches, also in drier situations; soils typically sandy; not in dense shade.

DESCRIPTION
Frond 15-25 cm long , with a single, simple blade; fertile branch (sporophore) arising near base of blade.

Sterile blade Sessile, smooth, entire, attached near middle of stalk; broadly lance-shaped to ovate, broadest near the middle; 4-10 cm long, 1.5-3 cm wide.

Sporophore Stalked spike extending 5-10 cm above blade, bearing 10-40 pairs of sporangia.

SYNONYMS
- *Ophioglossum vulgatum* auct. non L.
- *Ophioglossum vulgatum* L. var. *alaskanum* (E.G. Britton) C. Chr.
- *Ophioglossum vulgatum* L. var. *pseudopodum* (S.F. Blake) Farw.

SIMILAR SPECIES
- **Southern adder's-tongue** (*Ophioglossum vulgatum*) similar but blade widest near base and darker green; in northern adder's-tongue, blade widest at middle and pale green.

NAME
From Latin, *pusillus*, very small, perhaps alluding to its concealment by surrounding vegetation.

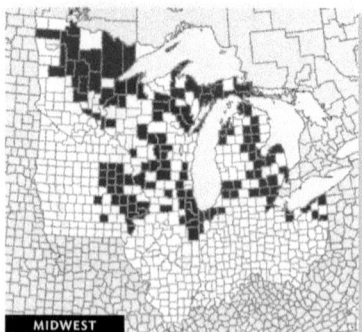

NORTH AMERICA

MIDWEST

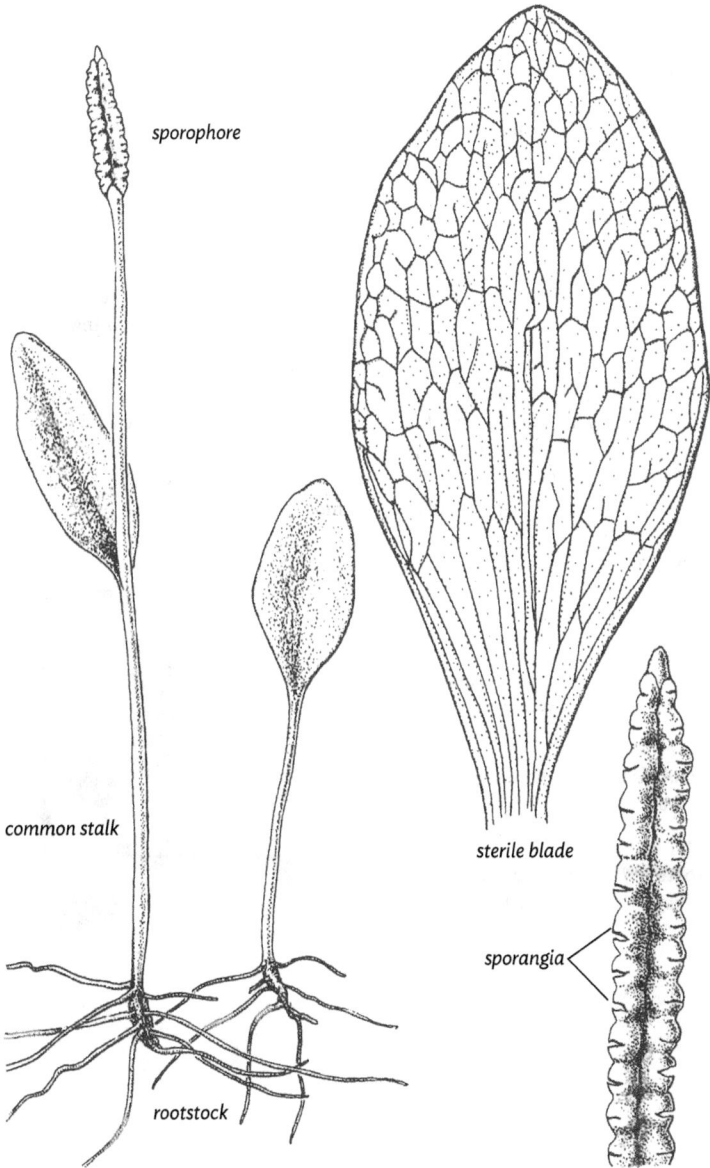

sporophore

common stalk

rootstock

sterile blade

sporangia

Ophioglossum vulgatum L.
SOUTHERN ADDER'S-TONGUE

FIELD TIPS

• small fern of moist woods and fields

• sterile blade dark green, widest near base

• sporangia in 2 rows on fertile branch

MIDWEST RANGE

Ill, Ind, se Iowa, s Mich (endangered), Ohio.

HABITAT

Moist deciduous forests, moist meadows, thickets and clearings.

DESCRIPTION

Fronds Scattered from rhizome, to about 25 cm long. divided into lower sterile and upper fertile portions; withering by midsummer.

Sterile blade Simple, oval to ovate, widest near the stalkless base, rounded at tip, to 10 cm long and 4 cm wide, smooth and fleshy, dark green; veins netlike.

Sporophore Stalked spike above sterile blade, with 10-35 pairs of sporangia; spores spherical, yellow.

SYNONYMS

• *Ophioglossum pycnostichum* (Fernald) Á. Löve & D. Löve

SIMILAR SPECIES

• **Northern adder's-tongue** (*Ophioglossum pusillum*), but blade in that species is paler green and widest at middle.

NOTES

Traditional European folk use of leaves and rhizomes as a poultice for wounds. This remedy was sometimes called the 'Green Oil of Charity.' A tea made from the leaves was used as a traditional European folk remedy for internal bleeding and vomiting.

NAME

Named from European specimens by Linnaeus in 1753; American plants are identical with European plants.

From Latin, *vulgaris*, common; the most common species in Europe and North America.

NORTH AMERICA

MIDWEST

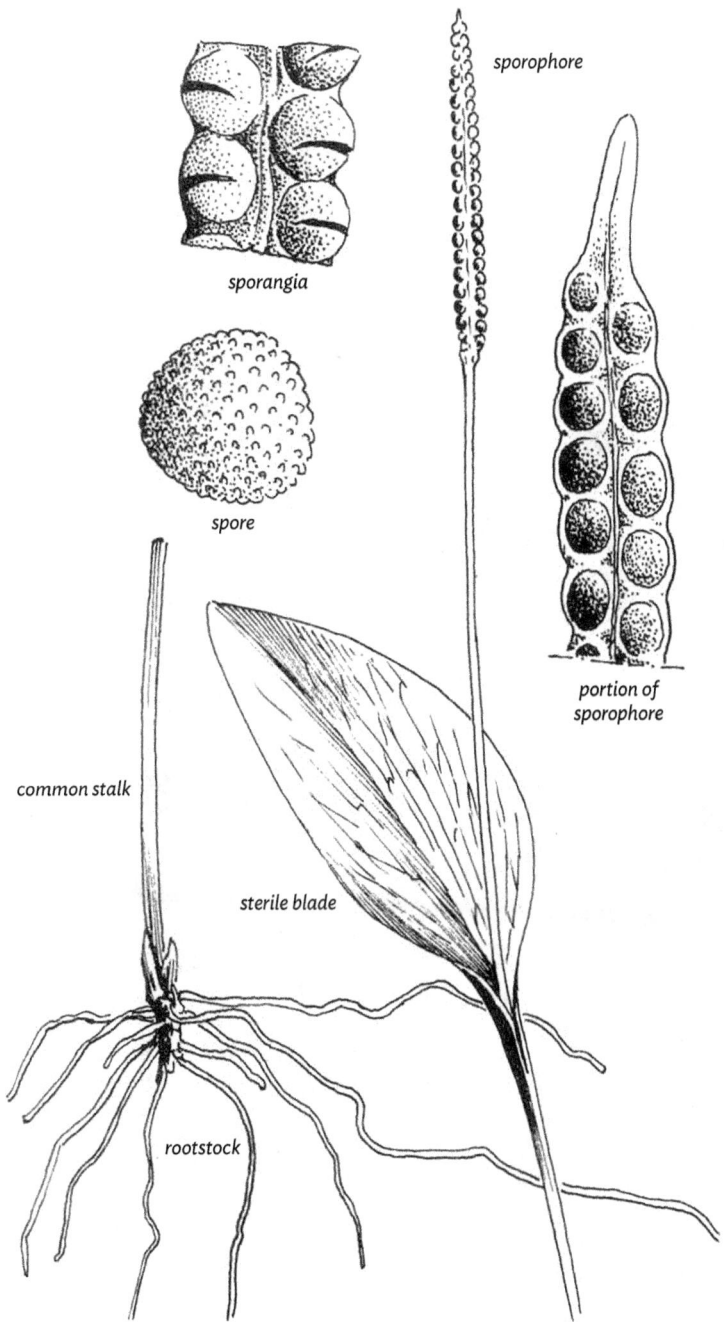

sporangia

spore

sporophore

portion of sporophore

common stalk

sterile blade

rootstock

Limestone adder's-tongue (*Ophioglossum engelmannii*).

Sceptridium
GRAPE FERN

Small to medium leathery ferns found in a variety of moist to dry, open to shaded habitats, often where sandy. **Sterile blades** dissected, winter-green, **sporophore** short-lived, withering by late summer. Sterile blade and sporophore joined near or below ground level. Worldwide, 10 species recognized, with 7 species known from North America; 5 in our flora.

KEY CHARACTERS

• Plants emerge in early summer, sterile blade persisting over winter.

• **Sterile blade** dissected, often appearing 3-parted; leathery or herbaceous.

• **Sporophore** short-lived, not overwintering; bearing yellow **sporangia** in a branched cluster

SIMILAR GENERA

Grape ferns (*Sceptridium*) distinguished from moonworts (*Botrychium*) by their generally larger size, usually leathery texture, the sterile blade nearly parallel to ground surface, and the joining of sterile blade and sporophore at or below ground surface.

Sceptridium and *Botrypus* separated from one another by different points of attachment of their sporophores:

• In **rattlesnake fern** (*Botrypus*), sterile blade (trophophore) and sporophore joined well above ground level, sterile blade not stalked;

• In **grape fern**s (*Sceptridiu*m), sterile blade and sporophore joined near ground level, sterile blade stalked.

NAME

From Greek, *sceptrum*, from resemblance of sporophore to a scepter or staff. Common name, grape fern, from resemblance of sporangia to small clusters of grapes.

KEY TO SCEPTRIDIUM | GRAPE FERN

1   Sterile blade segments deeply cut more than half way to the midvein, the entire blade lacerate **2**

1   Sterile blade segments finely to coarsely toothed **3**

2   Sterile blade mostly 2-pinnate, the segments sharply serrulate; rare in s Ill, s Ind and Ohio   **1. Sceptridium biternatum**
         SPARSE-LOBE GRAPE FERN, page 272

2   Sterile blade mostly 3-pinnate (or more divided, those forms keyed above), the segments entire to obscurely serrulate or crenulate; regionwide   **2. Sceptridium dissectum**
         CUT-LEAF GRAPE FERN, page 274

3   Ultimate segments of blade ± uniform in size; sterile blade segments finely toothed to ± entire; dissection of blade into segments extending to within 1 cm of apex at tips of blades **4**

3   Ultimate segments of blades variable in size, the apical segments much longer than the laterals; sterile blade segments coarsely and ± irregularly toothed or cut; dissection of blade into segments stopping at ca. 1–2.5 cm from apex at tips of blades. **5**

4   Segments of sterile blade rounded at base; symmetrically tapered to an often ± blunt or even rounded apex; larger segments mostly 9–17 mm long; margins nearly entire or finely and inconspicuously toothed; n Ill, e Iowa, Mich, Minn, Ohio, Wisc
         3. Sceptridium multifidum
         LEATHERY GRAPE FERN, page 276

4   Segments of sterile blade usually (obliquely) asymmetrical and angular, cuneate to the apex; larger segments mostly 4–9 mm long; margins clearly finely dentate, especially visible in immature leaves; Mich, Minn, Wisc   **4. Sceptridium rugulosum**
         TERNATE GRAPE FERN, page 280

5   Overwintering leaves green, not bronze; larger (terminal) segments of vegetative blades narrowly to broadly ovate, obtuse to rounded at apex; ± symmetrical at base; margins toothed but never lacerate; n Ill, Ind, Mich, Minn, Ohio, Wisc   **5. Sceptridium oneidense**
         BLUNT-LOBE GRAPE FERN, page 278

5   Overwintering leaves bronze-colored (or green if covered by leaves); larger (terminal) segments of sterile blades lance-shaped, acute, and strongly asymmetric at base; margins toothed to irregularly cut **6**

6   Sterile blade mostly 2-pinnate, the segments sharply serrulate; rare in s Ill, s Ind and Ohio   **1. Sceptridium biternatum**
         SPARSE-LOBE GRAPE FERN, page 272

6 Sterile blade mostly 3-pinnate (or more divided, those forms keyed above), the segments entire to obscurely serrulate or crenulate; regionwide **2. Sceptridium dissectum**
CUT-LEAF GRAPE FERN, page 274

Sceptridium biternatum
SPARSE-LOBE GRAPE FERN

Sceptridium dissectum
CUT-LEAF GRAPE FERN

Sceptridium multifidum
LEATHERY GRAPE FERN

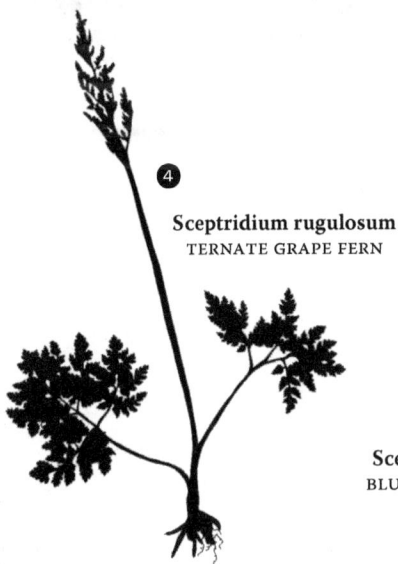

Sceptridium rugulosum
TERNATE GRAPE FERN

Sceptridium oneidense
BLUNT-LOBE GRAPE FERN

Sceptridium biternatum (Sav.) Lyon
SPARSE-LOBE GRAPE FERN

FIELD TIPS

• sterile blade evergreen, 3-parted, pinnae less divided than our other grape ferns

• uncommon in southern portions of region

MIDWEST RANGE

s Ill (threatened), s Ind, Ohio (endangered); more common in southeastern USA.

HABITAT

A variety of moist, shaded situations; low woods, ravines, thickets, forested floodplains.

DESCRIPTION

Sterile blade Dark green (and remaining green over winter), herbaceous, horizontal; 3-parted, 1-2-pinnate; pinnules few, elongate, sparsely cleft; margins sharp-toothed; veins forked.

Common stalk Short, mostly underground.

Sporophore Deciduous, several times longer than sterile blade, branched; **sporangia** yellow, in tight clusters.

SYNONYMS

• *Botrychium biternatum* (Sav.) Underw.
• *Botrychium dissectum* Spreng. var. *tenuifolium* (Underw.) Farw.

NOTES

Usually less dissected than **cut-leaf grape fern** (*Sceptridium dissectum*), and the pinnae trowel-shaped.

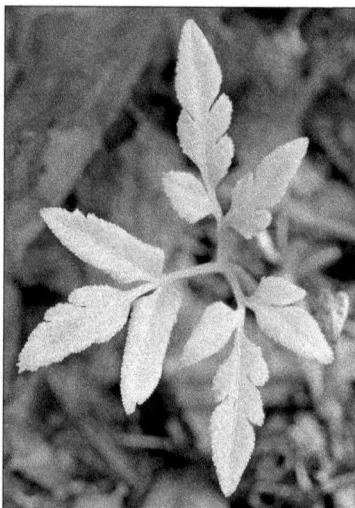

Sparse-lobe grape fern (*Sceptridium biternatum*), leaf blade of trophophore.

NORTH AMERICA

MIDWEST

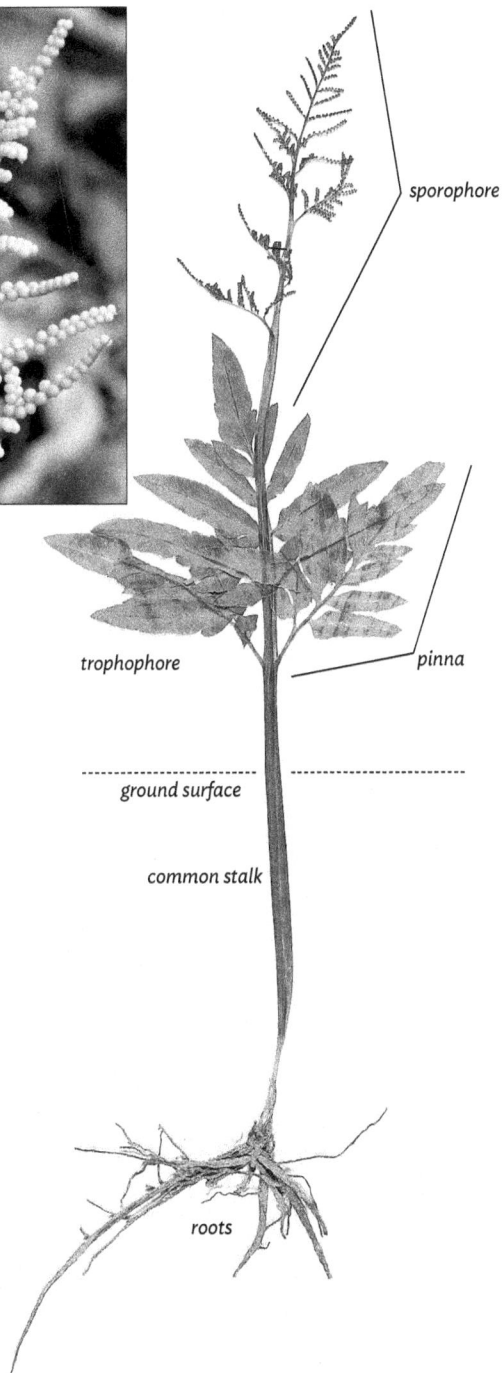

sporophore

trophophore

pinna

ground surface

common stalk

roots

Sceptridium dissectum (Spreng.) Lyon
CUT-LEAF GRAPE FERN

FIELD TIPS
- common grape fern of moist to dry, usually open and sandy places
- blade somewhat leathery, over-wintering
- blade color and dissection highly variable

MIDWEST RANGE
Regionwide but absent from far western areas.

HABITAT
Dry hilltops, dry to moist sandy fields, pastures and open woods; sometimes along old trails and roads.

DESCRIPTION
Frond Sterile blade diverges from sporophore at or below ground level; the single leaf is held close to the ground surface and overwinters, often turning a bronze or reddish color late in growing season following frosts.

Sterile blade Long-stalked, broadly triangular in outline, about 10-30 cm long and as wide, 3-parted; ultimate divisions of blade cut in linear segments.

Sporophore Stalked, panicle-like, with clusters of light yellow **sporangia**; **spores** mature in fall.

SYNONYMS
- *Botrychium dissectum* Spreng.
- *Botrychium dissectum* Spreng. var. *obliquum* (Muhl. ex Willd.) Clute
- *Botrychium dissectum* Spreng. var. *oblongifolium* (Graves) Broun
- *Botrychium obliquum* Muhl. ex Willd.

SIMILAR SPECIES
- Leaves less leathery than the similar **leathery grape fern** (*Sceptridium multifidum*).
- **Rattlesnake fern** (*Botrypus virginianus*) has somewhat similar dissected blade but is larger, and the sterile blade and sporophore are attached well above the ground.

NOTES
Extremely dissected and lacerated plants appear almost skeletonized and are very distinctive, and sometimes considered separate varieties:

- **var. *dissectum*** - pinnules deeply cut and margins sharply toothed.

- **var. *obliquum*** - pinnules less deeply dissected, and may be confused with **blunt-lobe grape fern** (*Sceptridium oneidense*) which has blunter pinnules, and blunt or small teeth.

NORTH AMERICA

MIDWEST

However, a wide range of forms may be present, even within a population.

NAME
This fern was discovered in Virginia and sent to Sprengel for identification, being named by him in 1804.

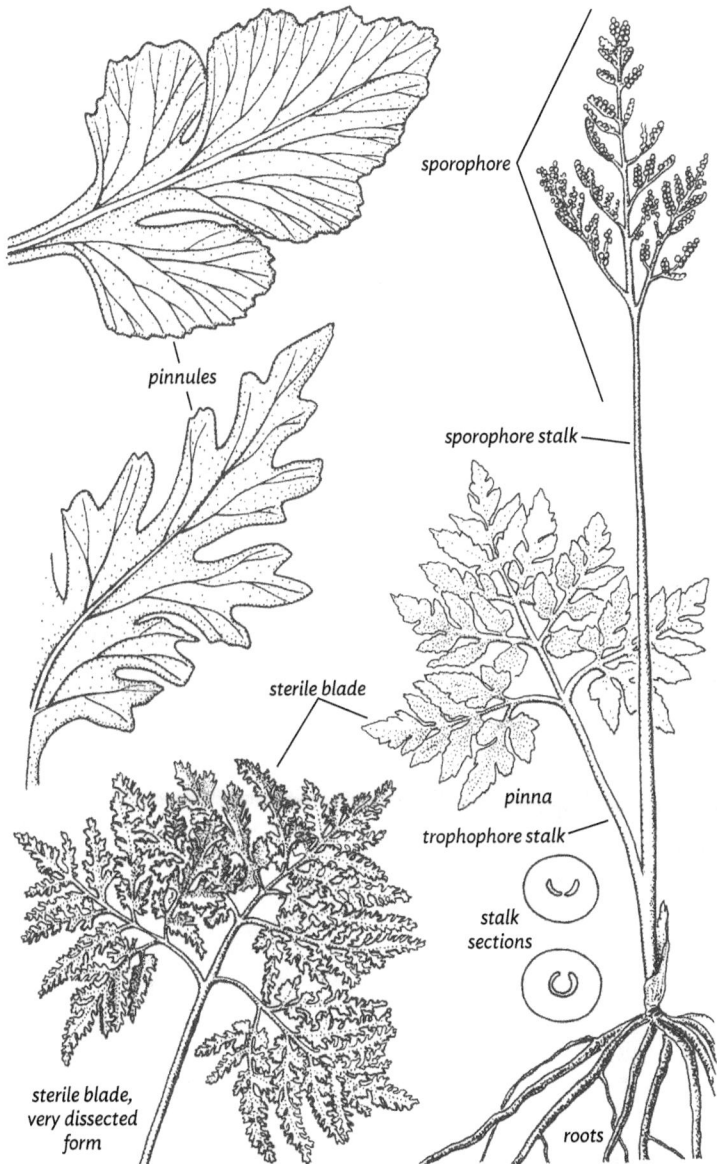

sporophore

pinnules

sporophore stalk

sterile blade

pinna

trophophore stalk

stalk sections

sterile blade, very dissected form

roots

Sceptridium multifidum
(Gmel.) Nishida ex Tagawa
LEATHERY GRAPE FERN

FIELD TIPS
- region's largest grape fern
- blade leathery, succulent, winter-green, not turning bronze-colored in winter
- sporophore large, branched

MIDWEST RANGE
n Ill (endangered), e Iowa (threatened), Mich, Minn, Ohio (endangered), Wisc; historical records from nw Ind.

HABITAT
Dry to moist sandy barrens, openings in oak and pine woods, old fields and clearings, along trails and old roads, grassy meadows and lawns; soils acidic; sterile plants sometimes found in shaded deciduous woods.

DESCRIPTION
Frond Up to 20 cm long, our largest *Sceptridium;* sterile blade and stalk very leathery and succulent; blade and sporophore attached near base of plant.

Sterile blade Long-stalked, broadly triangular in outline, 3-parted, shiny green to gray-green, evergreen, to 20 cm long and wide; held horizontally or bent backwards. Pinnae long-stalked, variable, ultimate segments crowded, sometimes overlapping.

Sporophore (if present) large and panicle-like; **sporangia** numerous,

yellow, maturing in late-summer.

SYNONYMS
- *Botrychium californicum* Underw.
- *Botrychium coulteri* Underw.
- *Botrychium matricariae* (Schrank) Spreng.
- *Botrychium multifidum* (S.G. Gmel.) Trevis.
- *Botrychium silaifolium* C. Presl
- *Sceptridium silaifolium* (C. Presl) Lyon

NOTES
Previous year's blade remains green through winter until a new leaf appears in spring, then sometimes yellowing and withering but persisting through much of the following summer.

NAME
In 1766 Gmelin applied the name *Osmunda multifida* to a Russian species; this changed to *Botrychium multifidum* by Ruprecht in 1859. The relative occupying northern North America was named *B. silaifolium* by Presl in 1825, but is considered only varietally distinct from those in Europe. In 1916, Farwell proposed *Botrychium multifidum* for North American plants.

NORTH AMERICA

MIDWEST

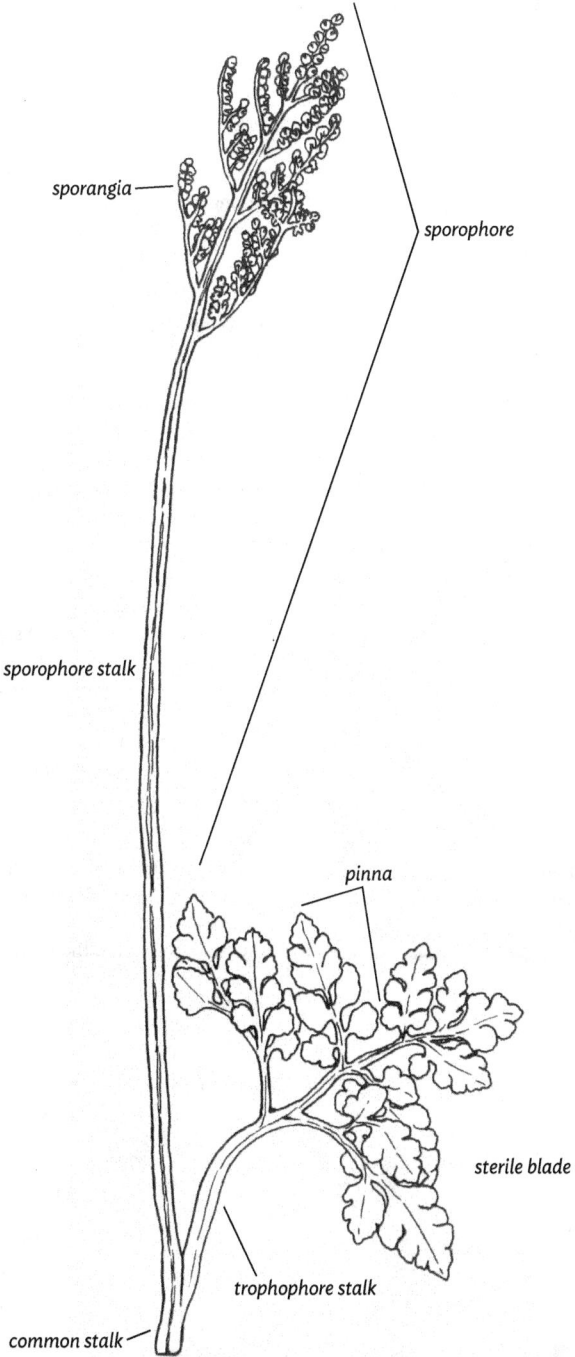

sporangia

sporophore

sporophore stalk

pinna

sterile blade

trophophore stalk

common stalk

Sceptridium oneidense (Gilbert) Holub

BLUNT-LOBE GRAPE FERN

FIELD TIPS

- sterile blade bright or bluish green, remaining green through winter
- sporophore large, sporangia yellow
- moist to wet, shaded woods

MIDWEST RANGE

n Ill, Ind, Mich, Minn (endangered), Ohio, Wisc.

HABITAT

Shaded, wet to dry, hardwood or mixed forests, occasionally in conifer swamps.

In Minnesota, typically in depressions and near vernal pools in forests of sugar maple, yellow birch, black ash, northern red oak and basswood; less commonly in wetter forests of northern white cedar and red maple.

DESCRIPTION

Frond To 40 cm long or longer; stem and sterile blade somewhat leathery; sterile blade diverges from the common stalk about 2.5-5 cm above the ground.

Sterile blade Winter-green, bluish green, more or less flat, broadly triangular in outline, 3-parted, to 20 cm long and 15 cm wide, borne near the ground; pinnules blunt at tip, margins with fine teeth.

Sporophore Panicle-like, not always produced each year; **sporangia** yellow, maturing in fall.

SYNONYMS

- *Botrychium oneidense* (Gilbert) House
- *Botrychium dissectum* Spreng. var. *oneidense* (Gilbert) Farw.
- *Botrychium multifidum* (S.G. Gmel.) Trevis. var. *oneidense* (Gilbert) Farw.

SIMILAR SPECIES

- **Leathery grape fern** (*Sceptridium multifidum*) similar in size and shape, but blunt-lobe grape fern has more unequally divided pinnae segments (the upper segments longer and less divided than the basal segments), and margins usually somewhat fine- toothed. In *S. multifidum*, segments usually wavy and lobed.

- **Cut-leaf grape fern** (*Sceptridium dissectum*) similar, but pinnae segments of blunt-lobe grape fern are blunt tipped, whereas those of *S. dissectum* somewhat pointed; *S. oneidense* leaves mostly green, those of cut-leaf grape fern reddish when young, and often bronzy in winter.

NOTES

Previously treated as a variety, form, or hybrid of both *Sceptridium dissectum* and *S. multifidum*. The broader, more rounded divisions and the more shaded habitats of blunt-lobe grape fern are characteristic.

NORTH AMERICA

MIDWEST

NAME
Oneidense refers to Oneida County,
New York.

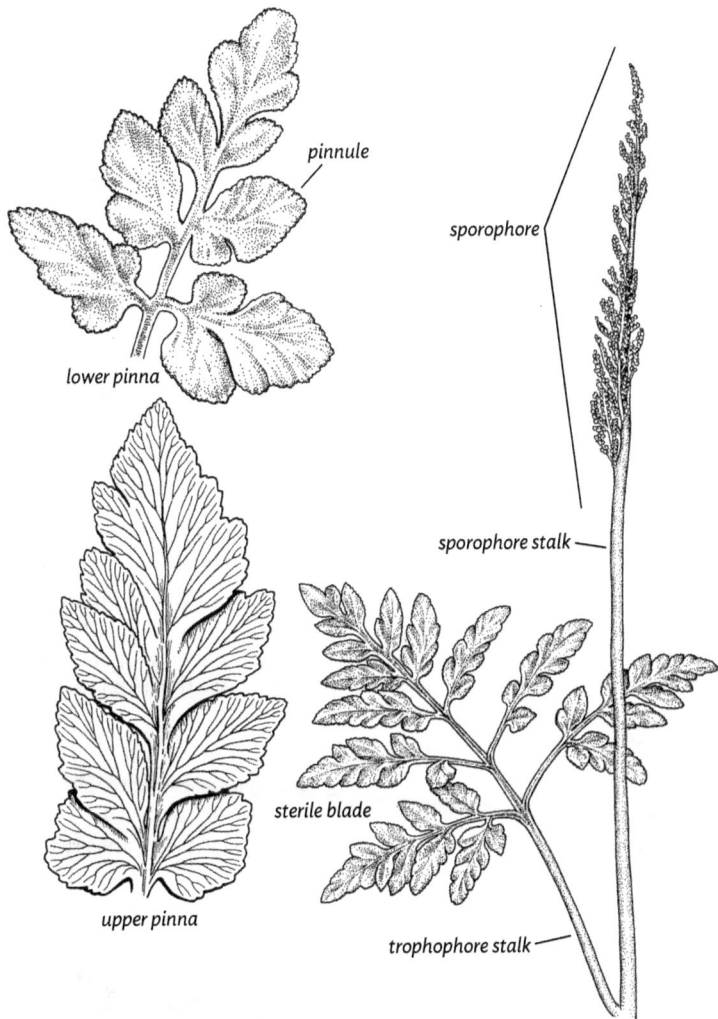

pinnule

lower pinna

sporophore

sporophore stalk

sterile blade

upper pinna

trophophore stalk

Sceptridium rugulosum
(W.H. Wagner) Skoda & Holub
TERNATE GRAPE FERN

FIELD TIPS
- sterile blade herbaceous, not leathery
- blade segments toothed, sometimes turned downward along margins
- sandy fields and open woods, sometimes where wet

MIDWEST RANGE
Mich, Minn (threatened), Wisc.

HABITAT
Dry to moist brushy or grassy meadows, sandy open woods, mossy places in forests of jack pine and red pine; sometimes in wet, swampy woods and along shaded streambanks.

DESCRIPTION
Frond 25 cm long or longer, thin and herbacous and not leathery.

Sterile blade Stalked, borne near the ground, broadly triangular in outline; 3-parted, the 3 major divisions stalked, the terminal portion not usually fully divided; to about 20 cm long and as wide; pinnae margins coarsely toothed, sometimes reflexed (turned downward).

Sporophore Panicle-like, atop a fleshy stalk; sporangia yellow, maturing in late summer to fall.

SYNONYMS
- *Botrychium rugulosum* W.H. Wagner

- *Botrychium ternatum* auct. non (Thunb.) Sw.

SIMILAR SPECIES
- **Cut-leaf grape fern** (*Sceptridium dissectum*) similar but its leaves often turn a drab reddish color, and in fall, often turn bronze-colored. In ternate grape fern, the leaf tends to remain green. Also, the margin teeth of cut-leaf grape fern are serrate (sharp and forward-pointing); in *S. rugulosum*, the teeth are dentate (squarish and outward-pointing)
- **Leathery grape fern** (*Sceptridium multifidum*) similar, but instead of the rounded segment lobes of ternate grape fern, the sterile blade angular and with margins mostly coarsely and irregularly toothed.

NOTES
Reported to occur with **cut-leaf grape fern** (*Sceptridium dissectum*), **leathery grape fern** (*S. multifidum*), and rarely **blunt-lobe grape fern** (*S. oneidense*).

NAME
Rugulosum refers to tendency of segments to become more or less wrinkled and convex.

'Ternate' refers to the 3-parted sterile blade.

NORTH AMERICA

MIDWEST

Also known as **St. Lawrence grape fern** due to its presence in St. Lawrence River valley and Great Lakes region.

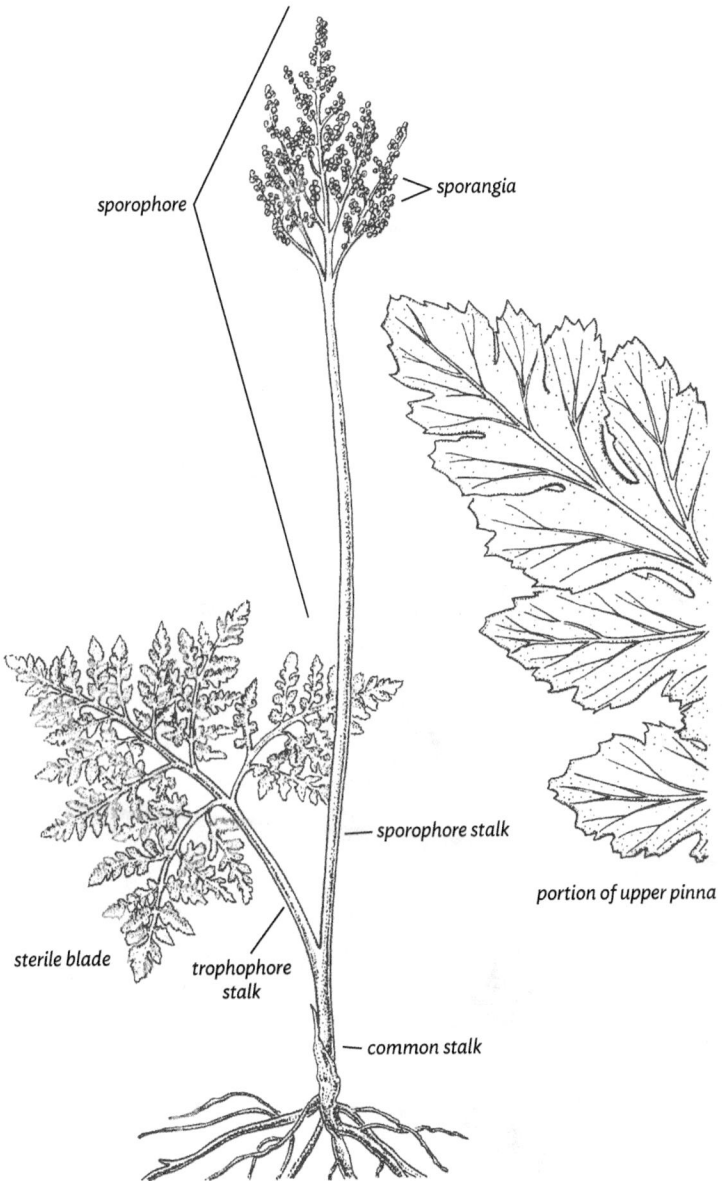

sporophore

sporangia

sporophore stalk

portion of upper pinna

sterile blade

trophophore stalk

common stalk

Cut-leaf grape fern (*Sceptridium dissectum*), extremely dissected form.

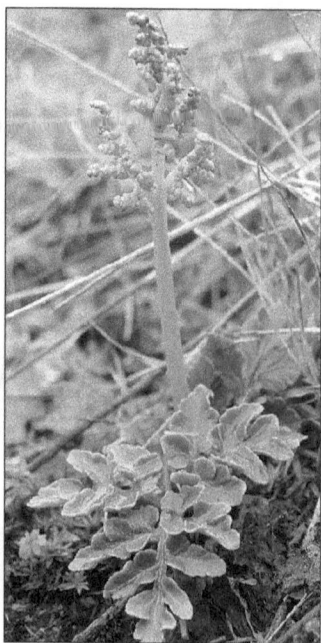

Leathery grape fern (*Sceptridium multifidum*), (l) sterile trophophore, (r) plant with trophophore (lower) and sporophore (upper).

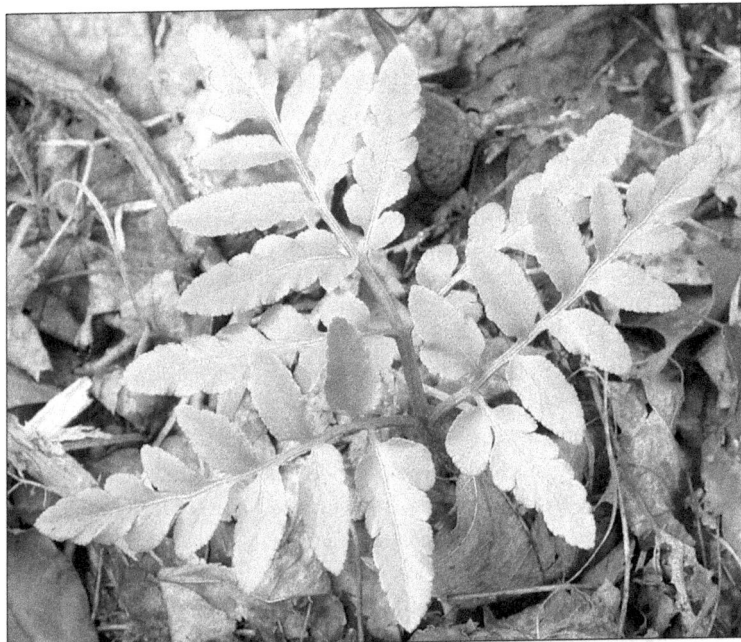

Blunt-lobe grape fern (*Sceptridium oneidense*), trophophore (sterile blade).

Ternate grape fern (*Sceptridium rugulosum*), trophophore (sterile blade).

OSMUNDACEAE | Royal Fern Family
Osmunda
ROYAL FERN

Large clumped ferns of wet to moist places. Fertile and sterile leaves different (*Osmunda cinnamomea*) or the fertile portions different than the sterile portions (*O. claytoniana, O. regalis*). **Rhizomes** stout, covered with persistent leaf bases but lacking scales; **fronds** variously compound; **sporangia** large, in dense clusters on reduced, stalklike pinnae, these on separate leaves from those bearing sterile pinnae, or on different portions of the leaf. A small family of 3 cosmopolitan genera; *Osmunda* includes 9 species, 3 in our flora.

KEY CHARACTERS
• Large deciduous ferns forming circular clumps.
• **Sterile fronds** surround inner, more erect **fertile fronds**.
• **Sporangia** large, round, in clusters on modified pinnae of fertile fronds.
• **Croziers** emerge early in spring, covered with wooly white or red-brown hairs.
• Found in swamps, shaded wet depressions, and wet to moist forests.

NOTES
Densely matted roots and rhizomes of *Osmunda* species may form an upright trunk, and are source of the fiber **osmundine**, used as a growing medium for orchids and other epiphytes.

NAME
From Saxon god Osmunder the Waterman, the Saxon equivalent of the Norse god Thor, who supposedly hid his family from danger in a clump of these ferns.

HYBRIDS
Osmunda × ruggii, uncommon hybrid between **interrrupted fern** (*O. claytoniana*) and **royal fern** (*O. regalis*), reported from Forest County in northern Wisconsin.

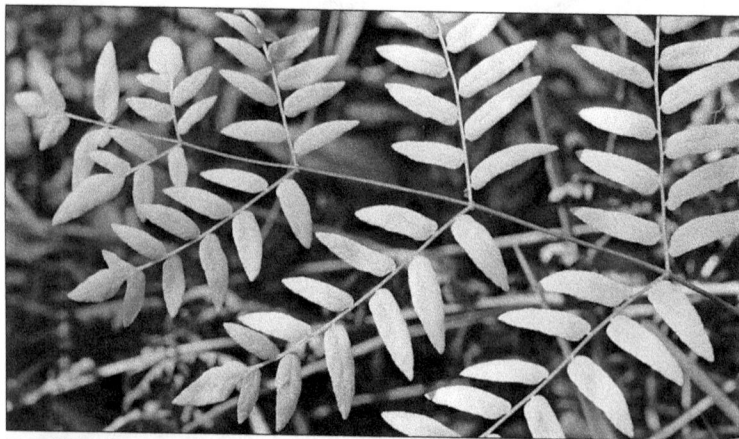

Royal fern (*Osmunda regalis*), portion of sterile frond.

KEY TO OSMUNDA | ROYAL FERN

1   Fronds of two types, sterile and fertile; sterile fronds broad, 1-pinnate, with wooly tufts at base of pinna-axes; fertile fronds slender, lacking leafy tissue; regionwide
1. Osmunda cinnamomea
CINNAMON FERN, page 286

1   Fronds alike, portions of fronds either sterile or fertile **2**

2   Blade 1-pinnate, lacking wooly tufts at base of pinna-axes; fertile pinnae small, in mid-portion of blade; regionwide
2. Osmunda claytoniana
INTERRUPTED FERN, page 288

2   Blade 2-pinnate, with simple oblong pinnules; fertile pinnae at tip of blade; Ill, Ind, e Iowa, Mich, Minn, Ohio, Wisc
3. Osmunda regalis
ROYAL FERN, page 290

fertile pinnae

fertile frond

Osmunda regalis
ROYAL FERN

Osmunda cinnamomea
CINNAMON FERN

fertile pinnae

Osmunda claytoniana
INTERRUPTED FERN

Osmunda cinnamomea L.
CINNAMON FERN

FIELD TIPS

- large, clumped, colony-forming fern of wet places
- sterile and fertile fronds different; linear fertile fronds surrounded by larger, leafy sterile fronds
- persistent tuft of wooly hairs at base of each pinna

MIDWEST RANGE

Ill, Ind, Iowa (endangered), Mich, Minn, Ohio, Wisc.

HABITAT

Bogs, acid swamp forests, wet shrub thickets, soils typically sandy or peaty and acidic; usually in partial shade, less common in heavy shade.

DESCRIPTION

Rootstock Erect, massive, forming a short trunk, occasionally branching, covered with old roots and winged stipe bases; scales absent; roots black, fibrous.

Crozier Circular, about 2 cm wide, densely covered with whitish hairs which turn cinnamon-brown.

Frond Deciduous, clustered; sterile and fertile fronds different; the outer ring of fronds sterile, arching, longer, to 100 cm long or more; inner fronds fertile, erect, green at first, turning cinnamon brown, developing earlier then withering in early summer and persisting as a hairy dried stalk.

Stipe With rusty wooly hairs when young, soon smooth; vascular bundle 1, U-shaped, the top of the arms continuing to curl inward.

Rachis Green; sparsely covered with cinnamon-colored wool when young, becoming smooth apart from a tuft of brownish hairs at base of each pinna.

Sterile blade Elliptic to oblong, arching, 1-pinnate-pinnatifid, pinnae broadly oblong.

Fertile blade Pinnae small, angled upward, bearing sporangia.

Pinnae 20 to 25 pairs, rotated to the horizontal; pinnules obtuse; costae and rachis shallowly grooved above; margins entire; veins free, forked 1-3 times; rusty wool on pinnae and rachis soon deciduous; tuft of tan hairs on underside of pinna base near rachis tends to persist.

Sori None; **indusium** absent; **sporangia** large, globose, greenish when young, tan or black when mature.

SYNONYMS

- *Osmundastrum cinnamomeum* (L.) C. Presl

NORTH AMERICA

MIDWEST

SIMILAR SPECIES

• **Interrupted fern** (*Osmunda claytoniana*), **ostrich fern** (*Matteuccia sruthiopteris*) and **Virginia chain fern** (*Woodwardia virginica*) are similar ferns of wet habitats, but all lack the tan wooly hairs at pinnae base.

NOTES

In early spring, young uncoiling sterile fronds (fiddleheads), were once considered edible but now known to be carcinogenic (the best edible fiddleheads are those of ostrich fern, page 220). The crisp, tender, central part of the crown, with its slightly nutty flavor, were also sometimes eaten raw.

NAME

Specimens of this fern were sent to Linnaeus from Maryland and named by him in 1753.

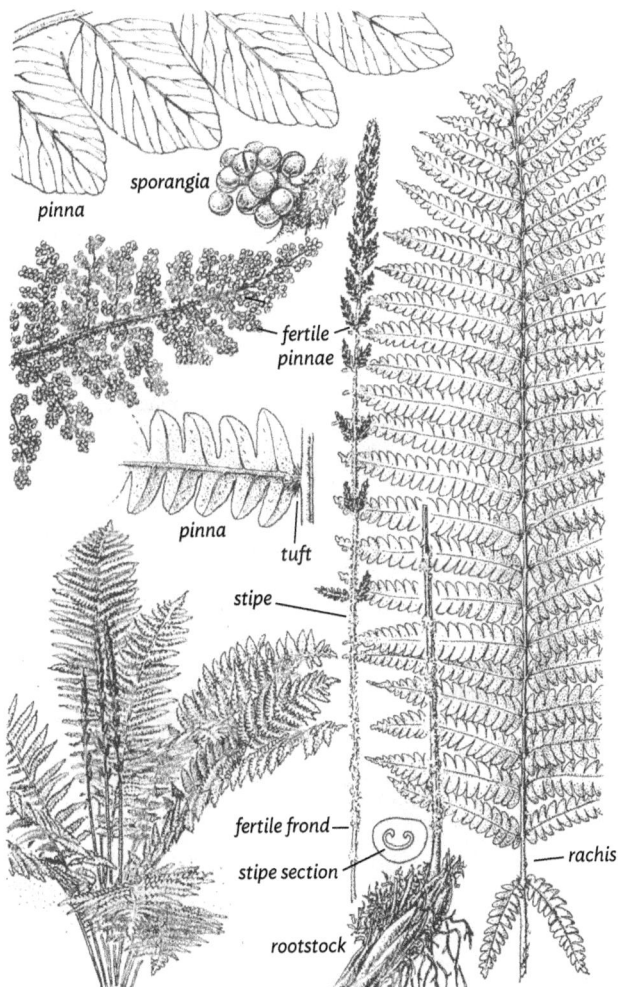

Osmunda claytoniana L.
INTERRUPTED FERN

FIELD TIPS
- large deciduous fern
- fronds with small fertile pinnae near middle of blade
- sterile fronds smaller, arching
- no tuft of hairs at base of pinnae as in cinnamon fern
- moist to wet forests

MIDWEST RANGE
Regionwide.

HABITAT
Low places in mesic to wet forests, on hummocks in swamps, roadsides, in shade to partial shade, less tolerant of wet conditions than cinnamon fern or royal fern.

DESCRIPTION
Rootstock Erect and also creeping and branching, large, covered with old roots and winged stipe bases; scales absent; roots black, fibrous.

Crozier Circular, about 2 cm wide, densely covered with whitish hairs when young, turning brownish; very similar to those of *O. cinnamomea*.

Frond Deciduous, clustered; sterile and fertile fronds different; outer ring of fronds sterile, arching, 40-180 cm long and longer than the fertile fronds; inner fronds fertile, erect, with widely spaced pinnae.

Stipe Hairy when young, soon smooth; vascular bundle one, U-shaped, the top of the arms continuing to curl.

Rachis Green; sparsely covered with reddish to light brown wooly hairs, at least when young, but no tufts of hairs at base of pinna as in cinnamon fern (*O. cinnamomea*).

Sterile blade Elliptic to oblong, arching, 1-pinnate-pinnatifid.

Fertile blade With 2-7 middle pinnae pairs reduced in size, these bearing the sporangia, and soon withering.

Pinnae Broadly oblong, 20 to 30 pairs, rotated to the horizontal; margins entire, tip rounded; veins free, once-forked.

Sori None; **indusium** absent; **sporangia** large, globose, greenish young, tan or black when mature.

NAME
John Clayton collected this fern in Virginia in the early 1700s, and it was named in his honor by Linnaeus in 1753.

NORTH AMERICA

MIDWEST

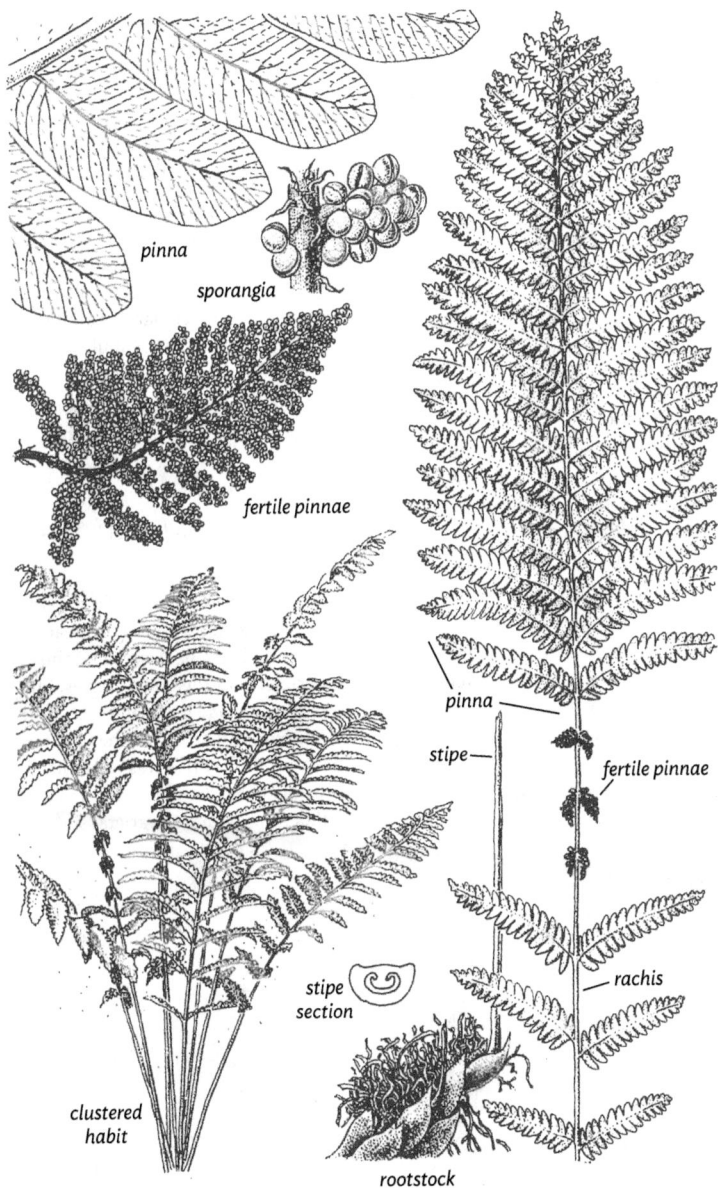

pinna

sporangia

fertile pinnae

pinna

stipe

fertile pinnae

rachis

stipe
section

clustered
habit

rootstock

Osmunda regalis L.
ROYAL FERN

FIELD TIPS

- large, clumped, colony-forming fern of wet places
- sterile fronds similar to compound leaves of locust tree
- fertile pinnae at tips of fronds

MIDWEST RANGE

Ill, Ind, e Iowa (threatened), Mich, Minn, Ohio, Wisc.

HABITAT

Swamps, low woods, marshy meadows, cedar bogs, on hummocks in very wet sites; tolerant of saturated soils.

DESCRIPTION

Rootstock Erect, large, forming a short hummocky, woody trunk, covered with old roots and winged stipe bases; scales absent.

Crozier Dark red, covered with brown hairs when very young but soon smooth.

Frond Deciduous, clustered; clustered; sterile and fertile fronds different; 50-180 (-300) cm long.

Stipe Hairy when young, soon smooth, pinkish to green; vascular bundle one, U-shaped, the top of the arms continuing to curl.

Rachis Grooved, pinkish to green, with reddish to light brown hairs, these soon deciduous.

Blade With sterile pinnae below and fertile pinnae above, broadly ovate; 2-pinnate.

Pinnae 5 to 9 pairs, not opposite; rotated to the horizontal; on fertile fronds usually the 2 to 4 lowest pairs sterile; pinnules 8-12 pairs plus a terminal pinnule; margins almost entire, the tip of the pinnules finely toothed; veins free, forked.

Sori none; **indusium** absent; **sporangia** large, globose, tan or black when mature.

SYNONYMS

- *Osmunda spectabilis* Willd.

NAME

In 1753, Linnaeus defined *Osmunda regalis* as including both European and Virginian plants; he designated the American plants as a subspecies, but Willdenow made it a species, *O. spectabilis*, in 1810. That they are too similar to justify this was recognized by Gray in 1857, when he transferred Willdenow's name to varietal status.

From Latin for royal.

NORTH AMERICA

MIDWEST

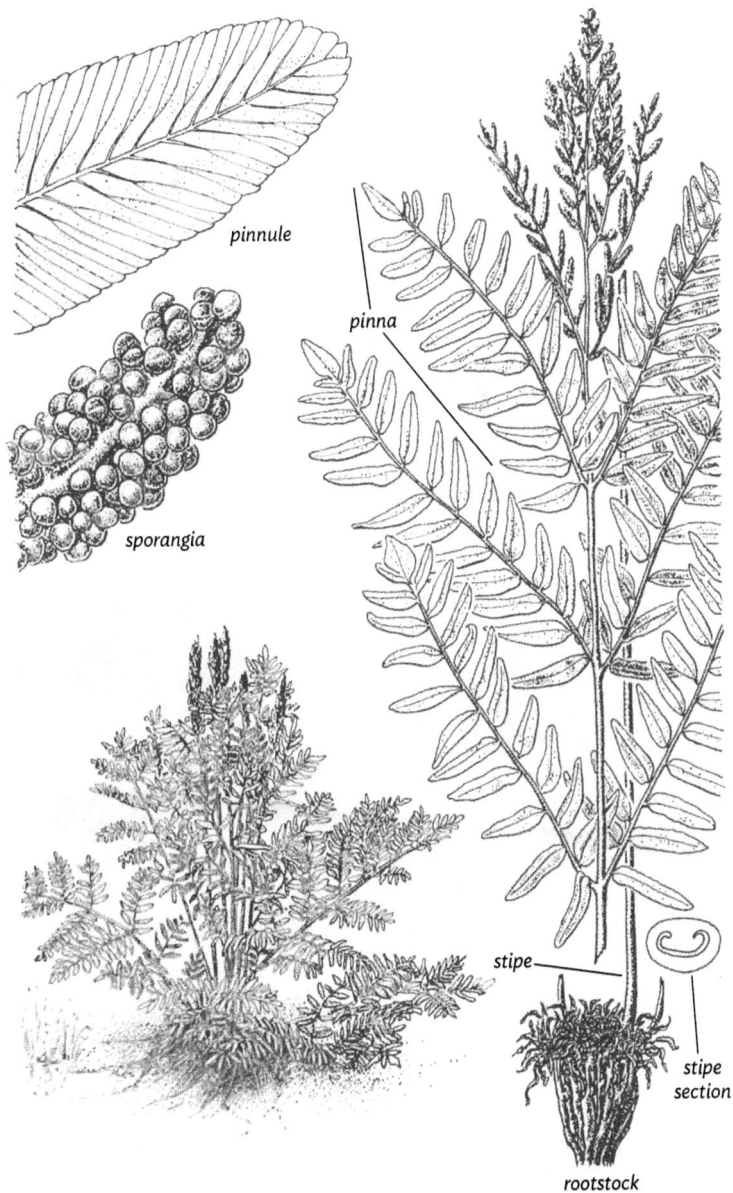

pinnule

pinna

sporangia

stipe

stipe section

rootstock

Cinnamon fern (*Osmunda cinnamomea*), upright fertile fronds surrounded by large, arching sterile fronds.

Cinnamon fern (*Osmunda cinnamomea*), sterile frond with conspicuous tufts of wooly hairs at base of pinnae.

Interrupted fern (*Osmunda claytoniana*), fertile pinnae in middle portion of frond.

Royal fern (*Osmunda regalis*), fertile frond with sporangia at tip of frond.

POLYPODIACEAE | Polypody Family

A large family of 89 genera and more than 1,110 mostly tropical species; 2 genera and 3 species in our region. Most ferns in the family grow on trees or rock rather than in soil.

KEY CHARACTERS

• Evergreen, colony-forming ferns of rock cliffs, boulders and sometimes on stumps or in trees (*Pleopeltis*).
• **Leaf blades** pinnatifid.
• Sterile and fertile **fronds** alike.
• **Sori** round, **indusia** absent.

NAME

From Greek, *polys*, many, and *pous*, foot, perhaps referring to the numerous stipe bases, or to the many knoblike branches of the rhizome.

KEY TO POLYPODIACEAE | POLYPODY FAMILY

1   Leaf blade densely scaly on lower surface; leaf segment margins entire; rhizome 1-2 mm in diameter; s Ill, s Ind, s Ohio
1. Pleopeltis polypodioides
RESURRECTION FERN, page 296

1   Leaf blade without scales on lower surface; leaf segment margins finely toothed; rhizome 3-6 mm in diameter; regionwide
2. Polypodium
ROCK POLYPODY, page 298

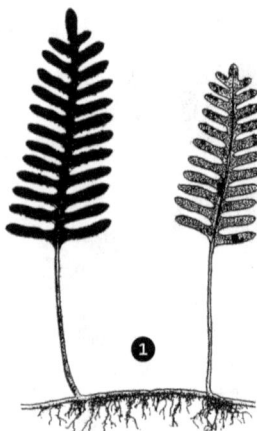

Pleopeltis polypodioides
RESURRECTION FERN

Polypodium
ROCK POLYPODY

Pleopeltis
SCALY-POLYPODY

Evergreen, colony-forming fern on rock or rarely (in our region) in trees or old stumps; genus includes ca. 56 species, mostly of New World tropics, 1 species in southern portion of Midwest.

KEY CHARACTERS
- Evergreen, colony-forming ferns of rock outcrops and sometimes on stumps or in trees (this characteristic more common south of our region).
- **Fronds** leathery, pinnatifid, underside densely scaly.
- Sterile and fertile **fronds** alike, from long-creeping rhizomes.
- **Sori** round, in single rows on each side of the midrib of the pinna lobes; **indusia** absent; young sori covered by peltate scales.
- Fronds curl up and appear dead in dry weather, reviving following rain.

NAME
From Greek, *pleos*, many, and *pelte*, shield, in reference to the peltate scales covering immature sori.

Pleopeltis formerly included within genus *Polypodium*.

Resurrection fern (*Pleopeltis polypodioides*), colony in typical rock crevice habitat.

Pleopeltis polypodioides
(L.) Andrews & Windham
RESURRECTION FERN

FIELD TIPS
- evergreen, creeping fern of rock outcrops, forms matlike colonies
- blade pinnatifid, leathery, underside with many gray scales
- in dry periods, fronds curl up and appear lifeless

MIDWEST RANGE
s Ill, s Ind, s Ohio (threatened); becoming common in se USA.

HABITAT
Outcrops of dolomite or sandstone, in sun to partial shade and where periodically moist; in southeast USA, also found on trees, stumps and fence posts.

DESCRIPTION
Rootstock Creeping, branching, cordlike and scaly.

Frond Evergreen, sterile and fertile fronds alike.

Stipe Jointed at base at rhizome, green to dark brown or black, flattened to round, often grooved below and winged above; vascular bundles 3.

Blade Ovate, pinnatifid, leathery, densely gray scaly on underside; segments linear to oblong, margins entire to serrate; veins free or netted.

Sori Circular to oval, discrete, borne at tips of single veins, in 1-3 rows on either side of midrib, often confined to upper portion of blade; **indusium** absent; **sporangia** transparent or yellowish.

SYNONYMS
- *Polypodium polypodioides* (L.) Watt

SIMILAR SPECIES
- Somewhat resembles **rock polypody** (*Polypodium virginianum*), but differing in the dense gray scales on frond underside.

NAME
Reached Linnaeus from Virginia and Jamaica, and he named it *Acrostichum polypodioides* in 1753; transferred to genus *Polypodium* by Watt in 1866; placed in *Marginaria* by Tidestrom in 1905.

Our plants var. *michauxiana* (Weath.) E.G.Andrews & Windham.

Common name refers to fronds curling up and appearing dead in dry weather, then reviving following rain.

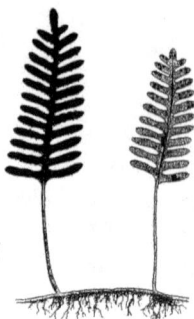

NORTH AMERICA

MIDWEST

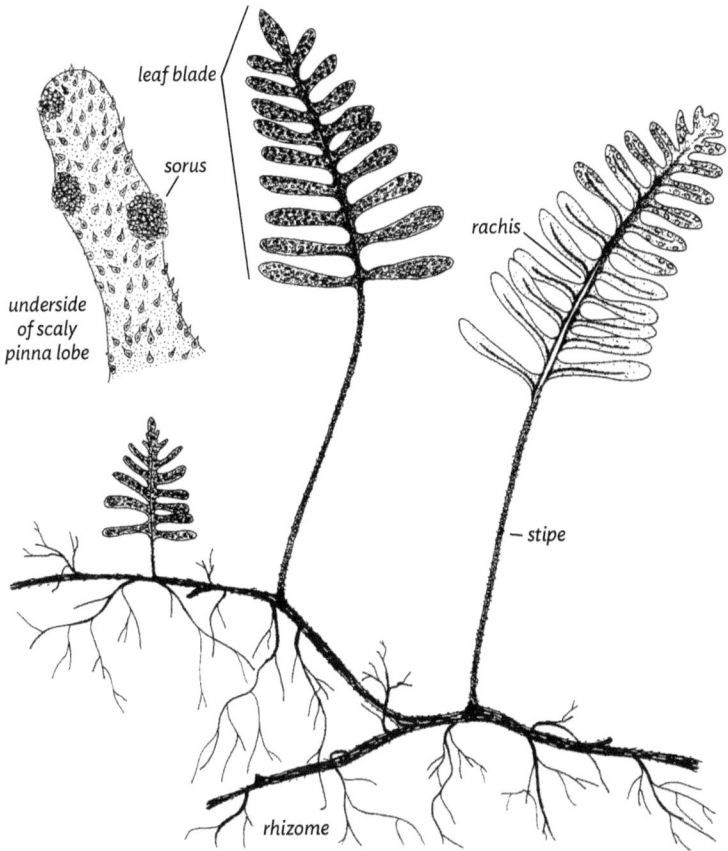

leaf blade

sorus

underside
of scaly
pinna lobe

rachis

stipe

rhizome

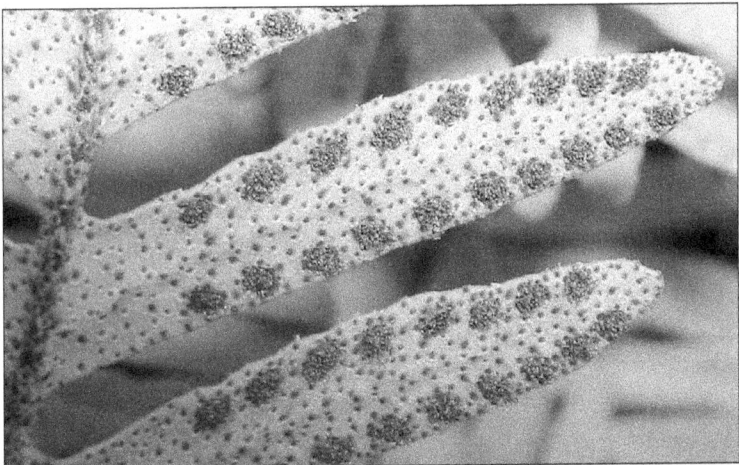

Resurrection fern (*Pleopeltis polypodioides*), underside of fertile lobes.

Polypodium
POLYPODY

Evergreen, colony-forming ferns of rock outcrops; **rhizome** creeping, scaly; **fronds** pinnatifid, the lobes rounded at tip, **stipes** scaly, with 3 vascular bundles; **sori round**, indusium absent. Genus includes ca. 160 species worldwide; 2 species in our flora.

KEY CHARACTERS

• Evergreen, colony-forming fern of rock outcrops.
• **Leaf blades** pinnatifid, leathery, without scales on underside.
• Sterile and fertile **fronds** alike.
• **Sori** round, **indusia** absent.
• Fronds wither in drought, quickly becoming green following rain.

NAME

From Greek, *polys*, many, and *pous*, foot, perhaps referring to the numerous stipe bases, or to the many knoblike branches of the rhizome.

KEY TO POLYPODIUM | POLYPODY

1 Leaf blade widest just above base, lobes narrowed at tip (especially lower lobes); rhizome scales a uniform golden brown; Ohio
1. Polypodium appalachianum
APPALACHIAN POLYPODY, page 299

1 Leaf blade widest near middle, the blades more evenly wide; lobes blunt-tipped; rhizome scales of 2 shades of brown; regionwide
2. Polypodium virginianum
ROCK POLYPODY, page 300

Polypodium appalachianum
APPALACHIAN POLYPODY

Polypodium virginianum
ROCK POLYPODY

Appalachian polypody (*Polypodium appalachianum* Haufler & Windham)

Evergreen fern of cliffs and rocky slopes; in our region, known from eastern Ohio, more common in eastern North America. Also called **rock cap fern** for its habit of densely covering rocks and boulders. **Rhizome** creeping, branching, with whitish covering; scales lanceolate, mostly uniformly golden brown. **Fronds** to 35 cm long, sterile and fertile fronds alike. **Stipe** jointed at base, straw-colored, smooth or with scattered light-brown scales; vascular bundles 3. **Blade** pinnatifid, oblong to narrowly lanceolate, usually widest near base, leathery; **rachis** sparsely scaly on underside, upperside smooth. **Pinnae** linear, entire to slightly toothed, broadest near their base; veins free. **Sori** round, midway between margin and midrib to nearly marginal; on all but lowest pinnae of fertile fronds, **indusium** absent, **sporangia** yellow, then brown when mature. Distinguished from **rock polypody** (*Polypodium virginianum*) by being widest near base of blade and narrower at tip, the pinnae tips tend to be more pointed than in *P. virginianum* (at least near base of blade), and rhizome scales uniformly golden brown.

Polypodium appalachianum
Appalachian polypody

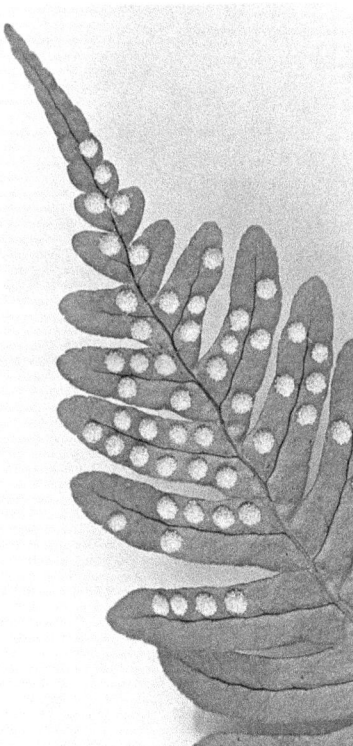

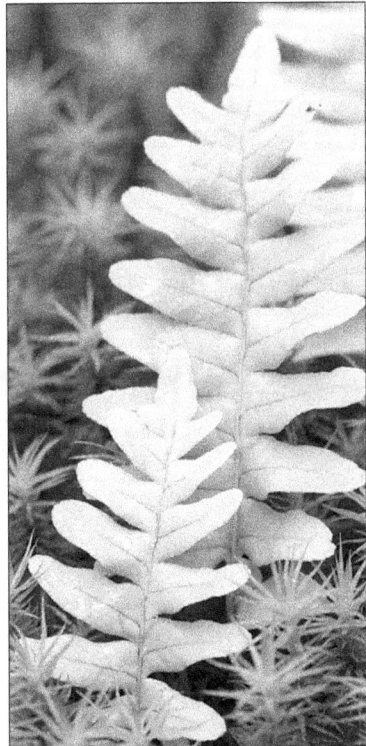

LEFT **Appalachian polypody** (*Polypodium appalachianum*), sori on underside of fertile blade. RIGHT Fronds of **rock polypody** (*Polypodium virginianum*).

Polypodium virginianum L.
ROCK POLYPODY

FIELD TIPS
- evergreen, colony-forming fern of rock outcrops
- blade pinnatifid, leathery
- sori round, indusia absent

MIDWEST RANGE
Regionwide.

HABITAT
Sandstone and calcareous bluffs, rock outcrops, ledges, boulders and mossy talus, rocky soil on steep slopes, and on decaying stumps and logs; usually in partial shade.

DESCRIPTION

Rootstock Long-creeping at ground surface, ropelike and spongy, occasionally branching, often partly exposed and mat-like, producing rows of fronds; densely covered with brown scales.

Frond Evergreen, last year's fronds withering as this year's appear in early summer, sterile and fertile fronds alike; mostly 10-40 cm long.

Stipe Jointed at base, straw-colored, smooth or with scattered light-brown scales; vascular bundles 3.

Rachis Straw-colored, smooth; sunken below plane of upper surface, somewhat raised on undersurface.

Blade Pinnatifid, cut almost to rachis, oblong to narrowly lance-shaped, usually widest near middle, occasionally at or near base, deep green and leathery, upper surface lustrous.

Pinnae 9 to 18 pairs, linear, tip rounded; margins entire to round-toothed; veins free, forked, obscure.

Sori Round, midway between margin and midrib to nearly marginal, at vein ends; on all but lowest pinnae of fertile fronds; **indusium** absent; **sporangia** yellow to brown at maturity; paraphyses (branching structures among the sporangia) present.

SYNONYMS
- *Polypodium vulgare* auct. non L. p.p.
- *Polypodium vulgare* L. var. *virginianum* (L.) Eaton

NOTES
Henry David Thoreau referred to the "fresh and cheerful communities" of polypody in early spring.

The rhizome has a strong licorice taste.

NAME
In 1753, Linnaeus recognized a European *Polypodium vulgare* and an

NORTH AMERICA

MIDWEST

American *P. virginianum.*
Virginianum describes the area of
Virginia.

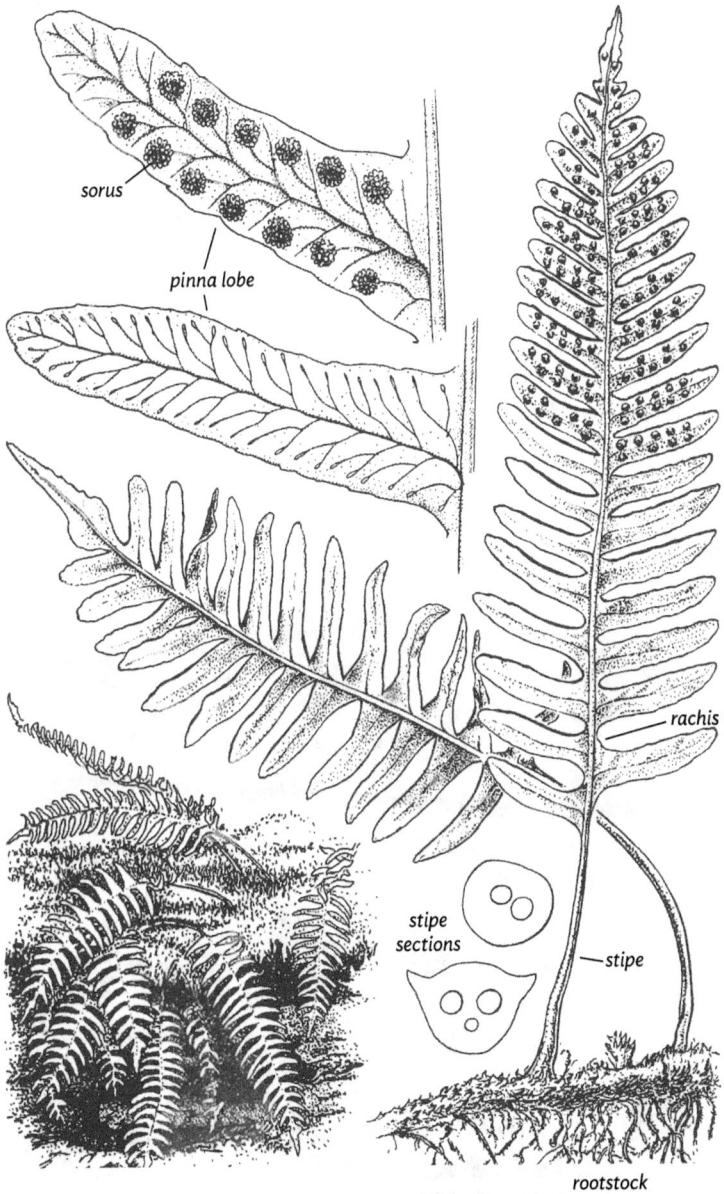

sorus

pinna lobe

rachis

stipe
sections

stipe

rootstock

PTERIDACEAE
Maidenhair Fern Family

Small to medium ferns, mostly growing on rock. **Fronds** compound, **sori** linear along or near pinnae margin, covered by false indusium formed by inrolled margin. Worldwide, the family includes about 50 genera and 950 species, most common in tropical and arid regions; 7 genera in our flora.

KEY CHARACTERS

- Typically growing on rocks (except for *Adiantum pedatum* which also grows in soil).
- **Rhizomes** scaly.
- **Sori** linear along pinnae margin, covered by **false indusium** formed by inrolled margin.

SUBFAMILIES

This large family sometimes subdivided into 5 subfamilies, 4 of which occur in the Midwest:

- **Cheilanthoideae** - *Argyrochosma, Cheilanthes*
- **Cryptogrammoideae** - *Cryptogramma*
- **Pteridoideae** - *Pellaea, Pteris*
- **Vittarioideae** - *Adiantum, Vittaria*

NAME

From Greek, *pteris*, fern, derived from *pteron*, wing or feather, for the closely spaced pinnae, the leaves resembling feathers.

Northern maidenhair (*Adiantum pedatum*), note dark, slender stipes and rachises.

ADDITIONAL MIDWEST SPECIES

• **Powdery cloak fern** [*Argyrochosma dealbata* (Pursh) Windham], west central Illinois, more common southwest of our region; prefers limestone rocks. **Rootstock** erect to short-creeping, with pale brown to black scales. **Frond** deciduous, monomorphic or somewhat dimorphic. **Stipe** ridged, scaly at base, vascular bundles 1 or more. **Leaf blade** 1-4 pinnate; segments sessile to short-stalked, veins conspicuous, free. **Sori** continuous, near margin; **indusium** false. Synonyms *Cheilanthes dealbata, Notholaena dealbata,* and *Pellaea dealbata.*

• **Spider brake** (*Pteris multifida* Poir.), introduced in southeastern USA, in our region, known only from se Illinois (and reported for s Indiana); on soil or rock. **Rootstock** short-creeping, densely scaly, scales reddish brown. **Frond** deciduous in our region, the sterile fronds leafier than the fertile. **Stipe** dark brown to straw-colored at base, sparsely scaly at base, smooth above; vascular bundle 1. **Leaf blade** 1-pinnate, bright green, smooth; **pinnae** 3 to 7 pairs, linear, lower pinnae 1-pinnate, upper not divided, joined to the winged rachis; fertile pinnae contracted; veins conspicuous, free, simple or once-forked. **Sori** continuous, near margin; **indusium** false, formed by reflexed margin of pinnae. Synonym *Pycnodoria multifida* (Poiret) Small.

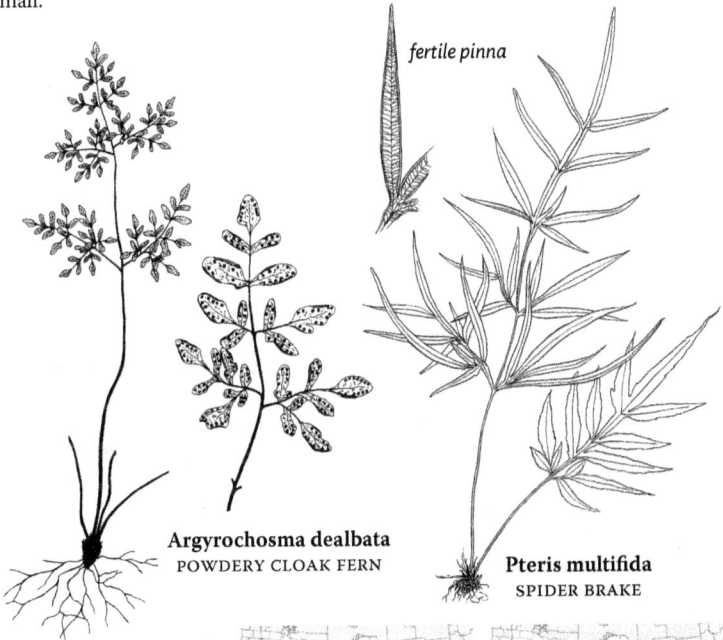

fertile pinna

Argyrochosma dealbata
POWDERY CLOAK FERN

Pteris multifida
SPIDER BRAKE

Argyrochosma dealbata
Powdery Cloak Fern

Pteris multifida
Spider Brake

KEY TO PTERIDACEAE | MAIDENHAIR FERN FAMILY

1 Plant a free-living gametophyte, resembling a thallose liverwort; rare in dark, moist recesses in rock overhangs in s Ind and Ohio
1. Vittaria appalachiana
SHOESTRING FERN, page 328

1 Plant a sporophyte, consisting of a stem, rhizome, corm, or crown producing well-developed leaves **2**

2 Plants of moist forests; sori distinct, short, mostly not merged together; regionwide
2. Adiantum pedatum
NORTHERN MAIDENHAIR, page 306

2 Plants of rock cliffs and boulders; sori usually merged as a marginal band . . **3**

3 Segments of frond hairy, sometimes bead-like
3. Cheilanthes
LIP FERN, page 308

3 Segments of frond without hairs, not bead-like **4**

4 Petioles green to straw-colored for at least the upper 1/3, rachis green; sterile and fertile fronds different, the fertile longer than the sterile and with narrower segments; n Ill, Iowa, n Mich, Minn, Wisc
4. Cryptogramma stelleri
FRAGILE ROCKBRAKE, page 318

4 Petioles and rachis dark brown to almost black throughout; leaves uniform
5. Pellaea
CLIFFBRAKE, page 320

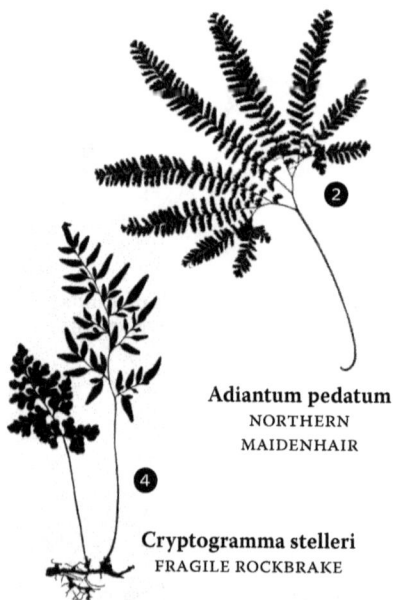

Cheilanthes
LIP FERN

Adiantum pedatum
NORTHERN
MAIDENHAIR

Cryptogramma stelleri
FRAGILE ROCKBRAKE

Pellaea
CLIFFBRAKE

Adiantum
MAIDENHAIR

Adiantum includes about 150 species worldwide, mostly of tropical regions. North America is home to 9 species; 2 species in our flora, one of which (*Adiantum aleuticum*) is disjunct in Michigan's Upper Peninsula from western North America (see below). Maidenhairs are deciduous, monomorphic ferns with slender purplish-black, glossy stipes and rachises; pinnules somewhat resemble the leaves of the *Ginkgo biloba* tree. Maidenhairs are terrestrial and typically found in shaded, rocky woods, often along streambanks and seepages.

KEY CHARACTERS

• Delicate fern of moist woods, blade fan-shaped to round in outline.

• **Pinnules** thin, midrib absent; main vein located along lower pinnule margin.

• **Stipe** and **rachis** glossy chestnut brown to purple-black, brittle.

• **Sori** along margins, covered by and attached to the inrolled edge of pinnule.

NAME

From Greek, *a*, not, and *diaine*, to wet, referring to the water-shedding properties of the leaves. Maidenhair may refer to either the slender black stipes or to the fine black fibrous roots.

ADDITIONAL MIDWEST SPECIES

• **Aleutian maidenhair** [*Adiantum aleuticum* (Rupr.) C.A. Paris], a fern mostly of western North America, is reported from Michigan's Upper Peninsula. Very similar to the much more common **northern maidenhair** (*A. pedatum*), and sometimes treated as a variety of that species. However, in *A. aleuticum,* pinnae sometimes oriented vertically; pinnae in *A. pedatum* always horizontal. Other differences include a middle pinna larger than the other pinnae; green, rather than red new growth; and presence of a single fan-shaped pinnule on the rachis between the fork and the first pinnae; this pinnule is rarely present on *A. pedatum*.

Aleutian maidenhair (Adiantum aleuticum), arrow indicates single pinnule on rachis between fork and first pinnae.

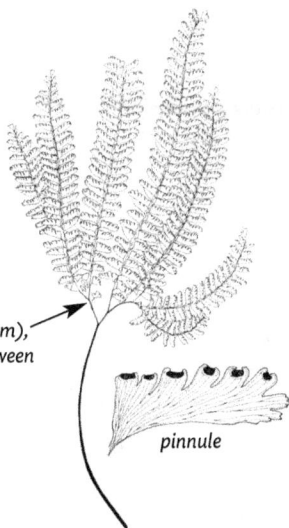

pinnule

Adiantum pedatum L.
NORTHERN MAIDENHAIR

FIELD TIPS
- delicate deciduous fern of moist woods, forming small colonies
- unique fan-shaped blade
- false indusium formed by inrolled pinnule margin
- stipe and rachis purplish black

MIDWEST RANGE
Regionwide.

HABITAT
Rich deciduous forests, often near seeps or streambanks; rocky ravines.

DESCRIPTION
Rhizome Short-creeping and branching, woody; scales bronze-colored.

Croziers Clustered, delicate, wine-red.

Frond Deciduous, separate but close together along the rhizome; sterile and fertile fronds alike; ca. 40 cm long by 30 cm wide.

Stipe Purplish black, slender, grooved above, smooth, occasionally waxy or with or a few scales at base; vascular bundle single, U- or V-shaped.

Rachis Purple-brown to black, smooth.

Blade Fan-shaped or nearly circular from the branched rachis, the primary divisions then pinnately divided; held horizontally, at 90° to the stipe; smooth.

Pinnae 6 to 10 pairs, oblong, those closest to the stipe longer, becoming smaller outwards; pinnules alternate, incised on upper margin; margins crenate; veins free, several times forked from main vein along lower margin, the midrib absent.

Sori Elongate, near upper margin of pinnule; **indusium** false, formed by reflexed margin of pinnule; **sporangia** yellow or yellowish brown.

NOTES
The dark, wiry stipes were used to accent larger tan-colored reeds in basket-making by Native Americans.

In earlier times, maidenhairs had many medicinal uses.

Southern maidenhair (*Adiantum capillus-veneris*), a fern of the southern USA, is not known from our region, but extends northward to nearby Kentucky and Missouri; unlike *A. pedatum*, the stipe is not forked.

NAME
Sent to Linnaeus from Canada and Virginia, and named by him in 1753.

From Latin, *pedatus*, having feet, referring to the branching of the blade.

NORTH AMERICA

MIDWEST

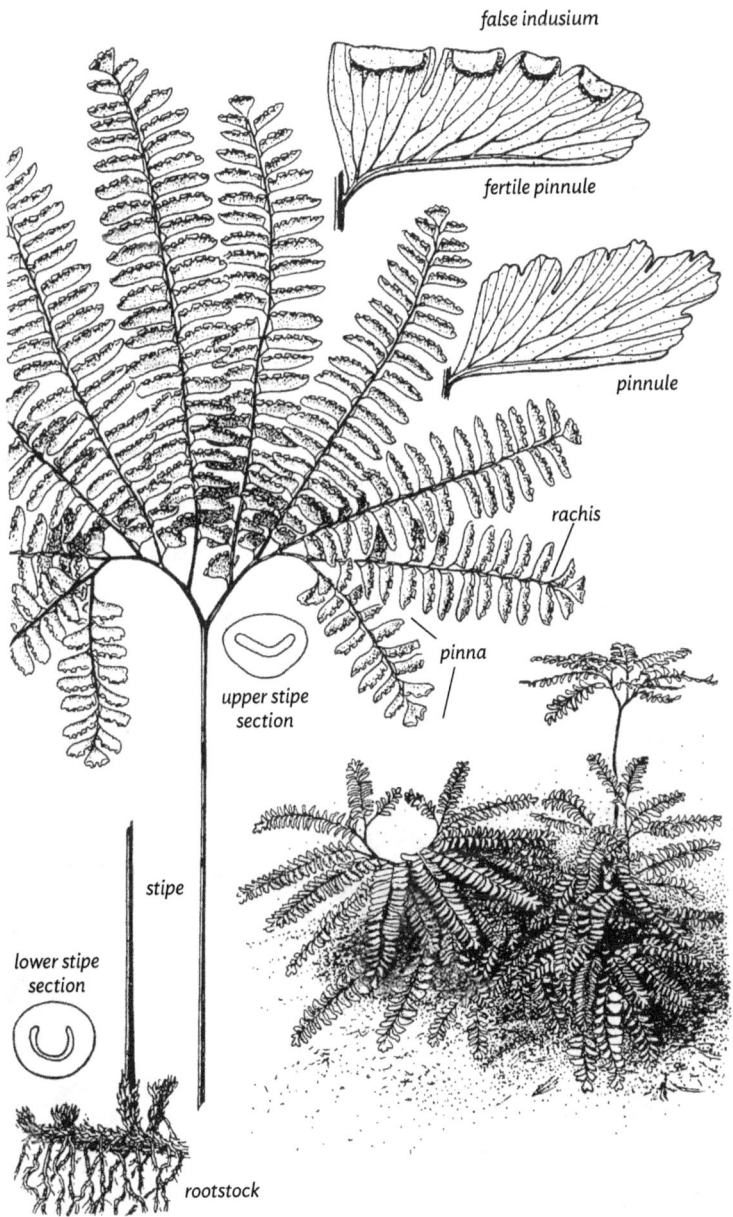

false indusium

fertile pinnule

pinnule

rachis

pinna

upper stipe
section

stipe

lower stipe
section

rootstock

Cheilanthes
LIP FERN

Small, evergreen (or nearly so), monomorphic ferns, usually wooly hairy, at least on underside of blade. Plants well-adapted to dry conditions of their typical rock cliff habitats, and during drought, may curl up and appear lifeless, then quickly greening after rain. *Cheilanthes* includes ca. 155 species, mostly in the Western Hemisphere; a few species occur in Europe, Asia, Africa, Australia and the Pacific Islands. There are 28 species in North America, mostly in deserts of the southwestern United States and northern Mexico; 2 species in our flora.

KEY CHARACTERS
• Small, nearly evergreen ferns of rock outcrops.
• **Stipe** brown to black.
• **Leaf blade** mostly 2-pinnate-pinnatifid, leathery, finely hairy and/or scaly below, finely hairy or mostly smooth above.
• **Pinnule lobes** sometimes rounded and beadlike.
• **Sporangia** at ends of veins in sori-like clusters, more or less continuous along margins of pinnules, covered by **false indusium** formed by inrolled margin.

NOTE
Separated from other small cliff-dwelling ferns by the small, hairy beadlike pinnule lobes and dark brown or black stipe and rachis.

Unlike most ferns, many *Cheilanthes* have noncircinate vernation, that is, the emerging bud is not at the center of a coil, in contrast to the more usual unfurling form of a crozier or fiddlehead form.

NAME
From Greek, *cheilos*, lip, and *anthos*, flower; a lip-like false indusium covers the sporangia. Our species now sometimes placed in genus *Myriopteris*.

Slender lip fern (*Cheilanthes feei*).

KEY TO CHEILANTHES | LIP FERN

1 Leaf blade mostly less than 10 cm long; pubescence sparse on stipe and upper side of blade, underside densely hairy, the hairs whitish, turning tan with age; plants usually on limestone; Ill, Iowa, Minn, Wisc **1. Cheilanthes feei**
SLENDER LIP FERN, page 310

1 Blade mostly more than 10 cm long; blade densely hairy, the hairs brownish; plants usually on sandstone; Ill, s Ind, Wisc
2. Cheilanthes lanosa
HAIRY LIP FERN, page 312

Cheilanthes feei
SLENDER LIP FERN

Cheilanthes lanosa
HAIRY LIP FERN

Cheilanthes feei T. Moore
SLENDER LIP FERN

FIELD TIPS
- small clumped fern of rock crevices, usually on limestone
- pinnae segments rounded and beadlike
- underside of blade with long wooly whitish hairs
- stipe dark brown to black
- plants dry up in drought

MIDWEST RANGE
Ill, e Iowa, Minn, Wisc; more common in w North America.

HABITAT
Cliffs and ledges of limestone or rarely sandstone.

DESCRIPTION
Rootstock Short-creeping, scales uniformly brown or some scales with darker central stripe.

Frond Nearly evergreen, in dense clusters, sterile and fertile fronds alike, 10-20 cm long by 3 cm wide.

Stipe Dark brown to black, the color extending into rachis and costae, white-hairy; vascular bundle 1.

Rachis Hairy; scales absent.

Blade Lance-shaped, 2-pinnate-pinnatifid or more divided at base; blue-green; underside with wooly white to tan hairs, upper surface with scattered hairs.

Pinnae 7 to 9 pairs, lowest pairs widely separated, becoming closer together above; smallest segments beadlike, cut to midrib; veins free.

Sori Linear, nearly continuous along segment margins, submarginal, **indusium** false, formed by incurved margin; **sporangia** black.

SYNONYMS
- *Myriopteris gracilis* Fée

SIMILAR SPECIES
- Distinguished from **hairy lip fern** (*Cheilanthes lanosa*) by forming tighter clumps, the fronds smaller and bluish green, and pinna segments dissected to midrib and more rounded.

NAME
Named for French botanist Antoine Fée (1789-1874), a student of ferns, lichens and fungi.

NORTH AMERICA

MIDWEST

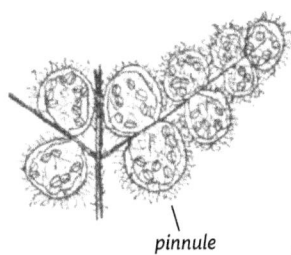

pinnule

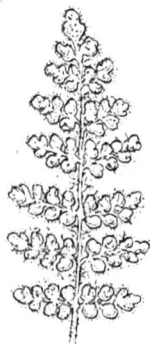

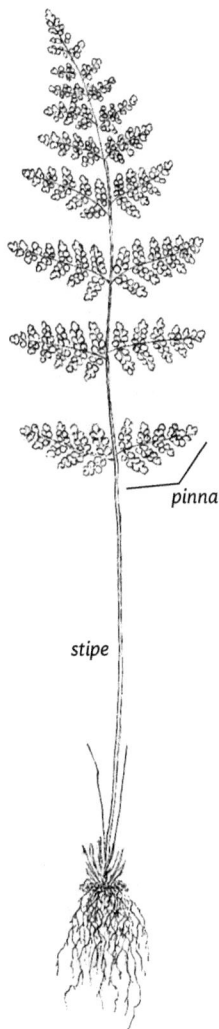

pinna

scale

stipe

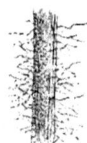

stipe hairy

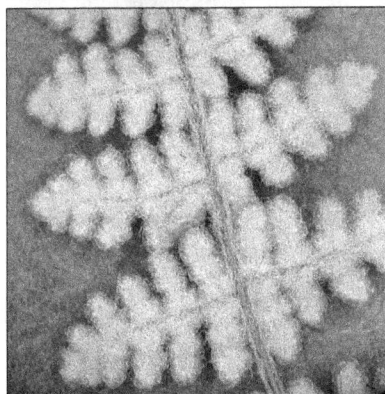

Cheilanthes lanosa (Michx.) D.C. Eaton
HAIRY LIP FERN

FIELD TIPS
- small loosely clumped fern in shallow soil on sandstone outcrops
- underside of blade with wooly tan-gray hairs
- stipe brown, hairy
- plants dry up in drought

MIDWEST RANGE
Ill, s Ind, Wisc.

HABITAT
Dry, exposed, sandstone bluffs and ledges.

DESCRIPTION
Rootstock Short-creeping, branched, without hairs; scales brown, narrowly lance-shaped.

Frond Nearly evergreen, becoming dry and dormant in droughts, sterile and fertile fronds alike; to 40 cm long by 5 cm wide.

Stipe Chestnut brown, the color extending into the rachis and costae, with many small white-grayish hairs; scales absent; vascular bundles 1.

Rachis Covered in white hairs, turning grayish when mature.

Blade Lance-shaped, 2-pinnate-pinnatifid at base, less cut above; underside wooly with jointed tan-gray hairs, upper surface less hairy.

Pinnae Gray-green; 12 to 14 pairs; pairs opposite and widely separated at base, closer together and sub-opposite to alternate above; pinnules opposite, pinnate at bottom of blade, pinnatifid above; underside sparsely reddish hairy; veins free, forked.

Sori Linear, discontinuous, near margin; **indusium** false, formed by the slightly inrolled margin; **sporangia** black.

SYNONYMS
- *Cheilanthes vestita* (Spreng.) Sw.
- *Myriopteris lanosa* (Michx.) Grusz & Windham

SIMILAR SPECIES
- Distinguished from **slender lip fern** (*Cheilanthes feei*) by its looser clumps, the larger, darker green fronds, and pinna segments less dissected and not beadlike.

NAME
Discovered by Michaux in the mountains of Tennessee and North Carolina and named *Nephrodium lanosum* in 1803; assigned to genus *Cheilanthes*, by Watt in 1874.

From Latin, *lana*, wooly or soft-hairy.

NORTH AMERICA

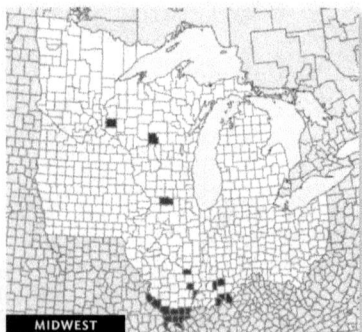

MIDWEST

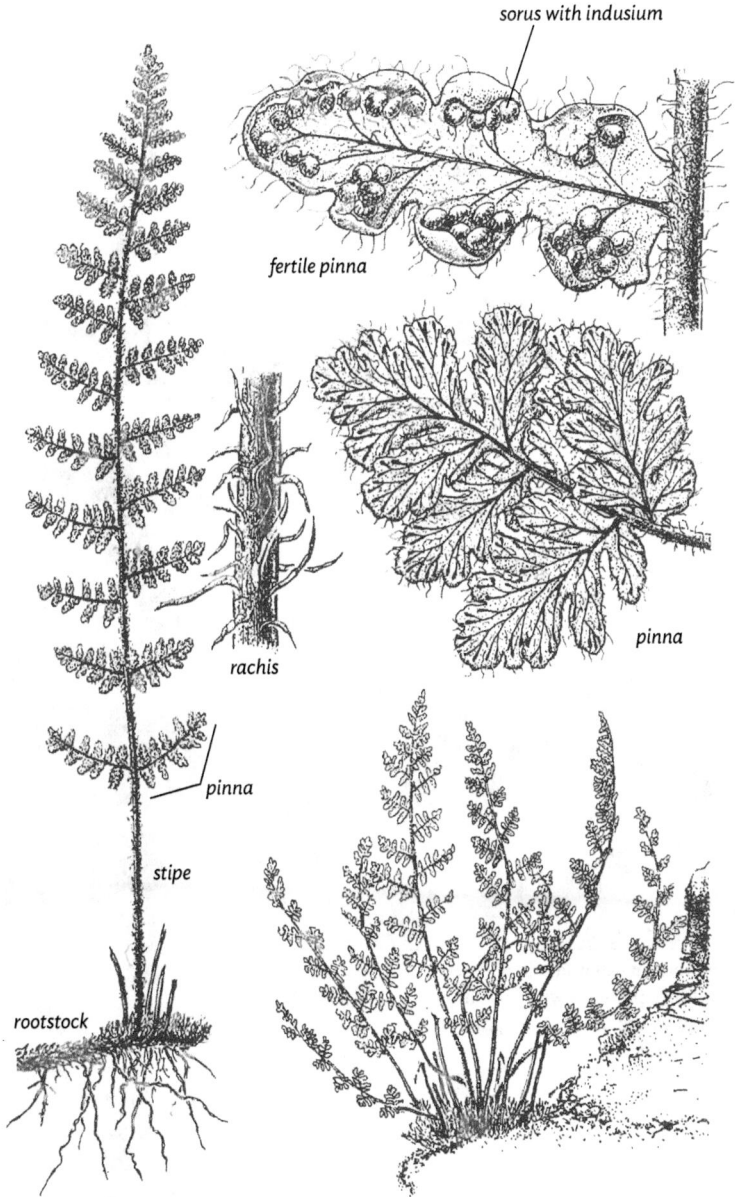

sorus with indusium

fertile pinna

rachis

pinna

pinna

stipe

rootstock

Northern maidenhair (*Adiantum pedatum*).

Northern maidenhair (*Adiantum pedatum*), pinnules with false indusia formed by the inrolled margins (above); frond (right).

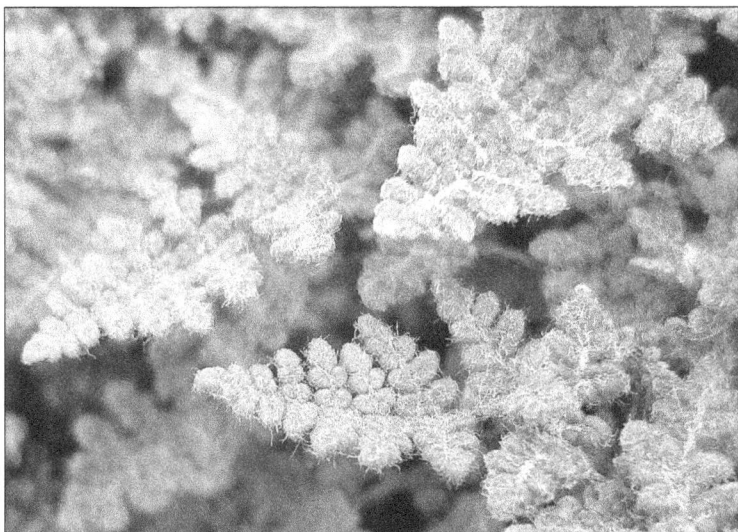

Slender lip fern (*Cheilanthes feei*).

Hairy lip fern (*Cheilanthes lanosa*); small colony (l); frond (r).

Cryptogramma
ROCKBRAKE

Small evergreen or deciduous, dimorphic ferns found on rock; fertile fronds larger than sterile. Genus includes about 11 species, mostly at higher elevations in North America, Europe, the Andes, and parts of Asia; 4 species in North America, 2 species in our flora.

KEY CHARACTERS

• Small evergreen or deciduous ferns of rock cliffs.

• **Fronds** dimorphic, fertile fronds taller than sterile fronds.

• **Leaf blades** 2-pinnate.

• **Stipes** dark brown at base, light brown to green upwards, grooved on upperside.

• **Sori** near margins of fertile pinnules, the margins inrolled and forming a false indusium.

NAME

From Greek, *kryptos*, hidden, and *gramme*, line; referring to hidden lines of sori under the reflexed leaf margin. **Rockbrake** refers to the typical rocky habitat, brake is used as a general term for fern. Also known as **parsely fern** as sterile fronds of some species resemble parsley.

ADDITIONAL MIDWEST SPECIES

• **American rockbrake** (*Cryptogramma acrostichoides* R. Br.), reported for Michigan Upper Peninsula (threatened), and possibly ne Minnesota (Cook County), on noncalcareous cliff crevices, rock outcrops, and talus, often where relatively dry. **Rootstock** short-creeping, branching, covered with old stipe bases, scales many, bicolored. **Fronds** tufted, 25 cm high by 4 cm wide, dimorphic, the sterile fronds evergreen and much shorter than the deciduous fertile fronds. **Stipe** brown at base, turning straw-colored, then green upwards; grooved; scaly at base; vascular bundles 2. **Leaf blade** 2-pinnate, bright green, with 5-6 pairs of pinnae. **Sori** elongate, near margins of narrowed segments; **indusium** false, strongly inrolled and covering the sori.

Separated from our more common *Cryptogramma stelleri* by its densely clumped habit and evergreen sterile fronds vs. the fronds along a rhizome and deciduous fronds of *C. stelleri*.

Acrostichoides means resembling *Acrostichum*, a genus of tropical ferns where the sori completely cover the undersurface.

Cryptogramma acrostichoides
AMERICAN ROCKBRAKE

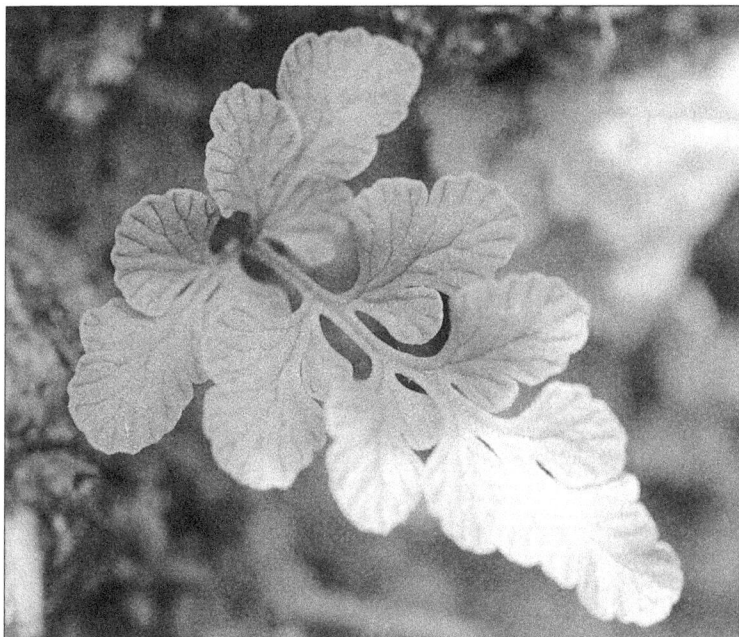

Fragile rockbrake (*Cryptogramma stelleri*), sterile frond.

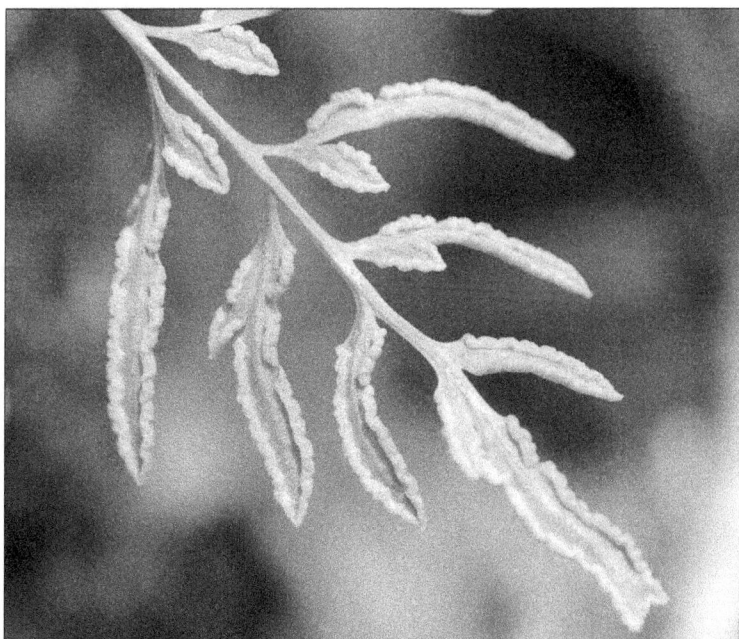

Fragile rockbrake (*Cryptogramma stelleri*), fertile frond.

Cryptogramma stelleri (Gmel.) Prantl
FRAGILE ROCKBRAKE

FIELD TIPS
• small fern of rock cliffs and ledges
• sterile and fertile fronds deciduous, much different in form
• margins inrolled over sori to form false indusium

MIDWEST RANGE
n Ill, Iowa, n Mich, Minn, Wisc.

HABITAT
Moist, shaded, rocks and cliffs; wet or seepy crevices and ledges of limestone, calcareous shale and sandstone.

DESCRIPTION
Rootstock Long-creeping, slender (to 1 mm wide), seldom branching, hairy, with a few transparent scales.

Frond Deciduous, dying by late summer; sterile and fertile fronds unalike, the sterile fronds shorter than the fertile; to about 18 cm long.

Stipe Brown at base, green upwards, smooth or with a few hairs at base; vascular bundles 2.

Rachis Green, smooth.

Blade Fertile fronds lance-shaped, 2-pinnate, the sterile less divided, pinnate-pinnatifid; bright green, thin-textured and nearly translucent, smooth.

Pinnae 5-6 pairs, the fertile pinnae smaller; margins entire, inrolled on fertile segments; veins free, forked.

Sori Elongate, near margins; **indusium** false, formed by the strongly inrolled margin; **sporangia** yellow.

SIMILAR SPECIES
• Somewhat like a small **cliffbrake** (*Pellaea*), but unlike that genus, the sterile and fertile fronds very different, and rachis greenish rather than dark brown.

NAME
First named *Pteris stelleri* by Gmelin based on plants from Siberia in 1768. Michaux named identical Canadian plants *Pteris gracilis* in 1803. In 1823, R. Brown proposed genus *Cryptogramma*, and Prantl placed the species in that genus in 1882.

Named in honor of German naturalist George Wilhelm Steller (1709-1746), who accompanied the Bering expedition of 1741 and who is considered the first European to set foot in what is now Alaska; his name is used with other species, including Steller's jay and Steller's sea lion.

NORTH AMERICA

MIDWEST

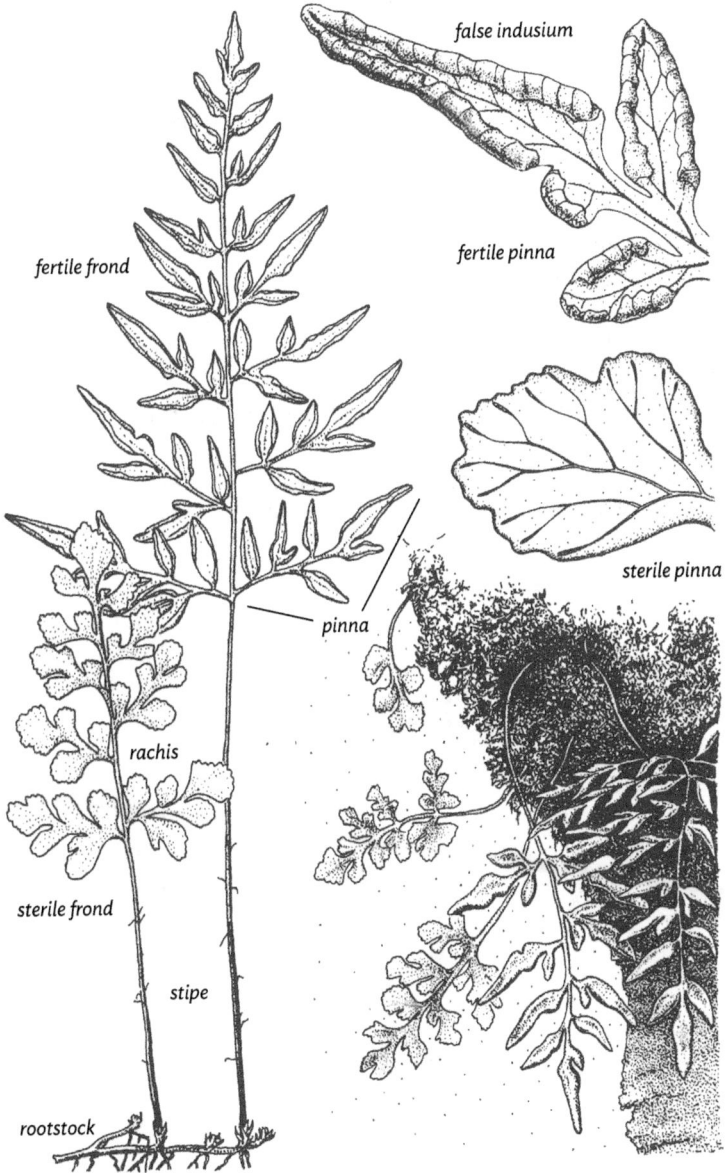

false indusium

fertile frond

fertile pinna

sterile pinna

pinna

rachis

sterile frond

stipe

rootstock

Pellaea
CLIFFBRAKE

Cliffbrakes are small evergreen ferns of rocky, often calcareous habitats. Sterile and fertile **fronds** similar (*Pellaea glabella*) or somewhat different (*P. atropurpurea*). **Sori** near pinna margin and covered by inrolled edge forming a false indusium. *Pellaea* includes ca. 40 species, mostly in the Western Hemisphere; 15 species in North America; 2 in our flora. In the United States, most species of *Pellaea* are found in arid regions of the Southwest.

KEY CHARACTERS

- Small evergreen ferns of rock cliffs and talus.
- **Rhizomes** short-creeping, scaly.
- Sterile and fertile **fronds** similar or slightly different.
- **Leaf blades** gray-green or blue-green, 2-pinnate at base, less so above; tipped with a terminal leaflet.
- **Stipes** shiny, dark brown or purple.
- **Sori** marginal, covered by **false indusium** formed by edge of inrolled pinna.

NAME

From Greek, *pellos*, dusky, referring to the dull, bluish-gray pinnae. Cliffbrake refers to the typical rock crevice habitat, brake is a general term for fern.

KEY TO PELLAEA | CLIFFBRAKE

1 Stipe and rachis without short, incurved hairs; lower pinnule surface smooth or with a few hairlike scales; Ill, Ind, Iowa, n Mich, Minn, Ohio, Wisc **1. Pellaea glabella**
SMOOTH CLIFFBRAKE, page 322

1 Stipe and rachis with short, incurved hairs on upper surfaces; lower pinnule surface with scattered hairlike scales; Ill, Ind, e Iowa, n Mich, s Minn, Ohio, Wisc **2. Pellaea atropurpurea**
PURPLE-STEM CLIFFBRAKE, page 324

Pellaea glabella
SMOOTH CLIFFBRAKE

Pellaea atropurpurea
PURPLE-STEM CLIFFBRAKE

Pellaea atropurpurea (L.) Link
PURPLE-STEM CLIFFBRAKE

FIELD TIPS
- small evergreen fern of rock and talus
- sterile and fertile fronds not alike, leathery, dull gray-green
- stipe and rachis dark purple-brown to nearly black, hairy
- sori covered by inrolled pinnule margin

MIDWEST RANGE
Ill, Ind, e Iowa (endangered), n Mich (threatened), s Minn, Ohio, Wisc.

HABITAT
Dry, exposed limestone outcrops, ledges and talus slopes; also found on shale and sandstone; less common on exposed vertical cliffs (that habitat more typical of *Pellaea glabella*).

DESCRIPTION
Rootstock Short-creeping, branched; densely covered with brown scales.

Frond Evergreen, somewhat dimorphic, sterile shorter and less divided than fertile fronds; 50 cm long by 18 cm wide.

Stipe Dark purple to black, shiny, with incurved hairs; base with scales, these uniformly brown; vascular bundle 1.

Rachis Dark purplish brown; usually covered with whitish hairs.

Blade Long triangular in outline, leathery, gray-green; usually 2-pinnate at base, but sometimes more or less.

Pinnae 5 to 9 pairs, lower pinnae stalked, upper sessile, with a terminal pinna like the upper lateral ones; pinnules sessile or nearly so, sometimes with 1 or 2 basal lobes; veins obscure.

Sori Elongate or linearly joined, submarginal; **indusium** false, margins inrolled; **sporangia** brown.

SIMILAR SPECIES
- Similar to **smooth cliffbrake** (*Pellaea glabella*), but differs in having hairs on rachis and costae; fronds also taller, more upright and more divided.
- Distinguished from **fragile rockbrake** (*Cryptogramma stelleri)* by its leathery pinnules and hairy stipe and rachis.

NAME
Discovered by John Clayton in Virginia and described by Gronovius in about 1743; named by Linnaeus *Pteris atropurpurea* in 1753. When Link founded genus *Pellaea* in 1841, he included this species.

From Latin, *atropurpureus*, dark purple, referring to color of stipe and rachis.

NORTH AMERICA

MIDWEST

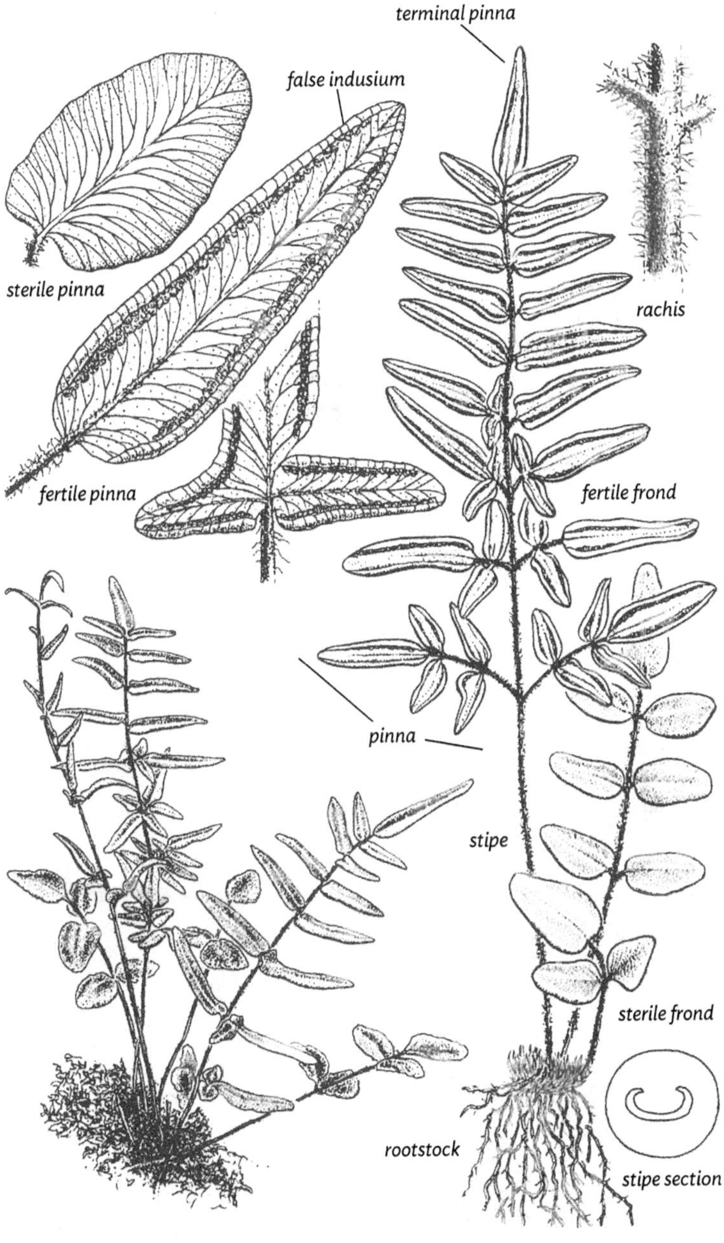

terminal pinna

false indusium

sterile pinna

rachis

fertile pinna

fertile frond

pinna

stipe

fertile frond

sterile frond

rootstock

stipe section

Pellaea glabella Mett. ex Kuhn
SMOOTH CLIFFBRAKE

FIELD TIPS
- small evergreen fern of rock cliffs
- sterile and fertile fronds similar, leathery, pale bluish green
- stipe dark reddish brown, shiny, without hairs
- sori covered by inrolled pinnule margin

MIDWEST RANGE
Ill, Ind, Iowa, n Mich, Minn, Ohio, Wisc.

HABITAT
Ledges and crevices of cliffs of limestone, dolomite and sandstone; in sun to light shade.

DESCRIPTION
Rootstock short-creeping or erect, branched; scales orange-brown.

Crozier Sparsely hairy at emergence, then smooth.

Frond Evergreen, clustered, sterile and fertile fronds similar but sterile fronds typically shorter and less divided; to 35 cm long by 8 cm wide.

Stipe Dark reddish brown, shiny, smooth or with a few hairs, vascular bundle 1.

Rachis Brown, smooth or very sparsely hairy.

Blade Usually 2-pinnate at the base, less above, leathery, pale bluish green,

Pinnae 4 to 9 pairs, lower pinnae stalked, upper sessile, with a terminal pinna like the upper lateral ones; pinnules sessile or nearly so, sometimes with 1 or 2 basal lobes; margins entire; veins obscure, forked.

Sori Elongate or linearly joined, submarginal; **indusium** false, margins inrolled to cover sori; **sporangia** pale brown.

SYNONYMS
- *Pellaea atropurpurea* (L.) Link var. *bushii*

SIMILAR SPECIES
- Similar to **purple-stem cliffbrake** (*Pellaea atropurpurea*), but smaller, the sterile and fertile fronds alike, and stipe and rachis without hairs.

NAME
For many years this species mistaken for depauperate specimens of *Pellaea atropurpurea*. Plants sent from Missouri to Mettenius were recognized by him to be distinct, and his suggested name, *Pellaea glabella*, was published by Kuhn in 1869.

Latin, *glabella*, hairless or smooth.

NORTH AMERICA

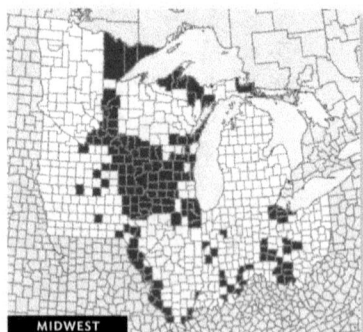

MIDWEST

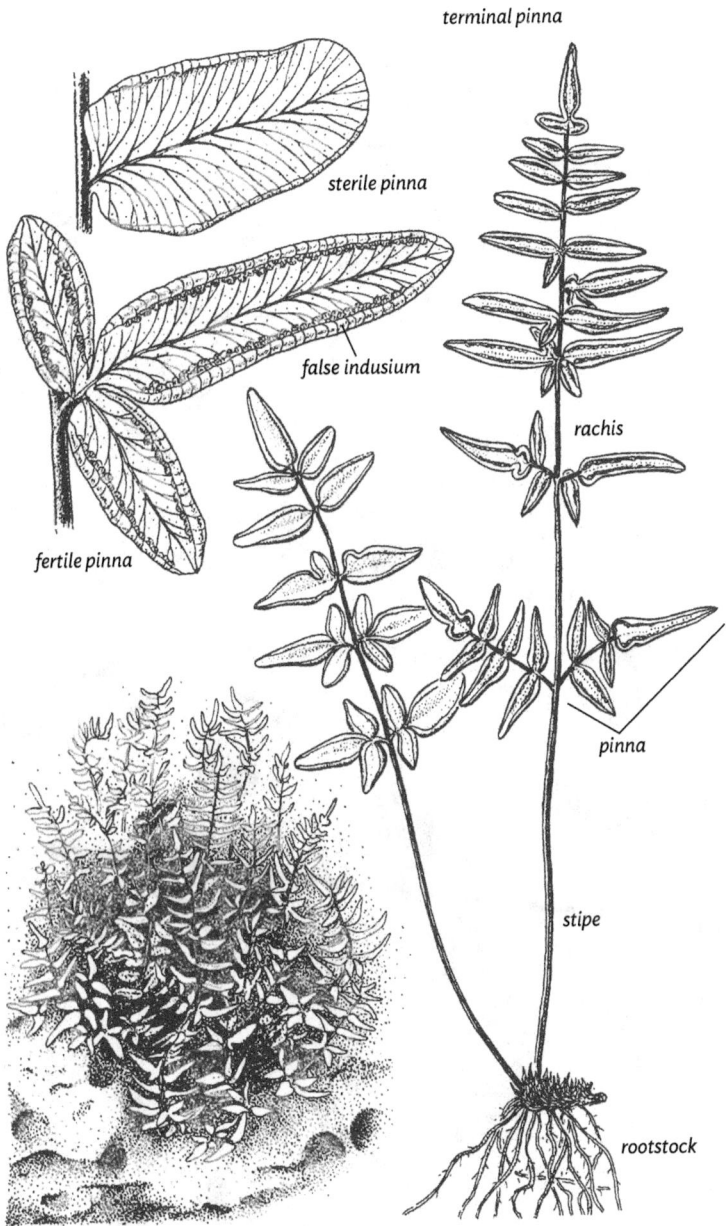

terminal pinna

sterile pinna

false indusium

fertile pinna

rachis

pinna

stipe

rootstock

PTERIDACEAE | MAIDENHAIR FERN FAMILY • *Pellaea atropurpurea*

Purple-stem cliffbrake (*Pellaea atropurpurea*).

Purple-stem cliffbrake (*Pellaea atropurpurea*), underside of fertile pinnae.

Smooth cliffbrake (*Pellaea glabella*), the sterile and fertile fronds alike.

Smooth cliffbrake (*Pellaea glabella*).

Vittaria appalachiana Farrar & Mickel
APPALACHIAN SHOESTRING FERN

gametophyte

FIELD TIPS

- restricted to dark, moist rock grottoes and small caves
- resembles a thallose liverwort
- may form dense colonies
- only found in gametophyte form
- in Midwest, only s Ind and Ohio

MIDWEST RANGE

s Ind, Ohio; more widespread south of Midwest region in the Appalachian Mountains.

HABITAT

On ceilings, walls and floors of cool, perennially moist, heavily shaded or dark cavities, rock shelters or overhangs in non-calcareous rock (sandstone or conglomerate); sometimes requiring a flashlight to see.

DESCRIPTION

Unique among our ferns in that the sporophyte generation (the typically fern-like stage of the life cycle) is essentially absent, and plants occur as asexually-reproducing gametophytes. Small (less than 1 cm long), abortive sporophytes have been recorded from Ohio, and have also been produced in culture.

Plants small, 1-3 cm long, with branches at tips bearing small gemmae (several-celled plantlets); gemmae shed to form new plants, and colonies up to several square meters may form in suitable habitat.

NOTES

Formerly included within Vittariaceae (Shoestring Fern Family), but that family no longer considered valid.

Worldwide, 29 species but taxonomy in a flux; 2 species in continental USA, the other species (*V. lineata*) occurs in both sporophyte and gametophyte forms, and occurs in Florida, s Alabama and Georgia. It is a common epiphyte on palm trees in Florida.

NAME

Genus named by James Edward Smith in 1793. *Vittaria* from Latin, *vitta*, a band or ribbon. Our plants first recognized as the gametophyte of a fern by A.J. Sharp in 1930, and assigned to genus *Vittaria* by Wagner and Sharp in 1963. Farrar and Mickel described the species in 1991. The common name refers to the hanging, long-linear form of other species of *Vittaria*.

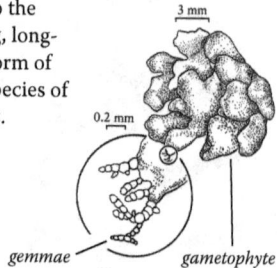

3 mm

0.2 mm

gemmae — *gametophyte*

NORTH AMERICA

MIDWEST

Appalachian shoestring fern (*Vittaria appalachiana*), typical habitat on ceiling of small cave.

Appalachian shoestring fern (*Vittaria appalachiana*).

SALVINIACEAE | Water Fern Family
Azolla
MOSQUITO FERN

Small, free-floating aquatic ferns of quiet or stagnant water of ponds, lakes, marshes and streams, may form large matlike colonies covering the water surface. Small family of 2 genera and ca. 16 species; within *Azolla*, 2 American species are recognized: *Azolla cristata* Kaulfuss (eastern and midwestern plants) and *Azolla filiculoides* Lam. (western USA from Washington to Arizona).

KEY CHARACTERS

• World's smallest fern and most important economically.

• Mat-forming, floating **aquatic fern** of quiet water.

• Plants small, reddish when in full sun to bright green in shade, the segments easily broken or dispersed to form new colonies.

• **Pinnae** small, scalelike.

• **Spores** of 2 types (but apparently rarely produced): female, egg-bearing megaspores; and male, sperm-bearing microspores; the 2 types of spores deveoping in separate capsules (sporocarps).

NOTES

Azolla sometimes placed in separate family, Azollaceae, with Salviniaceae including *Salvinia*, a genus of introduced aquatic ferns present in se USA and California.

Agriculturally, *Azolla* is important for its symbiosis with the nitrogen-fixing *Anabaena azollae* Strasburger, a cyanobacterium (blue-green alga). Because the plants fix nitrogen, they are often used as a green fertilizer and mixed with livestock feed as a nutritional supplement. *Azolla pinnata* has been cultivated for many centuries in rice paddies of Vietnam and China, where it acts as a fertilizer after it decomposes.

NAME

First described as a genus by Lamarck in 1783; family name in honor of Italian Antonio Maria Salvini (1633-1721).

Azolla from Greek, *azo*, to dry, and *ollupi*, to kill, referring to rapid death of plants when removed from water.

Crested mosquito fern (*Azolla cristata*).

Azolla cristata Kaulfuss
CRESTED MOSQUITO FERN

FIELD TIPS
- small, mosslike, free-floating, aquatic fern
- may form matlike colonies
- reddish to bright green

MIDWEST RANGE
Ill, Ind (threatened), Iowa, s Mich, se Minn, Ohio, Wisc.

HABITAT
Floating on quiet water of river backwaters and ponds; sometimes exposed on wet muddy banks; plants easily broken to form new plants and also dispersed by waterfowl to new habitats.

DESCRIPTION
Rhizomes Small (less than 3 cm long), pale brown, branching, with tiny roots.

Pinnae Small, less than 1 cm wide; pinnules 2 lobed, the upper smaller lobe floating and lower larger lobe submersed; red when in full sun or under stress, to green when in shade.

Sporocarps (rarely seen) Of 2 kinds, in the leaf axils; the smaller sporocarps contain a single large female megaspore, the larger sporocarps contain many tiny male microspores.

SYNONYMS
- *Azolla caroliniana* Willd.
- *Azolla mexicana* Schltdl. & Cham. ex C. Presl
- *Azolla microphylla* Kaulfuss

SIMILAR SPECIES
The region's only other true aquatic fern is **water-clover** (*Marsilea*, page 212); however, members of that genus are rooted rather than floating, and the leaf blade is larger and 4-parted.

NOTES
The pale bluish green color of the upper lobes is due, in part, to the presence of a symbiotic blue-green bacterium, *Anabaena azollae*.

NAME
First described by Kaulfuss in 1824. Common name from supposed ability of dense mats of this plant to smother mosquito larvae.

Plants of **crested mosquito fern** on finger tips.

NORTH AMERICA

MIDWEST

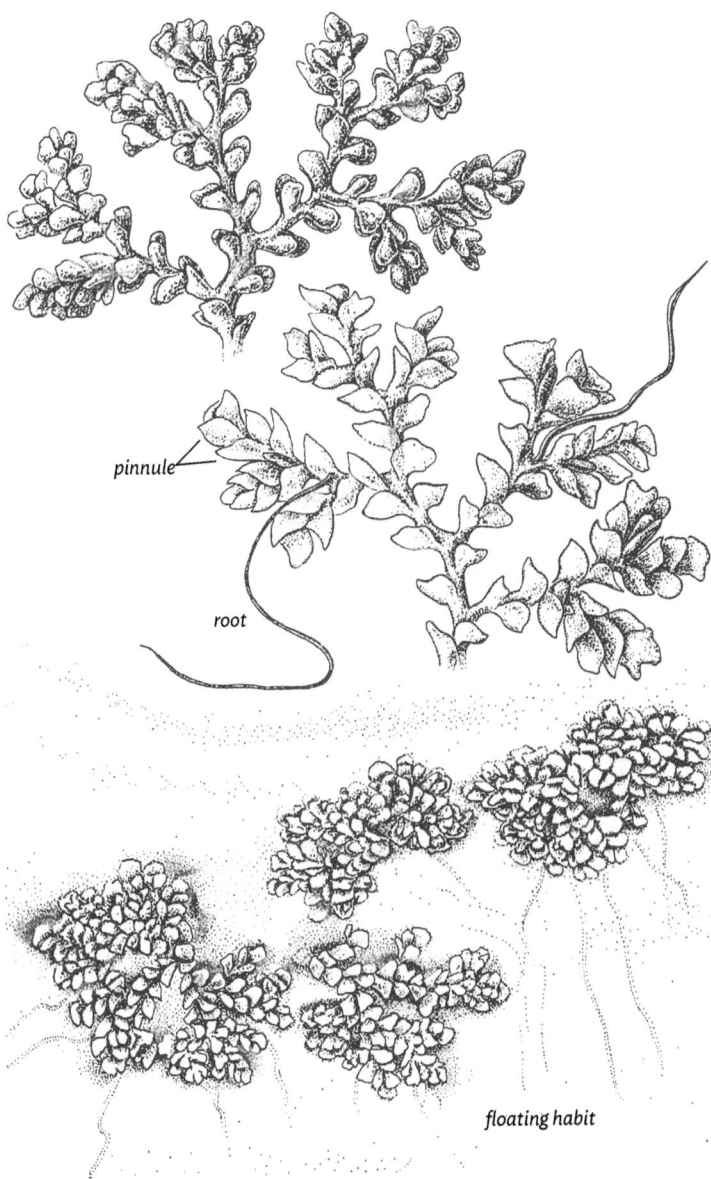

pinnule

root

floating habit

THELYPTERIDACEAE
Maiden Fern Family

Medium-sized deciduous ferns, spreading by rhizomes to form colonies; sterile and fertile **fronds** usually alike, 1-pinnate to pinnate-pinnatifid, with transparent needlelike hairs; **sori** usually on veins (but not marginal) on pinna underside; **indusia** present and often soon withering, or absent (*Phegopteris*). Worldwide, about 950 species in 5 (or more) mostly tropical genera; 4 genera in our flora, including 1 introduced species known from southern Illinois (below).

KEY CHARACTERS

• Deciduous, colony-forming ferns, of forests and sometimes in wetlands.

• Sterile and fertile **fronds** more or less alike, **blades** mostly 1-2 pinnate.

• Fronds with needlelike, transparent, usually 1-celled hairs; these also on rhizomes.

• **Sori** round, **indusia** present (often soon shed), kidney-shaped; or absent.

• **Vascular bundles** in stipe 2, these joined upwards.

NOTES

Taxonomic questions remain for this family. Our species previously included in genus *Thelypteris*, but are here separated into 3 genera; check synonyms listed for each species to see alternate scientific names.

NAME

Family name from *Thelypteris*, Greek *thelys*, female, and *pteris*, fern.

ADDITIONAL MIDWEST SPECIES

• **Mariana maiden fern** [*Macrothelypteris torresiana* (Gaud.) Ching], an introduced fern native to tropical and subtropical Africa and Asia, is reported in our region from Johnson County in southern Illinois; more common in southeastern USA. Distinguishing features are thick stipe (to 1 cm wide), rachis not winged; blades 2-pinnate or more divided; costae long-hairy; indusia very small, less than 0.3 mm wide. Previously known as *Thelypteris torresiana* (Gaud.) Alston.

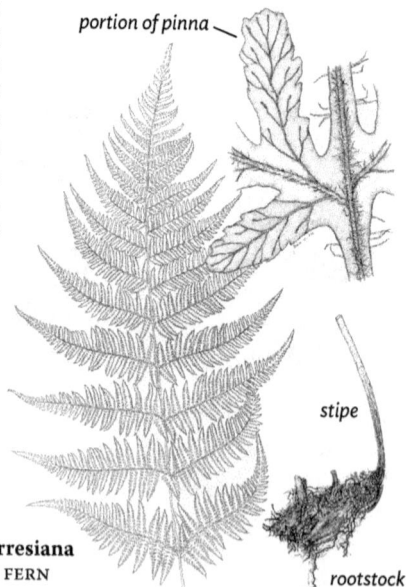

portion of pinna

stipe

Macrothelypteris torresiana
MARIANA MAIDEN FERN

rootstock

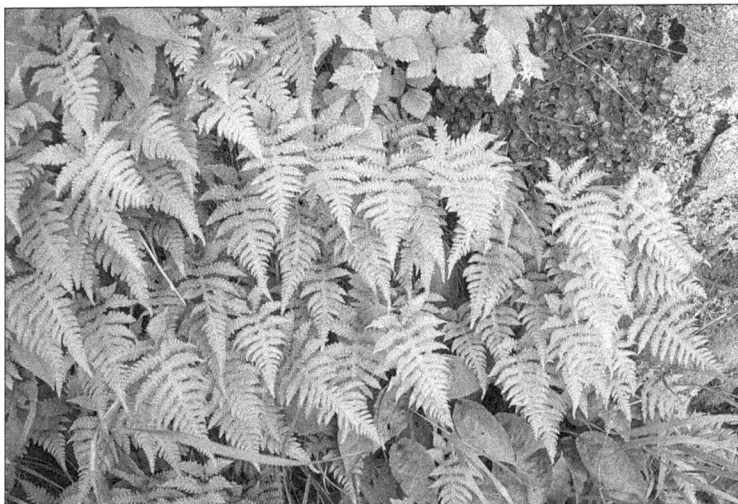

Narrow beech fern (*Phegopteris connectilis*), colony formed by creeping rhizomes.

Broad beech fern (*Phegopteris hexagonoptera*), note large basal pinnae, and winged rachis between two lowest pinna pairs.

KEY TO THELYPTERIDACEAE (ALL GENERA AND SPECIES)

1 Leaf blades 7-25 (-30) cm long, triangular, not more than 2x as long as wide; rachis with wings between the pinnae; sori without indusia; midribs of pinnae lacking a groove on upper surface **2**

1 Leaf blades (15-) 20-100 cm long, lance-shaped, oblong-lance-shaped, or triangular, more than 2x longer than wide; rachis without wings between the pinnae; sori with kidney-shaped indusia; midribs of pinnae with a groove on upper surface **3**

2 Rachis not winged between lowest two pinnae pairs; stipes finely hairy and usually also scaly in lower portion; blades usually longer than wide; Ill, e Iowa, Mich, Minn, Ohio, Wisc

1. Phegopteris connectilis
NARROW BEECH FERN, page 342

2 Rachis winged between all pinnae pairs; stipes becoming smooth just below base of blade (within 1 cm), apart from a few scales; blades usually wider than long; Ill, Ind, e Iowa, Mich, se Minn, Ohio, Wisc **2. Phegopteris hexagonoptera**
BROAD BEECH FERN, page 344

3 Leaf blade broadest near middle, gradually reduced to base; stipe less than 1/3 length of blade; plants of upland and wetland habitats; Ill, Ind, Mich, Ohio **3. Parathelypteris noveboracensis**
NEW YORK FERN, page 338

3 Leaf blade broadest near base, the pinnae stopping abruptly; stipe 2/3 to fully as long as blade; plants of wetlands **4**

4 Common regionwide; underside of blade without glands; lateral veins of sterile lobes once-forked between pinnule midvein and margin; lower surface of pinna midrib (costa) with tan, ovate scales; lobes of fertile leaves inrolled; indusia divided into hairlike segments (rarely smooth) **4. Thelypteris palustris**
EASTERN MARSH FERN, page 346

4 West-central and southwestern Wisc only (driftless area); underside of blade with small, stalkless, globular, golden to reddish glands; lateral veins of sterile lobes simple, not forked between the pinnule midvein and the margin; lower surface of pinna midrib lacking scales; lobes of fertile leaves flat to slightly revolute; indusia with minute glands along the margins

5. Parathelypteris simulata
BOG FERN, page 340

1

Phegopteris connectilis
NARROW BEECH FERN

2

Phegopteris hexagonoptera
BROAD BEECH FERN

3

**Parathelypteris
noveboracensis**
NEW YORK FERN

4

Thelypteris palustris
EASTERN MARSH FERN

5

Parathelypteris simulata
BOG FERN

Parathelypteris noveboracensis
(L.) Ching

NEW YORK FERN

FIELD TIPS
- medium, deciduous, colony-forming fern
- fronds yellow-green, in clumps of several from rhizome
- blade tapering to both ends, lowest pinnae very small, near base of frond

MIDWEST RANGE
Ill, Ind, Mich, Ohio.

HABITAT
Moist shady woods, especially near swamps, streams, and seeps in ravines, often in slightly disturbed secondary forests, may form dense colonies, especially in openings; soils moist, moderately acidic.

DESCRIPTION
Rootstock Long-creeping, cord-like, branching; scales few or absent.

Frond Deciduous, in tufts of 3 or more fronds along the rhizome; sterile and fertile fronds somewhat different; the ferile fronds larger, more erect and narrower than sterile fronds; to about 70 cm long by 15 cm wide; becoming whitish when dying.

Stipe Straw-colored, shading to green; scales tan to reddish brown; vascular bundles 2 at stipe base, merging above into a U-shape.

Rachis Pale green to straw-colored, covered with whitish hair.

Blade Elliptic, tapering to both ends, lowest pinnae very small (less than 1 cm long) and near base of frond; 1-pinnate-pinnatifid, the pinnule divisions falling short of meeting the costa by 1 mm; yellow-green; underside with transparent needlelike hairs.

Pinnae 20 to 24 pairs, lance-shaped, long tapering, less cut near the ends; costae grooved above, discontinuous with the rachis; margins entire or crenate, hairy; veins free, forked or not.

Sori Round, small, nearer the margin than the costule, absent from ends of the pinnules (or pinnae); **indusium** kidney-shaped, tan, withering.

SYNONYMS
- *Dryopteris noveboracensis* (L.) A. Gray
- *Polypodium noveboracense* L.
- *Thelypteris noveboracensis* (L.) Nieuwl.
- *Thelypteris thelypterioides* (Michx.) Holub

SIMILAR SPECIES
- **Hay-scented fern** (*Dennstaedtia punctilobula*) is larger, hairy, and with

NORTH AMERICA

MIDWEST

fronds mostly single in lines along rhizome; lowest pinnae larger and higher on frond.

• **Eastern marsh fern** (*Thelypteris palustris*) usually found in wetter habitats.

NOTES
Sent to Linnaeus from "Canada" (probably ne USA) and named by him *Polypodium noveboracense* in 1753. Transferred to *Thelypteris* by Nieuwland in 1910.

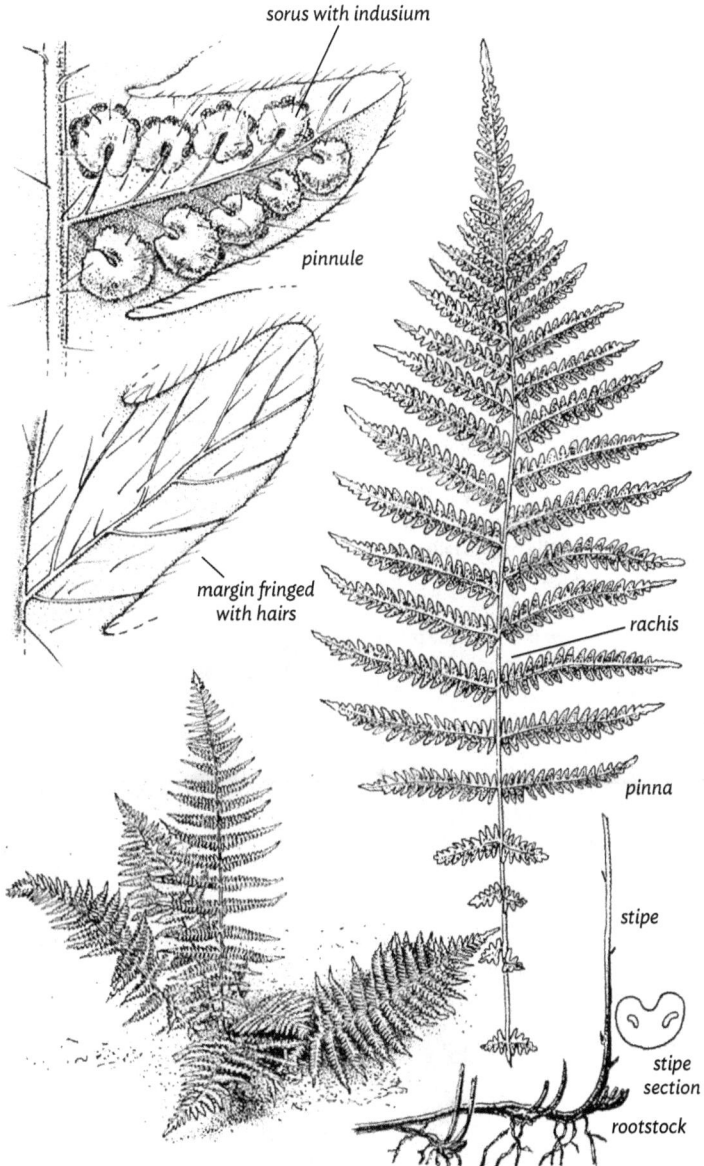

sorus with indusium

pinnule

margin fringed with hairs

rachis

pinna

stipe

stipe section

rootstock

Parathelypteris simulata (Davenport) Holttum
BOG FERN

FIELD TIPS
• medium deciduous fern of wet, acidic habitats
• pinnae narrow, separated along rachis
• in Midwest region, known only from sw Wisc

MIDWEST RANGE
Disjunct in unglaciated area of s Wisc; occasional in ne USA and Canada.

HABITAT
Acidic swamps and bogs, often growing with sphagnum mosses; in moist to wet shaded woodlands; peaty meadows.

DESCRIPTION
Rootstock Long-creeping, branching, some stipes distantly spaced, some closely spaced; scales sparse, pale brown.

Frond Deciduous, to 80 cm long by 15 cm wide; sterile and fertile fronds nearly alike, though sterile fronds arching, and fertile fronds taller, more erect, with longer stipes.

Stipe Brown at base, shading to yellow-green, with a few tan scales near base; vascular bundles 2, merging above to a U-shape.

Rachis Green; slightly hairy on upperside, smooth below.

Blade Lance-shaped, 1-pinnate-pinnatifid; pinnae widely separated, tapering to the tip, widest at or just above lowest pinna, lowest pair of pinnae bending down and forward; pale green; with short, whitish hairs, these soon falling, and with reddish to orangish, resinous, shiny, round glands.

Pinnae 16 to 18 pairs, lance-shaped, long tapering, less incised near ends, sessile; lower pinnae narrowed near rachis; fertile pinnules slightly inrolled; costae grooved above, discontinuous with the rachis; margins entire; veins free, simple.

Sori Round, midway between the margin than the costule, merging into each other; **indusium** narrowly kidney-shaped, whitish, maturing to tan, with tiny orange glands (use hand lens to see); **sporangia** brown.

SIMILAR SPECIES
• **New York fern** (*Parathelypteris noveboracensis),* but lowest pinnae in that species are very small.

• Separated from the much more common **eastern marsh fern** (*Thelypteris palustris*) by the veins never forking, the presence of small

NORTH AMERICA

MIDWEST

glands on the pinna underside, and the bending forward and down of the lowest pinna pair; in eastern marsh fern, the veins often forked and the glands absent.

SYNONYMS

• *Dryopteris simulata* Davenport
• *Thelypteris simulata* (Davenport) Nieuwl.

NAME

First collected by Dodge in New Hampshire, in 1880, and named by Davenport *Aspidium simulatum* in 1894. Transferred to *Thelypteris* by Nieuwland in 1910.

Simulata means resembling, in this case, perhaps to forms of lady fern, (*Athyrium filix-femina*), *Parathelypteris noveboracensis* and *Thelypteris palustris*.

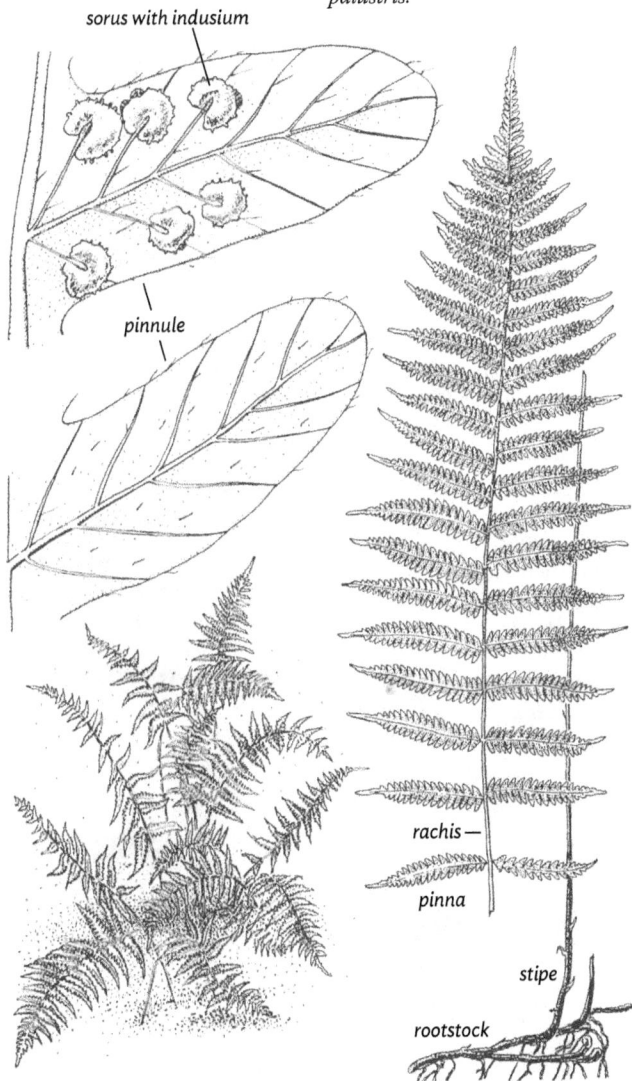

sorus with indusium

pinnule

rachis —

pinna

stipe

rootstock

Phegopteris connectilis (Michx.) Watt
NARROW BEECH FERN

FIELD TIPS

- deciduous creeping fern of moist woods and rocky places, may form colonies
- blade triangular, longer than wide
- lowest pinnae drooping down, not connected to other pinnae by wing along rachis
- indusia absent

MIDWEST RANGE

Ill (endangered), e Iowa (endangered), Mich, Minn, Ohio, Wisc.

HABITAT

Shaded moist to wet forests, streambanks, rocky ravines; sometimes on cliffs and near waterfalls.

DESCRIPTION

Rootstock Long-creeping, branching, cord-like, densely hairy and scaly when young, becoming smoother with age.

Frond Deciduous, new fronds all summer, sterile and fertile fronds alike, arching, to 50 cm long by 15 cm wide.

Stipe Straw-colored at base, with tan, lance-shaped scales and pointed hairs; vascular bundles 2, elongate, sometimes joined at top of stipe.

Rachis Green, with transparent needlelike hairs and spreading, lance-shaped scales, not winged between lowest pinnae.

Blade Triangular, widest at base, usually somewhat longer than wide, 1-pinnate-pinnatifid; light green, often finely hairy on both sides.

Pinnae 12 to 15 pairs, ovate-lance-shaped, base fused to rachis, except lowest pair unconnected by wings to the upper ones, and also typically drooping down and out; costae not grooved; veins free, simple or forked.

Sori Round, near margin; **indusium** absent; **sporangia** tan.

SYNONYMS

- *Dryopteris phegopteris* (L.) C. Chr.
- *Lastrea phegopteris* (L.) Bory
- *Phegopteris polypodioides* Fée
- *Thelypteris phegopteris* (L.) Slosson

SIMILAR SPECIES

- **Broad beech fern** (*Phegopteris hexagonoptera*) has a winged rachis between the 2 lowermost pinnae pairs.

NAME

Known to Linnaeus in Europe, and named by him *Polypodium dryopteris* in 1753; included in *Phegopteris* by Fée nearly 100 years later.

NORTH AMERICA

MIDWEST

Genus name from *phegos*, beech, and *pteris*, fern, in reference to its growing under beech trees. Epithet from Latin, *conecto*, to fasten together, referring to the upper joined pinnae.

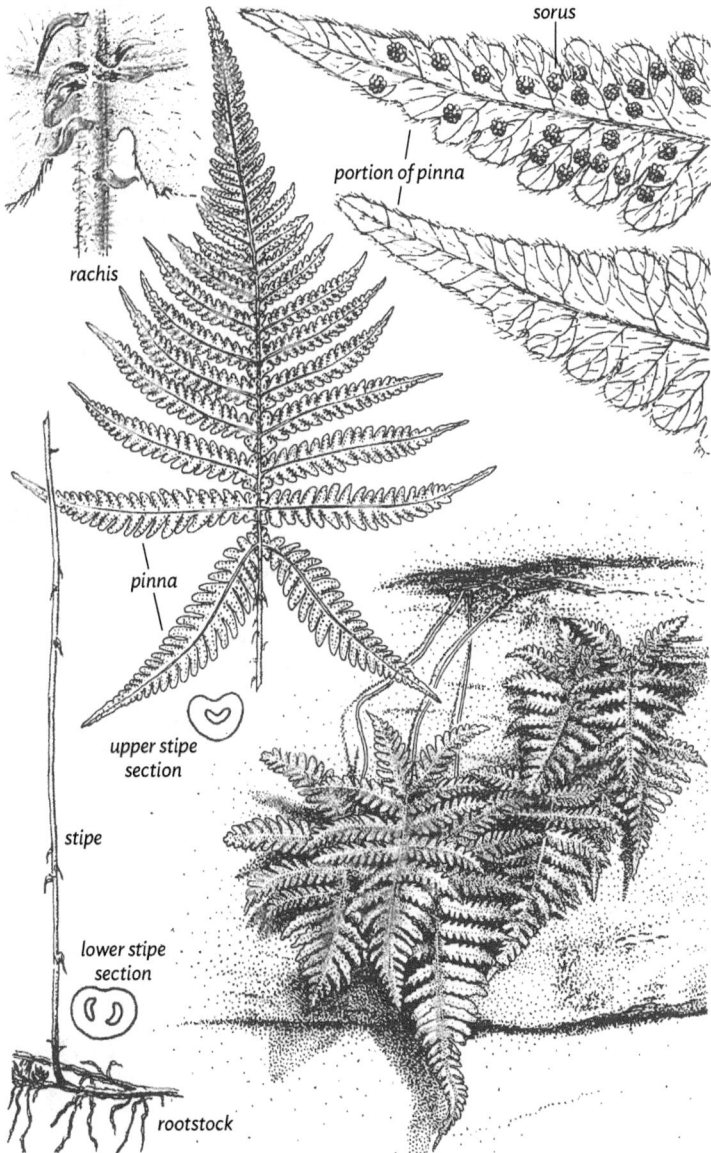

sorus

portion of pinna

rachis

pinna

upper stipe section

stipe

lower stipe section

rootstock

Phegopteris hexagonoptera (Michx.) Fée
BROAD BEECH FERN

FIELD TIPS

- deciduous fern of moist woods, forming small colonies
- blade about as wide as long
- all pinnae pairs connected by wing along rachis, lowest pair largest, spreading
- indusia absent

MIDWEST RANGE

Ill, Ind, Iowa, Mich, se Minn (threatened), Ohio, Wisc.

HABITAT

Rich, moist deciduous forests, often at bases of slopes, edges of seeps, and along streams; soils moderately acidic to alkaline.

DESCRIPTION

Rootstock Long-creeping, branching, cord-like, densely scaly and with a few hairs.

Frond Deciduous, arching, sterile and fertile fronds alike, to 60 cm long by 30 cm wide.

Stipe Reddish-brown at base, straw-colored above; scales tan, lance-shaped, and sometimes with hairs, vascular bundles 2, elongate, at 90°, merging or not by the top of the stipe to an open u-shape.

Rachis Green to straw-colored, with whitish scales; winged throughout.

Blade Broadly triangular, usually as wide as long, 1-pinnate-pinnatifid above to 2-pinnate-pinnatifid below; light green; underside with hairs along costae and veins, also with yellowish stalked glands.

Pinnae 12 to 15 pairs, ovate-lance-shaped, all connected by wing along rachis (including lowermost pinnae pair); basal pinnae largest, more spreading and more dissected than upper pinnae; costae not grooved; margins lobed, hairy; veins free, simple or forked.

Sori Round, sparse, near end of a vein; **indusium** absent; **sporangia** tan.

SYNONYMS

- *Dryopteris hexagonoptera* (Michx.) C. Chr.
- *Thelypteris hexagonoptera* (Michx.) Weath.

SIMILAR SPECIES

- **Narrow beech fern** (*Phegopteris connectilis*) lacks the winged rachis between the 2 lowermost pinnae pairs.

NORTH AMERICA

MIDWEST

NAME

Though early collected in Virginia and figured by Plukenet in 1691, this was confused with *P. connectilis* by Linnaeus. Michaux named it *Polypodium hexagonopterum* in 1803, and Fée placed it in his genus *Phegopteris* in about 1850.

Hex means six; *gona*, joint; *pterus* from *pteron*, Greek for wing, thus a six-jointed wing, referring to angular appearance of the winged rachis.

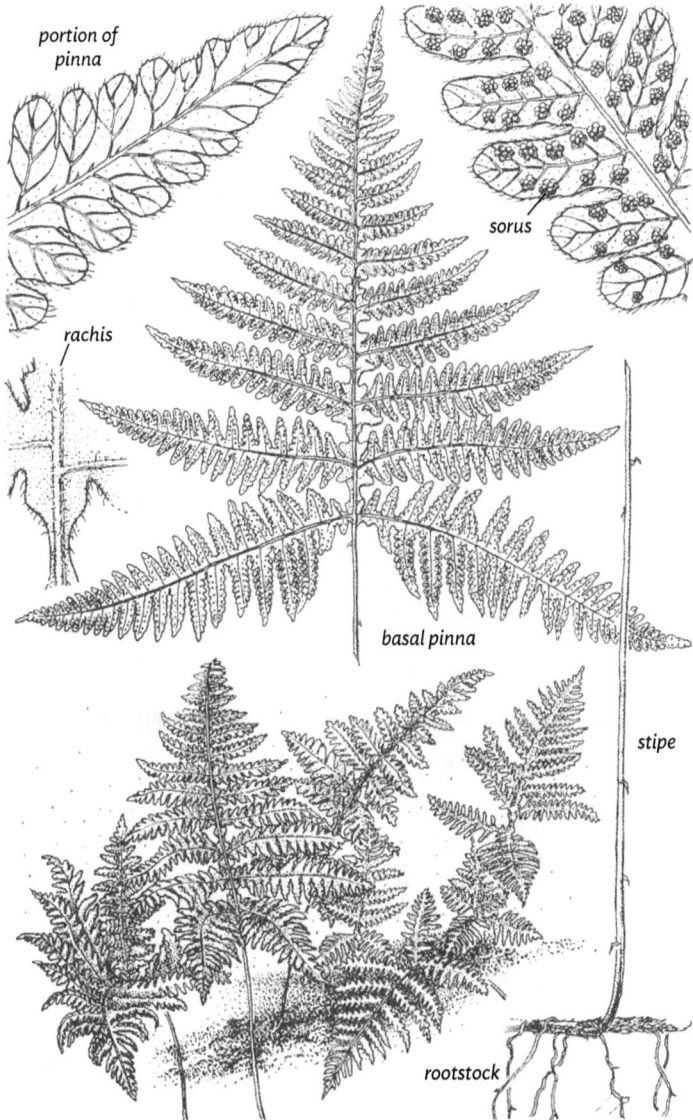

portion of pinna

sorus

rachis

basal pinna

stipe

rootstock

Thelypteris palustris Schott

EASTERN MARSH FERN

FIELD TIPS

• common deciduous fern of wet places
• blade widest below middle
• stipe long, brown
• veins once-forked

MIDWEST RANGE

Ill, Ind, Iowa, Mich, Minn, Ohio, Wisc; common.

HABITAT

Wet meadows, swamps, wet thickets, ditches, fens, bogs, wet shores, seeps, etc., in sun or partial shade, often on raised hummocks or sometimes in standing water.

DESCRIPTION

Rootstock Long-creeping, branching, black behind the white tip; scales sparse.

Frond Deciduous, often twisted; sterile and fertile fronds somewhat different; sterile fronds earlier in season, fertile fronds narrower, taller, developing about 6 weeks later, to 75 cm long by 20 cm wide.

Stipe Usually longer than blade, straw-colored or darker near base, shading upward to green, with a few scales near base, vascular bundles 2, oblong at stipe base, merging above to an open u-shape.

Rachis Pale green to purplish green; smooth and slender; grooved.

Blade Lance-shaped, 1-pinnate-pinnatifid, pinnae widely separated, tapering to tip, widest below the middle, narrowing somewhat to the lowest pinna, thin and delicate, pale green; with short, whitish hairs, these falling soon.

Pinnae 14 to 20 pairs, lance-shaped, long tapering, less dissected near the ends, sometimes twisted toward horizontal, sessile; costae grooved above, discontinuous with rachis; margins of sterile pinnae entire to slightly toothed; fertile pinnae triangular, margins folded downward to cover some of the sori; veins free, once-forked.

Sori Round, midway between the margin than the costule, merging into each other; **indusium** small, irregular or kidney-shaped, often hairy but without glands, soon shed.

SIMILAR SPECIES

• **Bog fern** (*Parathelypteris simulata*) similar but all veins unforked vs. 1-forked in *Thelypteris palustris*.

• **Hay-scented fern** (*Dennstaedtia punctilobula*) has hairy stipes shorter than the blade.

NORTH AMERICA

MIDWEST

• **Lady fern** (*Athyrium filix-femina*) is clumped, with true pinnules and toothed margins.

NAME

Linnaeus in 1753 gave the name *Acrostichum thelypteris* to a European fern, and 9 years later Schmidel made

Thelypteris a genus. Schott proposed T. palustris in 1834. In 1848 Gray made the combination *Dryopteris thelypteris*.

Greek *thelys*, female, and *pteris*, fern. *Palustris* means marshy or swampy.

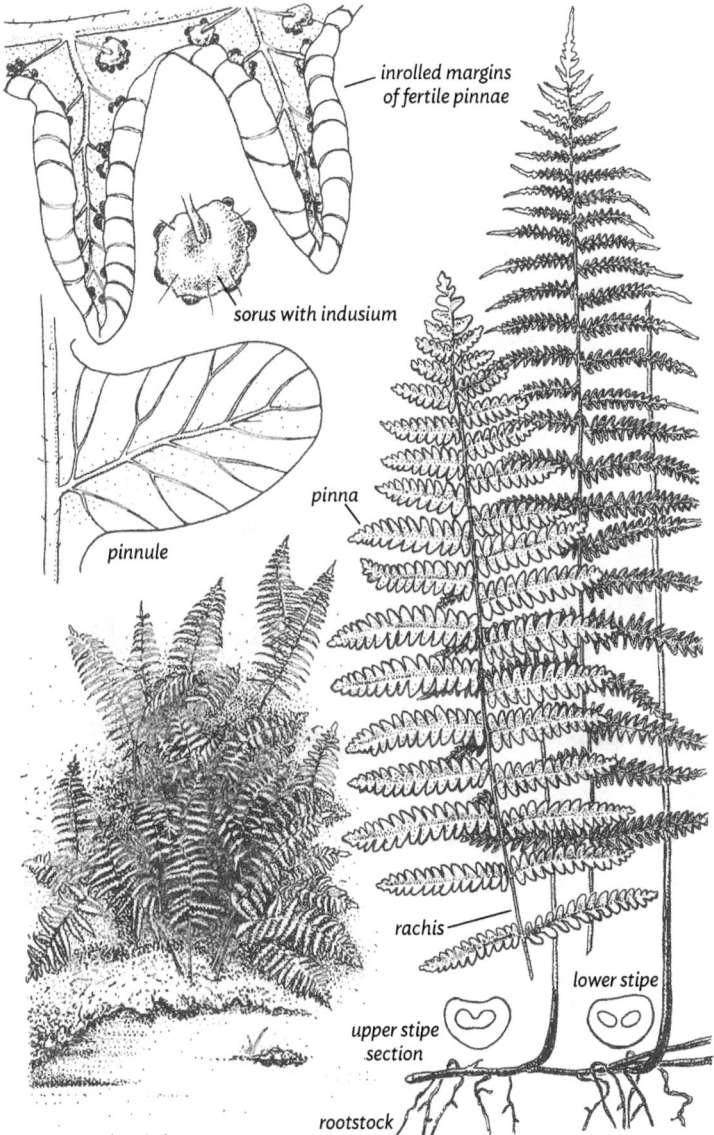

inrolled margins of fertile pinnae

sorus with indusium

pinna

pinnule

rachis

lower stipe

upper stipe section

rootstock

New York fern (*Parathelypteris noveboracensis*), blade (l), frond silhouette (r).

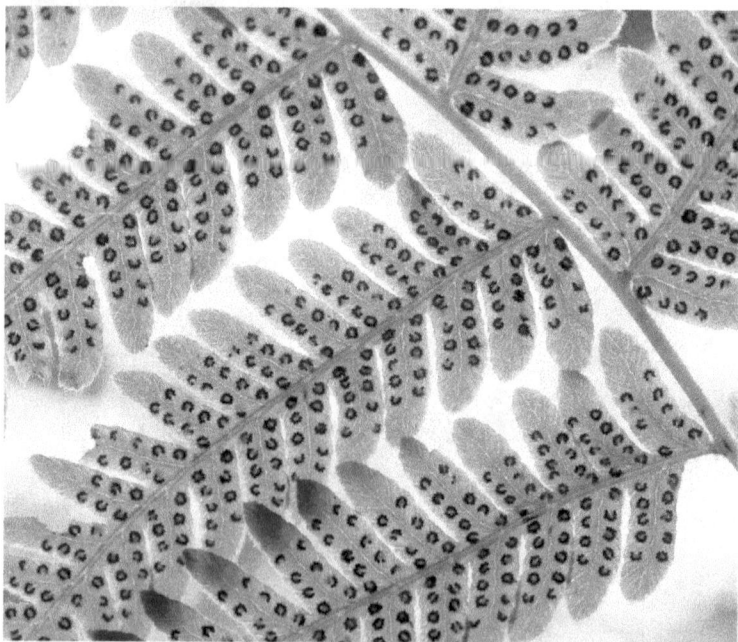

New York fern (*Parathelypteris noveboracensis*), underside of fertile frond.

Bog fern (*Parathelypteris simulata*), portion of blade.

Bog fern (*Parathelypteris simulata*), pinnae.

Narrow beech fern (*Phegopteris connectilis*), frond.

Broad beech fern (*Phegopteris hexagonoptera*), frond.

Eastern marsh fern (*Thelypteris palustris*), sterile fronds.

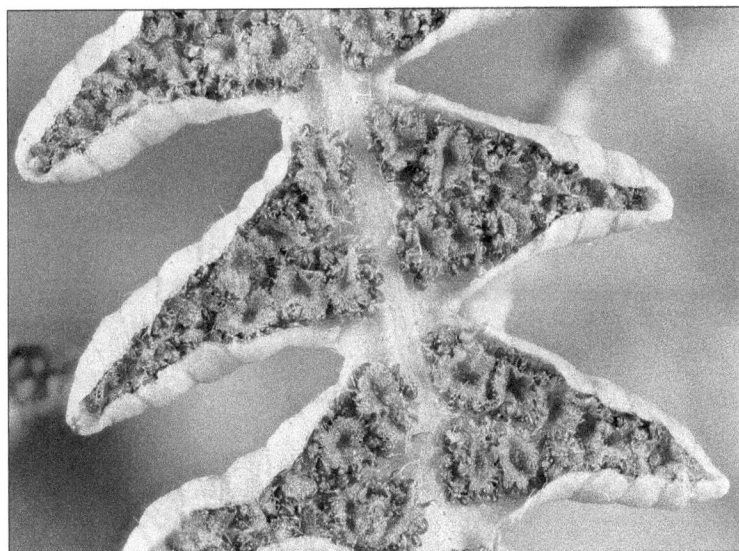

Eastern marsh fern (*Thelypteris palustris*), underside of fertile pinnae.

WOODSIACEAE | Cliff Fern Family
Woodsia
CLIFF FERN

Small deciduous ferns, mostly growing on rock cliffs and ledges; sterile and fertile **fronds** alike. Plants generally covered with hairs, scales and glands; also, in some species, the **stipes** are jointed, the frond eventually breaking at this point, and leaving a more or less even stubble. There are ca. 30 species of *Woodsia* occurring in northern temperate regions and at high elevations throughout the tropics; 10 species found in North America; 6 species in our flora, 3 of which occur only in the north portion of our region.

KEY CHARACTERS

• **Fronds** covered with hairs, scales or glands
• **Stipes** jointed in some species, the frond breaking at the joint to leave a more or less even stubble
• **Veins** ending before margin, sometimes the vein tip swollen.
• **Indusium** cup-like or dissected into several straplike or scalelike segments which encircle the round sorus.

NAME

Named for Joseph Woods, 1776-1864, an English botanical author.

SIMILAR SPECIES

Woodsia and *Cystopteris* are small ferns with thin-textured leaves, often growing together on or near rock outcrops, and often confused. *Woodsia* has the indusium divided into a series of scale-like or hair-like structures, attached below the sorus; *Cystopteris* has an undivided indusium, pocket-like or hood-like, attached around one side of the sorus. *Woodsia* has persistent dark stipe bases; in *Cystopteris* the petiole bases are deciduous; *Woodsia* stipes are scaly most of the way to the blade; *Cystopteris* stipes without scales except near the base. In *Woodsia*, veins do not reach the margin and tip of vein often enlarged; *Cystopteris* veins extend to margin.

HYBRIDS

Woodsia × abbeae Butters is a hybrid between **Woodsia ilvensis** and **Woodsia oregana**, local in northern Midwest region where known from ne and wc Minnesota, nw Wisconsin, and Michigan's Upper Peninsula.

Woodsia glabella
SMOOTH CLIFF FERN

Woodsia alpina
NORTHERN CLIFF FERN

Woodsia ilvensis
RUSTY CLIFF FERN

KEY TO WOODSIA | CLIFF FERN

1 Stipe jointed above base, the old stubble more or less uniform height **2**

1 Stipe not jointed, the stubble irregular **4**

2 Fronds delicate, smooth; stipe green; blade to only 1.5 cm wide, lower pinnae reduced to wings; indusium splitting into short hairs; rare in ne Minn   **1. Woodsia glabella**
SMOOTH CLIFF FERN, page 356

2 Fronds firm; stipe and lower rachis brown; indusium splitting into many long hairs **3**

3 Blade narrow, nearly smooth; rare in Mich Upper Peninsula and Minn   **2. Woodsia alpina**
NORTHERN CLIFF FERN, page 354

3 Blade wider, densely hairy; Ill, Iowa, Mich Upper Peninsula, Minn, Wisc   **3. Woodsia ilvensis**
RUSTY CLIFF FERN, page 358

4 Indusium shallowly split into wide segments; Ill, Ind, Iowa, Mich, Minn, Ohio, Wisc   **4. Woodsia obtusa**
BLUNT-LOBE CLIFF FERN, page 360

4 Indusium deeply split into narrow segments **5**

5 Pubescence a mix of gland-tipped and long white hairs; pinnae crowded; rare in ne Minn   **5. Woodsia scopulina**
ROCKY MOUNTAIN CLIFF FERN, page 364

5 Pubescence of gland-tipped hairs only, or hairs absent; pinnae spaced; Iowa, Mich Upper Peninsula, Minn, Wisc   **6. Woodsia oregana**
OREGON CLIFF FERN, page 362

Woodsia obtusa
BLUNT-LOBE CLIFF FERN

Woodsia scopulina
ROCKY MOUNTAIN CLIFF FERN

Woodsia oregana
OREGON CLIFF FERN

Woodsia alpina (Bolton) S.F. Gray
NORTHERN CLIFF FERN

FIELD TIPS

- very small, erect, clumped fern
- stipes dark brown or purple when mature, jointed, breaking to leave uniform stubble
- sori near margins, indusia of hairlike segments
- acidic rock habitats, rare in Midwest

MIDWEST RANGE

Mich Upper Peninsula (endangered), e Minn.

HABITAT

Crevices and ledges of moist, partially shaded cliffs of acidic rock.

DESCRIPTION

Rootstock Erect, compact, with numerous persistent stipe bases of nearly equal length; scales uniformly brown.

Frond Deciduous, in dense clusters; sterile and fertile fronds alike; 10-20 cm long.

Stipe Persistent base, red-brown or dark purple when mature, jointed above base at swollen node halfway up stipe, red-brown lance-shaped scales at base, fewer upwards; vascular bundles 2.

Rachis With scattered scales and red-brown hairs.

Blade Lance-shaped, broadest below middle; 1-pinnate-pinnatifid (barely so), bright green.

Pinnae 8 to 15 pairs, largest pinnae with 1-3 pairs of pinnules, the shorter ones merely fan-shaped; costae grooved above, grooves continuous from rachis to costae; margins nearly entire, merely lobed; veins free, simple or forked.

Sori Round, near the margin; **indusium** dissected into hairlike segments enveloping sorus, unravelling with maturity; **sporangia** brownish.

SYNONYMS

- *Woodsia glabella* R. Br. ex Richardson var. *bellii* (G. Lawson) G. Lawson

SIMILAR SPECIES

Distinguished from:

- **Rusty cliff fern** (*Woodsia ilvensis*) by having scattered hairs and scales, and the largest pinnae with only 1-3 pairs of pinnules. *W. ilvensis* has abundant hairs and scales; its largest pinnae have 4-9 pairs of pinnules; also a much more common species in our region.
- **Smooth cliff fern** (*Woodsia glabella*) by having a dark, hairy stipe.
- **Oregon cliff fern** (*Woodsia oregana*) and **Rocky Mountain cliff fern** (*W.*

NORTH AMERICA

MIDWEST

scopulina) by having a jointed stipe.

NOTES

Originated as a hybrid between *Woodsia ilvensis* and *W. glabella*.

NAME

Fiirst reference to this species was in John Ray's 1690 Synopsis, which recorded the discovery of a rare fern near the summit of Snowdon in Wales; however, the genus *Woodsia* was not established until 1810 by Robert Brown, who named it after Joseph Woods, an English botanist.

From Latin, *alpinus*, alpine.

sorus with indusium

sporangia

fertile pinna

indusium segment

rachis

pinna

stipe

stipe bases

rootstock

Woodsia glabella R. Br. ex Richards.
SMOOTH CLIFF FERN

FIELD TIPS
- very small, deciduous, clumped fern
- blades pale green, without hairs or scales
- stipes green to straw-colored, jointed
- rare in ne Minn in rock crevices

MIDWEST RANGE
ne Minn (threatened), more common northward.

HABITAT
Crevices in moist, north-facing cliffs, the rock basic to acidic.

DESCRIPTION
Rootstock Erect, compact, with abundant persistent petiole bases of nearly equal length; scales uniformly brown.

Frond Deciduous, in small clusters; sterile and fertile fronds alike; 5-15 cm long.

Stipe Green or straw-colored throughout, jointed above base at swollen node, to only ca. 3 cm long; scales broadly lance-shaped, to 4 mm, yellow-brown at base, smooth above the joint; vascular bundles 2, round or oblong.

Rachis Green, smooth.

Blade Linear to linear-lance-shaped, 1-pinnate-pinnatifid; pale green, without hairs.

Pinnae 7 to 9 pairs, sessile, largest pinnae with 1-3 pairs of pinnules, lower pinnae sometimes fan-shaped; costae grooved above, grooves continuous from rachis to costae; margins entire or crenate, often folding downwards; veins free, simple or forked.

Sori Small, near margin; **indusium** a saucer-shaped disk underneath sori, dissected into hairlike segments, these visible only while sori young; **sporangia** brownish.

SIMILAR SPECIES
- Distinguished from region's other *Woodsia* by its yellow-green stipe and lack of hairs.
- **Bright-green spleenwort** (*Asplenium viride*) has elongate sori, the indusium attached to side of sorus and not below it.

NAME
Latin, diminutive form of *glaber*, smooth or without hair.

NORTH AMERICA

MIDWEST

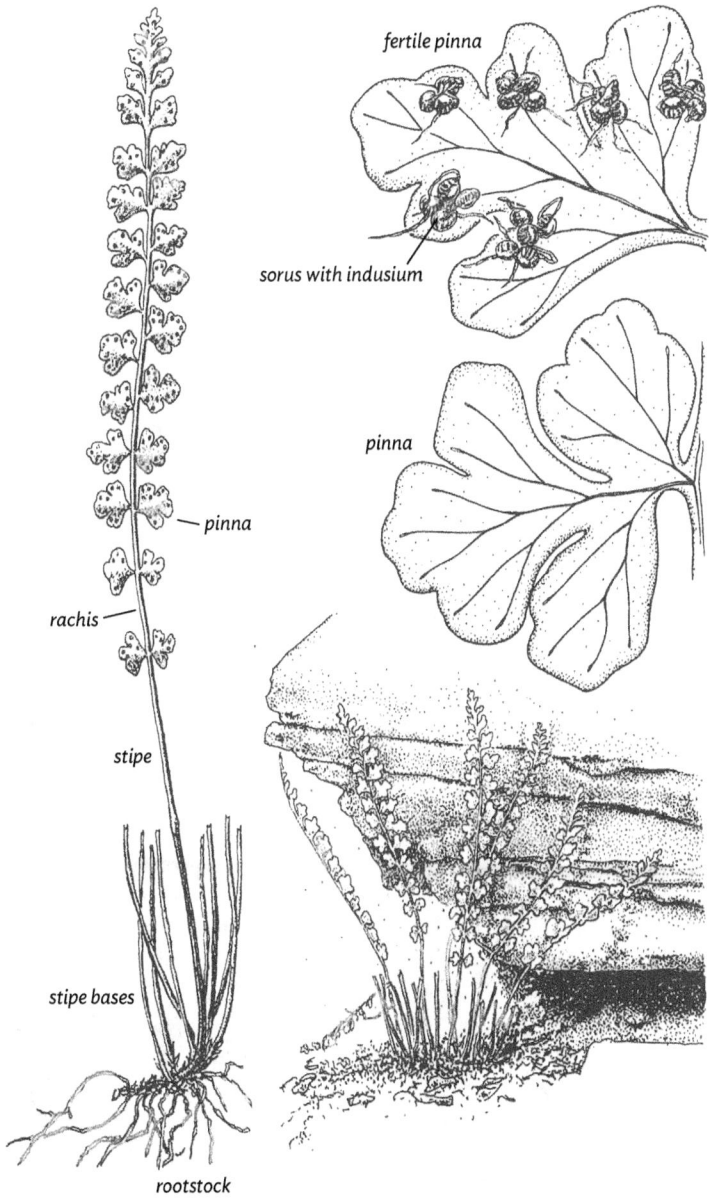

fertile pinna

sorus with indusium

pinna

pinna

rachis

stipe

stipe bases

rootstock

Woodsia ilvensis (L.) R. Br.
RUSTY CLIFF FERN

FIELD TIPS

• small, deciduous, clumped fern of acidic rock

• rachis and blade underside densely hairy, the hairs white or rusty brown

• mature stipe dark brown or purple, jointed above base

MIDWEST RANGE

n Ill (endangered), ne Iowa (endangered), Mich Upper Peninsula, Minn, Wisc.

HABITAT

Dry, exposed or partly shaded cliffs and ledges of non-calcareous rock

DESCRIPTION

Rootstock Short, often branched, dark brown to black, with abundant persistent petiole bases of nearly equal length; scales many, uniformly brown.

Crozier Covered with silvery white hairs, may be present all summer.

Frond Deciduous, in dense clusters; sterile and fertile fronds alike; 5-25 cm long.

Stipe Persistent base, brown or dark purple when mature, jointed above base at swollen node halfway up stipe; red-brown lance-shaped scales near base, a mixture of scales and hairs above and into the rachis; vascular bundles 2, oblong at stipe base,

merging above to an open U-shape.

Rachis Green, densely hairy and scaly.

Blade Lance-shaped, broadest below middle, 2-pinnate at base, less divided above; underside with dense silvery-gray hairs when young, turning rusty-brown later, with narrow scales below, and multicellular hairs along costa above.

Pinnae 10 to 20 pairs, lowest somewhat smaller, sessile or nearly so; with 4-9 pairs of rounded, variable lobes; costae grooved above, grooves continuous from rachis to costae; margins hairy, entire or crenate, often folding downwards; veins free, simple or forked.

Sori Small, near margin, often hidden by dense hairs and scales; **indusium** fringed with hairlike segments enveloping sorus, persistent but often obscure; **sporangia** brownish.

SIMILAR SPECIES

Distinguished from:

• **Northern cliff fern** (*Woodsia alpina*) by having abundant hairs and scales, and the greater number of pinnules (4-9 pairs on each pinna).

• **Smooth cliff fern** (*Woodsia glabella*) by having hairs and scales.

NORTH AMERICA

MIDWEST

• **Oregon cliff fern** (*Woodsia oregana*) and **Rocky Mountain cliff fern** (*W. scopulina*) by having a jointed stipe.

NOTES

Plants may dry and turn brown during drought, with green fronds formed following rain.

NAME

In 1753, Linnaeus named *Acrostichum ilvense*, a fern found in parts of Europe (the species name referring to the island of Elba). In 1810, Robert Brown placed these ferns in a new genus he named *Woodia*, correcting it to *Woodsia* 5 years later. Plants from America proved so similar to the European that the same names were applied.

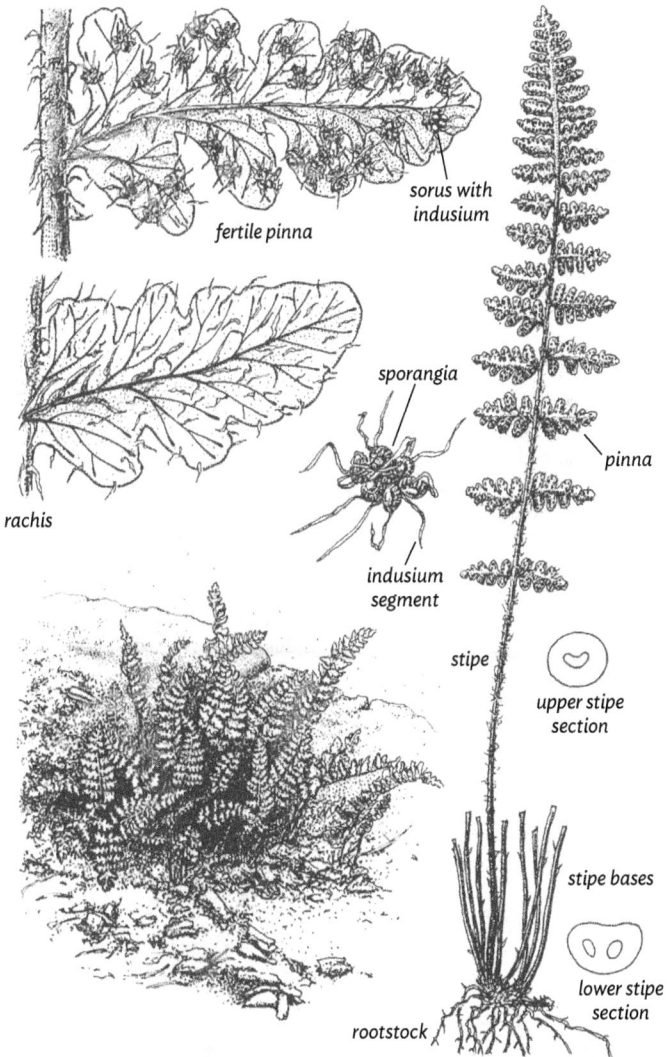

sorus with
indusium

fertile pinna

sporangia

pinna

rachis

indusium
segment

stipe

upper stipe
section

stipe bases

lower stipe
section

rootstock

Woodsia obtusa (Spreng.) Torr.
BLUNT-LOBE CLIFF FERN

FIELD TIPS

- nearly evergreen, loosely clumped fern, usually on calcareous rock
- stipe bases persistent, of unequal length, straw-colored to light brown when mature
- blade with glands on both sides
- region's largest cliff fern

MIDWEST RANGE

Ill, Ind, Iowa, Mich (threatened), Minn, Ohio, Wisc.

HABITAT

Open to partly shaded, outcrops and cliff ledges; the rock often (but not always) calcareous.

DESCRIPTION

Rootstock Erect to short-creeping, often branched, dark-brown to black, with stipe bases of unequal length and with persistent old fronds attached; scales few.

Frond Nearly evergreen, clustered; old fronds persistent; sterile and fertile fronds nearly alike, but the sterile fronds decline and remain green through winter; 25-40 cm long.

Stipe Not jointed, the persistent bases of unequal lengths, light brown or straw-colored when mature, occasionally darker at base, scales tan some with darker central stripe;

vascular bundles 2, oblong at stipe base, merging above to a U-shape.

Rachis Pale yellow to brown, with glandular hairs and scattered, narrow scales.

Blade Tapering to both ends, truncated and 2-pinnate-pinnatifid at base; herbaceous, gray-green, glands on both surfaces.

Pinnae 8 to 15 separated pairs, held nearly horizontally, lowest somewhat reduced, sessile or nearly so, pairs further apart the closer to the base; pinnules 3-8 pairs before changing to a pinnatifid tip; glandular hairs on both upper and lower surface and on costae; costae grooved above, grooves continuous from rachis to costae; margins dentate or lobed; veins free, simple or forked, ending before the margin, vein tips usually enlarged.

Sori Round, in one row between midrib and margin; **indusium** starlike, with 4-6 translucent lobes encircling sorus; **sporangia** brown, then black.

SIMILAR SPECIES

- Distinguished from **brittle bladder fern** (*Cystopteris fragilis*) by pinnae with glandular hairs, and the starlike indusium under sori.

NORTH AMERICA

MIDWEST

NAME

Discovered by Muhlenberg in Pennsylvania, and named by Sprengel in 1804 *Polypodium obtusum*, which was changed by Torrey in 1840 to *Woodsia obtusa*.

Obtusa refers to the blunt pinnae.

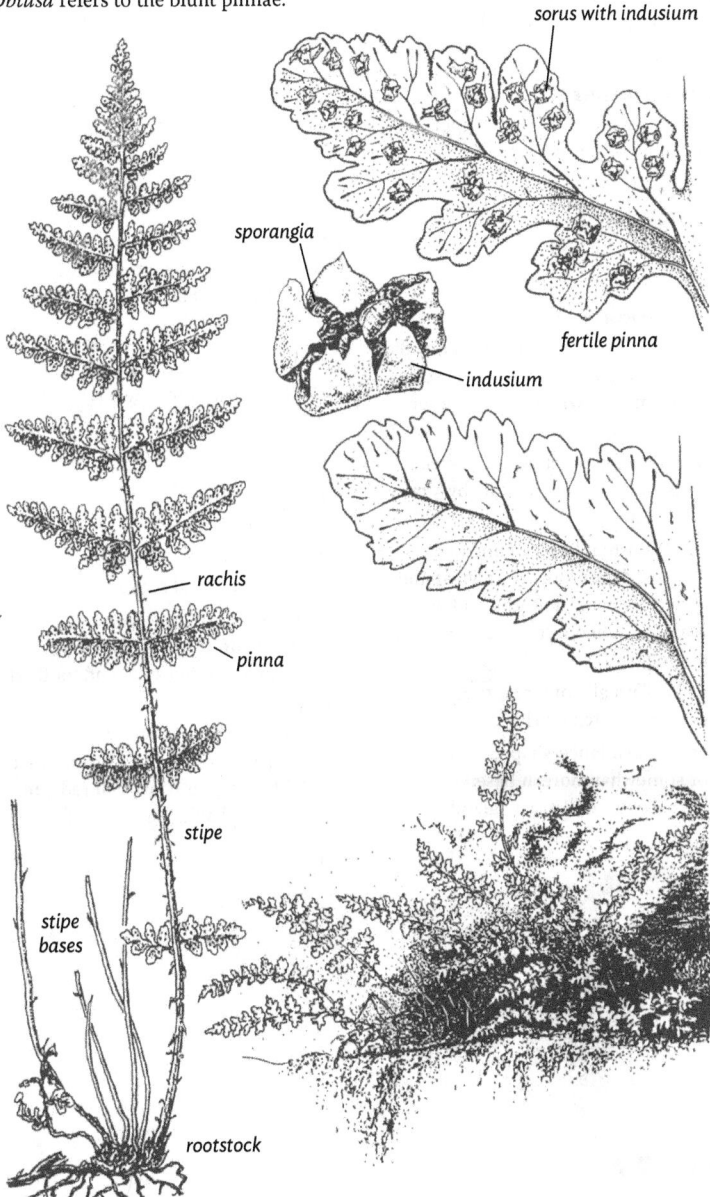

sorus with indusium

sporangia

fertile pinna

indusium

rachis

pinna

stipe

stipe bases

rootstock

Woodsia oregana D.C. Eaton
OREGON CLIFF FERN

FIELD TIPS
- small, deciduous, clumped fern
- blade with gland-tipped hairs
- base of mature stipe red-brown to dark purple, not jointed above base
- rock habitats

MIDWEST RANGE
n Iowa (threatened), Mich Upper Peninsula, Minn, Wisc.

HABITAT
Rock outcrops, ledges, crevices and talus slopes, on both acidic and calcareous rock.

DESCRIPTION

Rootstock Erect, with few to many old stipe bases of unequal length; scales uniformly brown or with a darker central stripe.

Frond Deciduous, sterile and fertile fronds alike; to 25 cm long by 5 cm wide,.

Stipe Persistent bases of unequal lengths, dark reddish-brown at and above the base, smooth or a few scales; vascular bundles 2, round or oblong.

Rachis With glandular hairs and scattered, often hairlike scales.

Blade Linear lance-shaped, 2-pinnate or somewhat more at base; herbaceous, dull green, gland-tipped hairs on upper and lower sides.

Pinnae 10 to 14 pairs, held nearly horizontally, triangular-oblong, pairs further apart the closer to the base, opposite; costae grooved above, grooves continuous from rachis to costae; margins crenate to serrate, often rolled under; veins free, simple or forked, ending before the margin.

Sori Round, in 1 row between midrib and margin; **indusium** narrow, of thread-like or hair-like segments; **sporangia** brown.

SYNONYMS
Our plants *Woodsia oregana* subsp. *cathartica*.
- *Woodsia cathcartiana* B. L. Rob.

SIMILAR SPECIES
Distinguished from:
- **Smooth cliff fern** (*Woodsia glabella*) by having hairs and scales.
- **Northern cliff fern** (*Woodsia alpina*) and **rusty cliff fern** (*W. ilvensis*) by having an unjointed stipe and toothed margins.
- **Rocky Mountain cliff fern** (*Woodsia scopulina*) by the mature stipes, which are light brown to straw-colored and much more pliable than the brown

NORTH AMERICA

MIDWEST

stipes of *W. scopulina*.

• **Brittle bladder fern** (*Cystopteris fragilis*) by pinnae with glandular hairs, and the starlike indusium.

NAME
First found in the Dalles of Oregon.

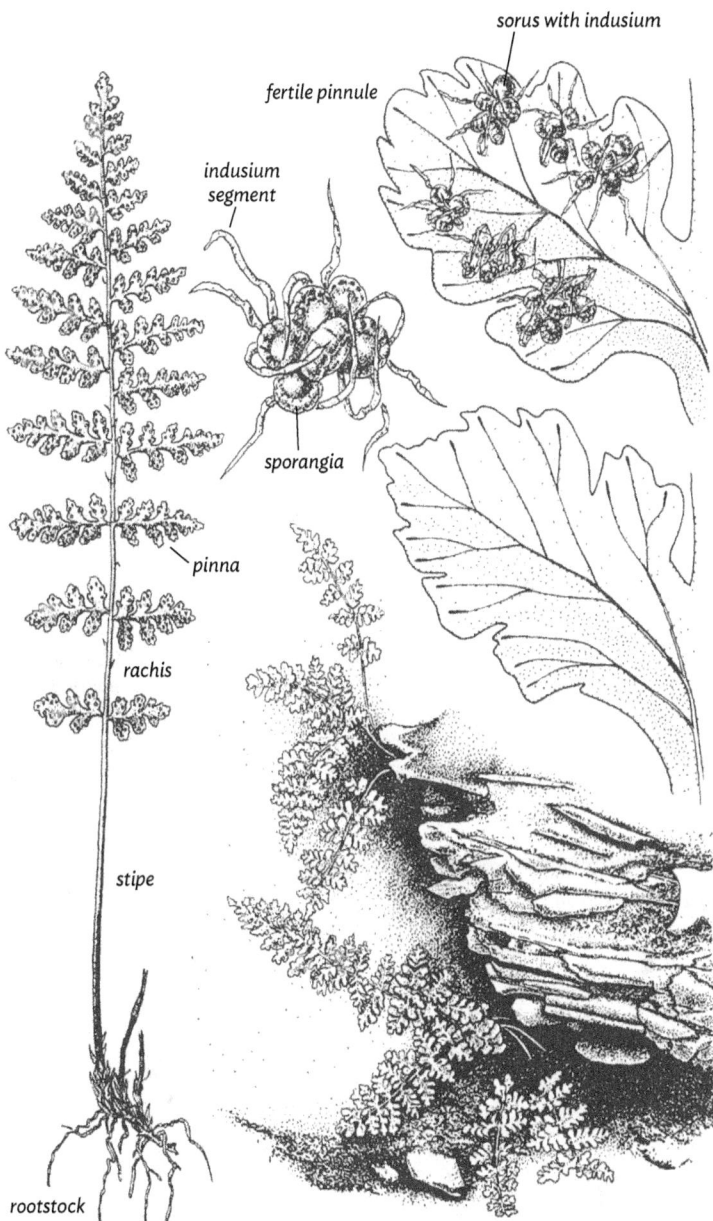

sorus with indusium

fertile pinnule

indusium segment

sporangia

pinna

rachis

stipe

rootstock

Woodsia scopulina D.C. Eaton
ROCKY MOUNTAIN CLIFF FERN

FIELD TIPS
- small, deciduous, clumped fern
- blade with glands and hairs
- mature stipe red-brown to dark purple, not jointed above base
- rare in Midwest (ne Minn only)

MIDWEST RANGE
ne Minn (threatened), more common in w North America.

HABITAT
North-facing cliffs of slate rock; sites cool, moist and shaded.

DESCRIPTION
Rootstock Erect, with few to many old stipe bases of unequal length; scales uniformly brown or with a darker central stripe.

Frond Deciduous, clumped, sterile and fertile fronds alike; to 30 cm long.

Stipe Shiny, chestnut-brown, with persistent bases of unequal lengths; scales tan; vascular bundles 2, round or oblong.

Rachis With glandular hairs and scattered, white, hairlike scales.

Blade Rhombic, tapering to both ends, truncated at base, mostly 2-pinnate, but variable, herbaceous, gray-green, glandular and sparsely hairy on both surfaces.

Pinnae 10 to 15 pairs, lowest distinctly reduced, sessile or nearly so, pairs further apart the closer to the base; costae grooved above, grooves continuous from rachis to costae; margins dentate or lobed; veins free, simple or forked, ending before margin.

Sori Round, near the margin; **indusium** of strap-like lobes encircling sorus, frayed at tips; **sporangia** brown, maturing to black.

SIMILAR SPECIES
Distinguished from:
- **Smooth cliff fern** (*Woodsia glabella*) by having hairs and scales.
- **Northern cliff fern** (*Woodsia alpina*) and **rusty cliff fern** (*W. ilvensis*) by having an unjointed stipe and toothed margins.
- **Oregon cliff fern** (*Woodsia oregana*) by having hairs concentrated along rachis on both surfaces, and by the mature stipes, typically a reddish brown to dark purple color and relatively brittle.

NAME
Named based on Rocky Mountain plants examined by D. C. Eaton in 1865.

From Greek, *skopelos*, peak, headland, or promontory, perhaps in reference to habitat.

NORTH AMERICA

MIDWEST

Oregon cliff fern (*Woodsia scopulina*) in typical rock crevice habitat.

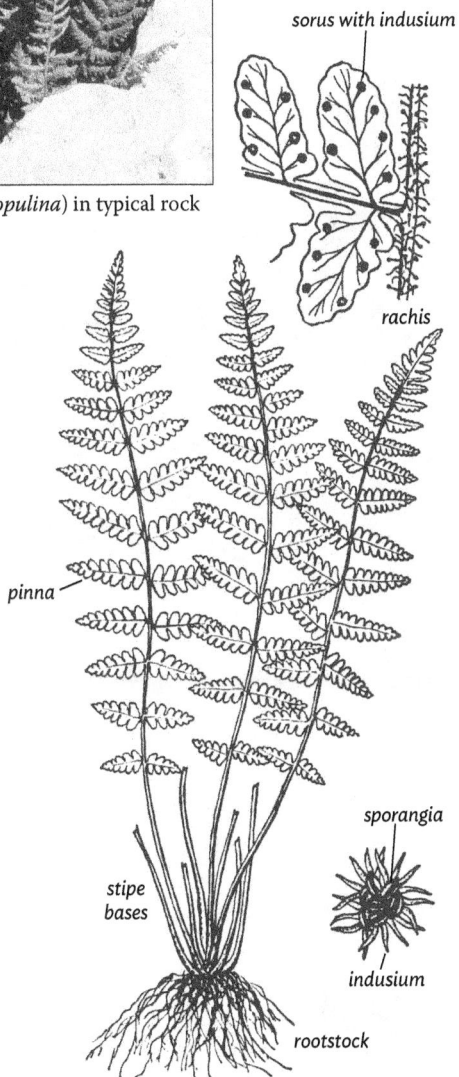

Rusty cliff fern (*Woodsia ilvensis*).

Rusty cliff fern (*Woodsia ilvensis*).

Blunt-lobe cliff fern (*Woodsia obtusa*), underside of fertile frond.

Oregon cliff fern (*Woodsia oregana*).

Colony of **broad beech fern** (*Phegopteris hexagonoptera*).

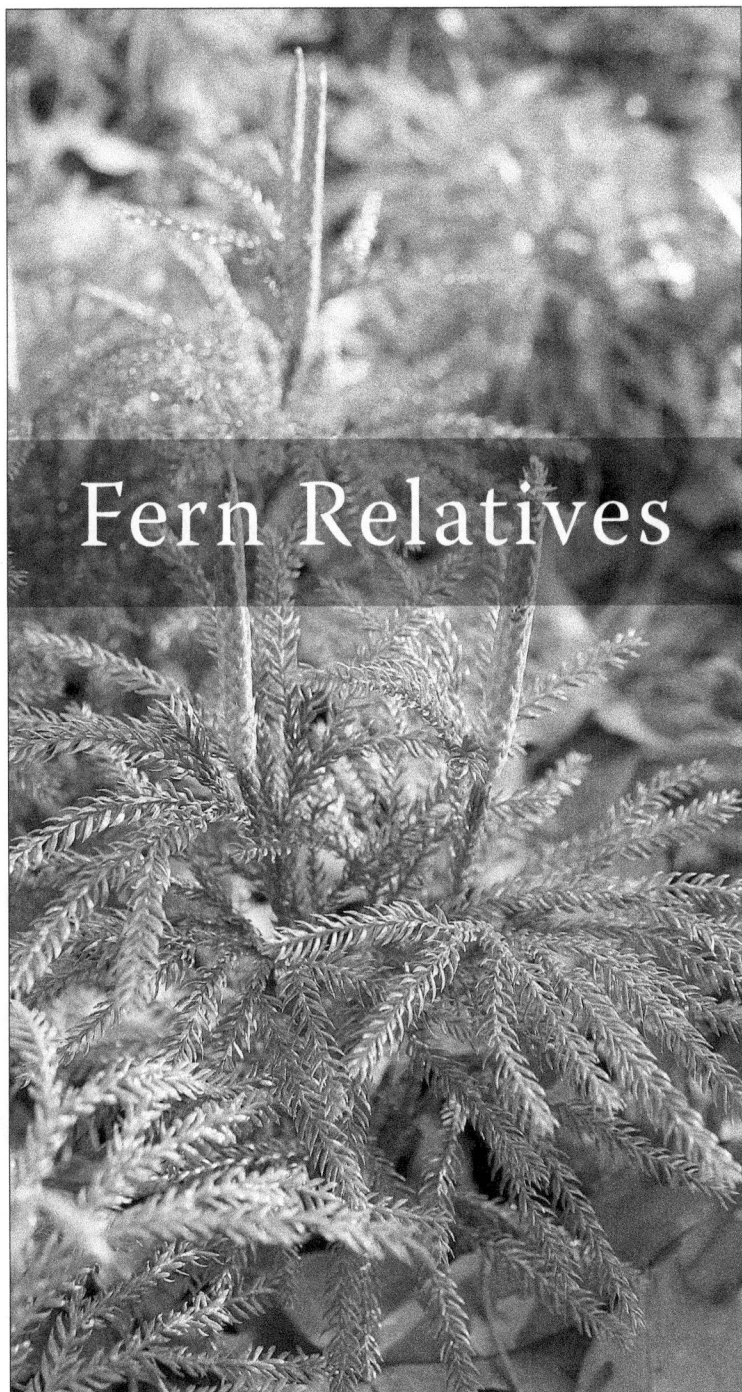

Princess-pine (*Dendrolycopodium obscurum*), a member of the Clubmoss Family.

ISOETACEAE | Quillwort Family
Isoetes
QUILLWORT

Worldwide, the Quillwort Family includes a single genus, *Isoetes*, of ca. 50 species, with 24 species present in North America, and 5 species in our flora. *Isoetes* are small, perennial, usually aquatic herbs. Plants often totally submerged or sometimes in drier situations along shores. **Leaves** grasslike, with internal air chambers, broadened and flattened at base, in clusters of few to many from a swollen, cormlike rhizome with numerous long, fleshy roots. Innermost leaves usually sterile; the next outer leaves usually with a small pocket at their expanded base containing the sporangia. The **sporangia** more or less covered by the thin edges of the velum, and with a small ligule situated above.

As in the spikemosses (*Selaginella*) and water ferns (*Azolla, Marsilea*), **spores** of *Isoetes* are of two types, numerous, and the surface is variously patterned; relatively large white **megaspores** (female) are borne in megasporangia (the case containing the spores); smaller **microspores** (male) are borne in microsporangia.

KEY CHARACTERS
- Small aquatic or wetland plants of lakes and shores.
- **Leaves** tufted, grass- or chive-like, swollen at base.
- **Sporangia** in a pocket at base of leaves.

NAME
From Greek, *isos*, equal, and *etos*, year, referring to evergreen habit of some species. The long pointed, hollow leaves suggest a quill; wort is an old English word to refer to any herbaceous plant.

NOTES
Identification of species is dependent on characters of the **megaspores**; these are about 0.5 mm wide, and a 20x hand lens (or more powerful microscope) may be needed. In the field, surface features are best seen when the megaspore is dry.

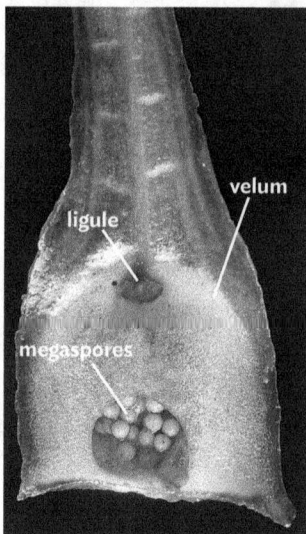

*Swollen base of **Isoetes** leaf.*

Quillworts can be distinguished from similar aquatic grasslike plants by the swollen leaf base which contains the megaspores.

Hybrids may also be present in habitats supporting several species of quillwort; hybrids often have megaspores of variable size, some of which may be deformed.

ADDITIONAL MIDWEST SPECIES
- **Limestone quillwort** (*Isoetes butleri* Engelm.); known in Midwest from ne Ill (endangered), more common southward in Missouri and Arkansas. Leaves narrow, triangular in section; megaspores covered with tiny bumps. Found in muddy depressions, temporary pools and northern white cedar glades; soils calcareous.

KEY TO ISOETES | QUILLWORT

1 Megaspores conspicuously covered with small spines; Mich, Minn, Wisc   **1. Isoetes echinospora**
SPINY-SPORED QUILLWORT, page 372

1 Megaspores not spiny. **2**

2 Plants normally underwater; leaves without outer fibrous strands; Mich, Minn, Wisc   **2. Isoetes lacustris**
LAKE QUILLWORT, page 374

2 Plants underwater or on exposed, drying shores; leaves with 4 or more fibrous strands **3**

3 Plants underwater or emergent; megaspores with a raised, netlike surface; rare in s Ind, s Mich and Ohio   **3. Isoetes engelmannii**
ENGELMANN'S QUILLWORT, page 373

3 Plants emergent and on drying shores; megaspores with numerous small bumps; Ill, s Ind, e Iowa, sw Minn   **4. Isoetes melanopoda**
BLACK-FOOT QUILLWORT, page 375

ISOETES MEGASPORES (ABOUT 1/2 MM IN DIAMETER)

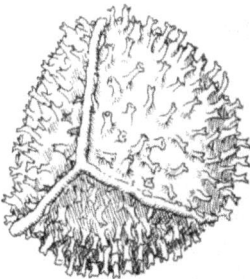

Isoetes echinospora
SPINY-SPORED QUILLWORT

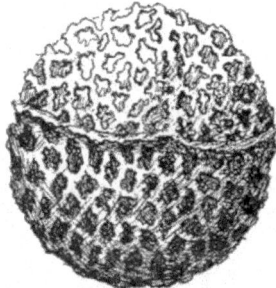

Isoetes lacustris
LAKE QUILLWORT

Isoetes engelmannii
ENGELMANN'S QUILLWORT

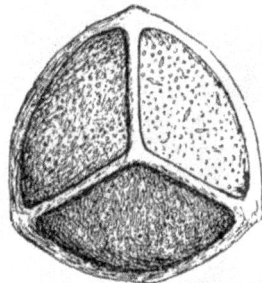

Isoetes melanopoda
BLACK-FOOT QUILLWORT

ISOETACEAE | QUILLWORT FAMILY

Isoetes echinospora Durieu
SPINY-SPORE QUILLWORT

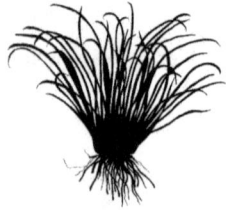

FIELD TIPS
• northern portions of region where our
  most common quillwort
• plants usually submerged, sometimes
  emergent in shallow water along
  shores
• leaves variable

MIDWEST RANGE
Mich, Minn, Wisc; historical record
from ne Ohio.

HABITAT
Shallow water and shores of lakes,
ponds and streams with nutrient poor,
acidic, sandy or gravelly bottoms.

DESCRIPTION
Leaves Bright green to reddish green,
  pale toward base, often deciduous, to
  about 30 cm long.
Megaspores White, covered with sharp
  spines; **spores** mature late summer.

SYNONYMS
• *Isoetes braunii* Durieu
• *Isoetes muricata* Durieu

NOTES
May hybridize with Midwest species
Isoetes engelmannii (*I.* × *eatonii* Dodge)
and *Isoetes lacustris* (*I.* × *hickeyi* Taylor
& Luebke).

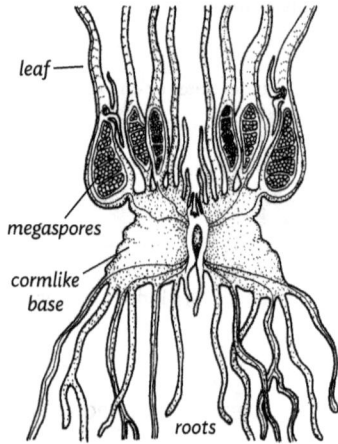

Isoetes echinospora, cross-section of
plant base.

NORTH AMERICA

MIDWEST

372 | FERN RELATIVES

Isoetes engelmannii A. Braun
ENGELMANN'S QUILLWORT

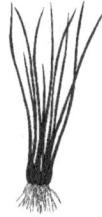

FIELD TIPS
• our largest quillwort
• local in se portions of region
• leaves bright green, evergreen

MIDWEST RANGE
s Ind (endangered), s Mich (endangered), Ohio (endangered); historical records from s Ill.

HABITAT
Swales between old sand dune ridges, pond margins, ditches; plants submerged or emergent.

DESCRIPTION
Leaves Bright green, pale toward base, evergreen, to about 60 cm long or sometimes longer, can be sprawling.

Megaspores White; surface ridged and honey-combed; **spores** mature in summer.

NAME
First collected in Missouri by Engelmann, and named in his honor by A. Braun in 1846.

Isoetes engelmannii, lower portion of plant with leaves and cormlike base.

NORTH AMERICA

MIDWEST

Isoetes lacustris L.
WESTERN LAKE QUILLWORT

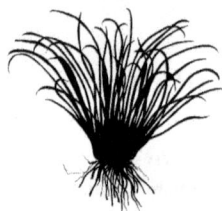

FIELD TIPS
• submerged in acidic lakes
• leaves evergreen, stiff

MIDWEST RANGE
Mich, n Minn, Wisc.

HABITAT
Submerged in nutrient poor, acidic lakes with sandy or gravelly bottoms; found at depths up to 3 m. Unlike *Isoetes echinospora*, **western lake quillwort** almost always found in deep cold water rather than emergent on shores.

DESCRIPTION
Leaves Dark green to reddish green, pale brown near base, evergreen, stiff, abruptly tapered to tip, to about 25 cm long.

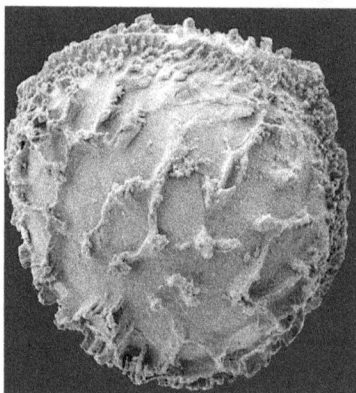

Isoetes lacustris, megaspore.

Megaspores White, sporadically covered with netlike ridges and with a distinct ridged girdle; **spores** mature late summer

SYNONYMS
• *Isoetes hieroglyphica* A.A. Eaton (sometimes treated as a separate species of northeastern USA and adjacent Canada, with a disjunct population reported from Wisconsin).

• *Isoetes macrospora* Durieu

NORTH AMERICA

MIDWEST

Isoetes melanopoda
Gay & Durieu ex Durieu
BLACK FOOT QUILLWORT

FIELD TIPS
• small plants at water's edge
• megaspores gray to black

MIDWEST RANGE
Ill, Ind (endangered), e Iowa
(endangered), sw Minn (endangered).

HABITAT
Wet depression, ditches. In Minnesota,
restricted to prairie region, specifically
where Sioux quartzite bedrock is
exposed at the surface. Vernal pools
may form between the outcrops or in
depressions in the rock itself. *Isoetes
melanopoda* occurs on margins of deep
rainwater pools and pools maintained
by groundwater seepage.

DESCRIPTION
Leaves Bright green, pale toward base,
deciduous, flexible, gradually tapered
to tip, 10-30 cm long.

Megaspores Gray to black, covered
with fine bumps; **spores** mature in
late spring.

NAME
'Black foot' refers to the sometimes
blackish base of the plant; the
megaspores are also blackish.

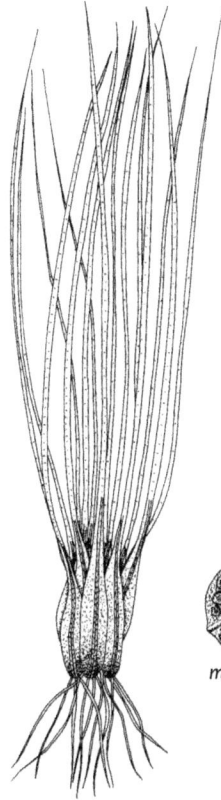

megaspore

NORTH AMERICA

MIDWEST

LYCOPODIACEAE | Clubmoss Family

Mostly evergreen, perennial herbs, somewhat mosslike (but larger) and some genera resembling miniature conifer trees. **Stems** upright, trailing or creeping, most species with stems lying flat on the ground and with upright shoots terminating in cylindrical, spore-producing **strobili** (cones, these absent in *Huperzia)*; covered by many small, linear to lance-shaped, 1-veined leaves. **Gametophytes** small, and in all but *Lycopodiella*, are produced underground, lack chlorophyll and are dependent upon mycorrhizal fungi; gametophytes of *Lycopodiella* develop on the surface and are photosynthetic. Earlier treatments of the family recognized just 2 genera: *Lycopodium* (North America) and *Phylloglossum* (Australia and New Zealand). Now, the family is separated into ca. 19 genera worldwide; 6 genera in our flora; see page 378 for a summary of old and new names.

NOTES

Dried clubmoss spores have water-repellent properties, and were sold under the name **lycopodium**. Uses included dusting wounds and chafed skin in infants, and dusting pills to prevent their sticking to one another. Spores were also used in fireworks; when placed in fire, the spores ignite to produce a bright light.

KEY TO LYCOPODIACEAE | CLUBMOSS FAMILY

1 Strobili absent, sporangia borne in axils of regular leaves; stems with flattened, two-lobed gemmae (tiny plantlets) that fall off and can produce new plants **1. Huperzia**
FIR-MOSS, page 396

1 Sporangia always borne in terminal, differentiated strobili; gemmae absent. **2**

2 Sporophylls (bracts subtending sporangia) green, leaf-like; erect branches with soft, herbaceous leaves, deciduous, unbranched (rarely forked) and always bearing terminal strobili; horizontal stems leafy, attached to substrate surface and rooting at frequent intervals **2. Lycopodiella**
CLUB-MOSS, page 406

2 Sporophylls yellowish, scale-like; sterile erect or ascending branches often present, these sometimes branched, all branches with evergreen, leathery leaves; horizontal stems either deeply buried or ± on the surface and rooting at irregular intervals **3**

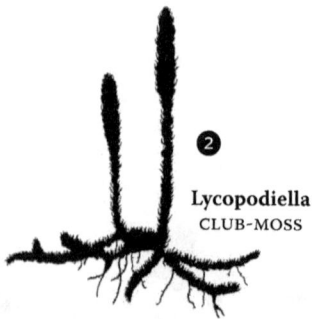

Huperzia
FIR-MOSS

Lycopodiella
CLUB-MOSS

3 Strobili sessile. **4**

3 Strobili elevated on distinct peduncles (stalks) **5**

4 Aerial stems simple or occasionally forked once with ascending branching; horizontal stems at the ground surface; strobilus one per upright shoot   **3. Spinulum**
CLUB-MOSS, page 416

4 Aerial stems much branched and "tree-like," with the branches often spreading; strobili usually several per upright shoot
4. Dendrolycopodium
TREE-CLUBMOSS, page 378

5 Leaves spreading, lance-shaped, tipped with a whitish, hairlike bristle   **5. Lycopodium**
GROUND-PINE, page 410

5 Leaves appressed, partially fused with stem, ± scalelike, not tipped with a whitish bristle   **6. Diphasiastrum**
CREEPING-CEDAR, page 386

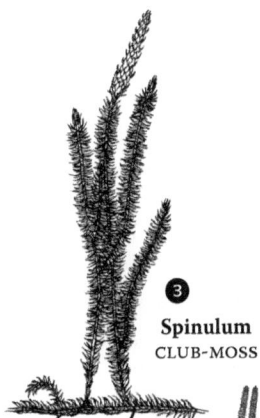

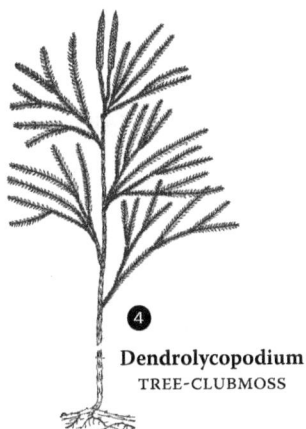

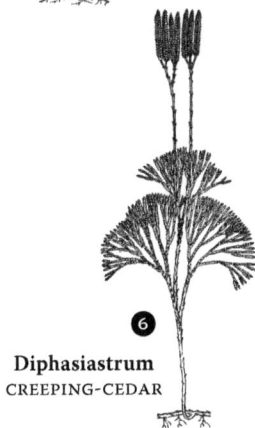

③
Spinulum
CLUB-MOSS

④
Dendrolycopodium
TREE-CLUBMOSS

⑤

⑥
Diphasiastrum
CREEPING-CEDAR

Lycopodium
GOUND-PINE

Dendrolycopodium
TREE-CLUBMOSS

Plants branched and resemble small trees; horizontal stems buried in soil. **Branches** round or compressed in cross-section. **Leaves** 6- to 8-ranked, tapered to a point at tip but not spine- or hair-tipped. **Strobili** (cones) stalkless, mostly 1-7 at ends of upright stem branches. Worldwide, 4 species in genus, 3 of which occur in mostly eastern North America and in our flora.

KEY CHARACTERS
• Plants treelike, the main axis with forked branches.

• **Branches** mostly round in cross-section.

• **Strobili** (cones) not stalked.

NOTES
Dendrolycopodium dendroideum and *D. hickeyi* long considered to be varieties of **princess-pine** (*D. obscurum*) and the 3 species can be difficult to distinguish, especially in dried specimens.

NAME
From Greek, *dendro*, tree, referring to treelike form of these clubmosses.

SUMMARY OF CLUBMOSS GENERA		
NEW GENUS NAME	**OLD NAME**	**COMMON NAME**
Dendrolycopodium	*L. obscurum* group	Tree-clubmoss
Diphasiastrum	*L. complanatum* group	Creeping Cedar
Huperzia	*L. selago* group	Fir-moss
Lycopodiella	*L. inundatum* group	Bog Clubmoss
Lycopodium	*L. clavatum* group	Ground-pine
Spinulum	*L. annotinum* group	Bristly Clubmoss

KEY TO DENDROLYCOPODIUM | TREE-CLUBMOSS

1 Branches somewhat flattened in cross-section; leaves on one of the flattened faces smaller than the others; leaves tightly appressed (not spreading) on the main stem below the branches; ne Ill, Ind, Mich, Minn, Ohio, Wisc   **1. Dendrolycopodium obscurum**
PRINCESS-PINE, page 384

1 Branches round in cross-section; leaves all the same size; leaves appressed or spreading on the main stem **2**

2 Leaves arranged in 2 ranks on each lateral edge of a branch, with 1 rank each on upper and lower surface; leaves on lower stem tightly appressed, dark green; Ill, n Ind, Mich, Minn, Ohio, Wisc
2. Dendrolycopodium hickeyi
PENNSYLVANIA TREE-CLUBMOSS, page 382

2 Leaves arranged in 1 rank on each lateral edge of a branch, with 2 ranks on upper and lower surfaces; leaves on lower stem spreading, pale green; Ill, s Ind, ne Iowa, Mich, Minn, Ohio, Wisc
3. Dendrolycopodium dendroideum
PRICKLY TREE-CLUBMOSS, page 380

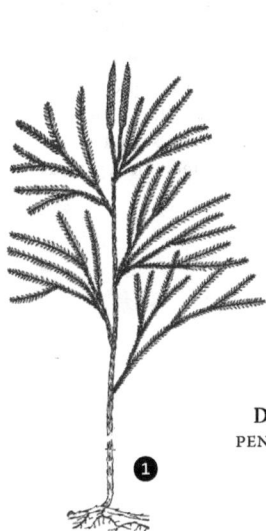

Dendrolycopodium hickeyi
PENNSYLVANIA TREE-CLUBMOSS

**Dendrolycopodium
obscurum**
PRINCESS-PINE

**Dendrolycopodium
dendroideum**
PRICKLY TREE-CLUBMOSS

Dendrolycopodium dendroideum
(Michx.) Haines

PRICKLY TREE-CLUBMOSS

FIELD TIPS
- evergreen, treelike clubmoss
- leaves spreading, stiff and prickly
- strobili (cones) stalkless, single at branch ends

MIDWEST RANGE
Ill (endangered), s Ind (endangered), ne Iowa (threatened), Mich, Minn, Ohio, Wisc.

HABITAT
Moist to dry deciduous and mixed forests; may form large colonies; soils sandy and acidic.

DESCRIPTION
Stems Spreading by underground horizontal stems, with branched, upright shoots to ca. 30 cm tall, appearing somewhat like a small tree; branches round in cross-section.

Leaves Spreading to ascending, linear; leaves of lower stems (below the branches) stiff and needlelike (prickly to touch); leaves of branches 6-ranked.

Strobili Stalkless, single at branch ends, with 1-7 or more on each upright stem.

SYNONYMS
- *Lycopodium dendroideum* Michx.
- *Lycopodium obscurum* L. var. *dendroideum* (Michx.) D.C. Eaton
- *Lycopodium obscurum* L. var. *hybridum* Farw.

SIMILAR SPECIES
The spreading, prickly leaves on lower stems distinguish **prickly tree-clubmoss** from our other 2 species. Also, the lateral branches are nearly round, and not conspicuously flattened as in **princess-pine** (*D. obscurum*).

NAME
When Michaux found this on his travels from "Canada to the Carolina mountains"' he considered it a new species and in 1803 named it *Lycopodium dendroideum*. It is so close to the one Linnaeus had previously named *L. obscurum*, that in 1890 D. C. Eaton made Michaux 's plant a variety. It is currently treated, once again, as a valid species.

NORTH AMERICA

MIDWEST

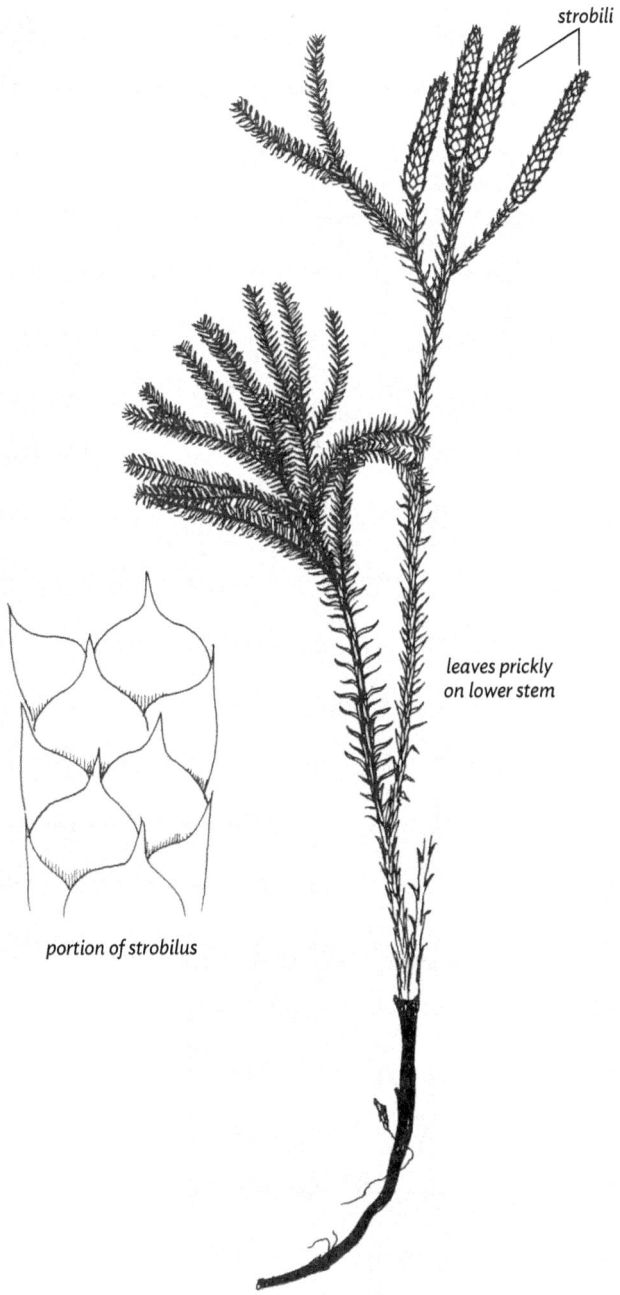

strobili

leaves prickly
on lower stem

portion of strobilus

Dendrolycopodium hickeyi

(W.H. Wagner, Beitel & R.C. Moran) A. Haines

PENNSYLVANIA TREE-CLUBMOSS

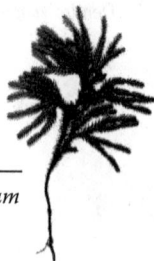

FIELD TIPS

- evergreen, treelike clubmoss
- base of upright stem not prickly
- leaves spreading and ascending
- strobili (cones) stalkless, 1-7 per each upright stem

MIDWEST RANGE

Ill, n Ind, Mich, Minn, Ohio, Wisc.

HABITAT

Drier deciduous, mixed, and conifer forests; may form colonies; soils sandy and acidic.

DESCRIPTION

Stems Spreading by underground horizontal stems, with branched, upright shoots to ca. 20 cm tall, appearing somewhat like a small tree; branches round to elliptic in cross-section.

Leaves All of about equal size, widest near middle, tapered to a pointed tip; margins entire; branch leaves 6-ranked, spreading; stem leaves below branches appressed.

Strobili Stalkless, 1-7 on each upright stem.

SYNONYMS

- *Lycopodium hickeyi* W.H. Wagner, Beitel & Moran

- *Lycopodium obscurum* L. var. *isophyllum* Hickey

SIMILAR SPECIES

- **Princess-pine** (*D. obscurum*) has branches flattened in cross-section, with twisted branch leaves of different sizes; in *D. hickeyi*, branches are nearly round, and the leaves are all similar size and not twisted.

NOTES

A recently recognized taxon (1977), first described as *Lycopodium obscurum* var. *isophyllum*, and now treated as a valid species.

NORTH AMERICA

MIDWEST

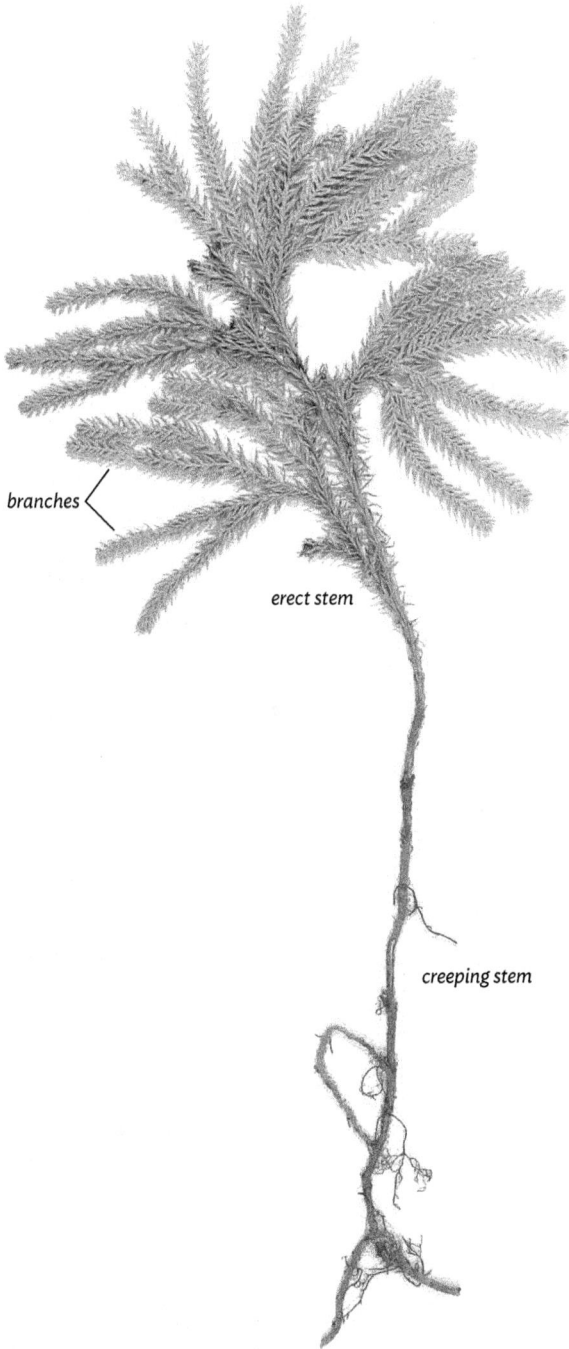

branches

erect stem

creeping stem

Dendrolycopodium obscurum
(L.) Haines

PRINCESS-PINE

FIELD TIPS
- evergreen, treelike clubmoss
- main stem not prickly
- leaves ascending, 6-ranked
- strobili (cones) stalkless, single at branch ends, 1-6 per each upright stem

MIDWEST RANGE
ne Ill, Ind, Iowa, Mich, Minn, Ohio, Wisc.

HABITAT
Moist to dry deciduous and mixed conifer-deciduous forests, and forest margins; may form large colonies; soils sandy and acidic.

DESCRIPTION
Stems Spreading by underground horizontal stems, with branched, upright shoots to ca. 30 cm tall, appearing somewhat like a small tree; branches flattened in cross-section or if viewed from the side.

Leaves Ascending, narrowly lance-shaped, tapered to a pointed tip; margins entire; branch leaves 6- to 8-ranked, lateral leaves twisted, leaves along lower surface of branch smaller than the others; stem leaves below branches appressed.

Strobili Stalkless, 1-6 on each upright, branched stem.

SYNONYMS
- *Lycopodium obscurum* L.

SIMILAR SPECIES
- **Prickly tree-clubmoss** (*D. dendroideum*) has lower stems (below the branches) prickly to the touch.
- **Pennsylvania tree-clubmoss** (*D. hickeyi*) has branches nearly round in cross-section, and branch leaves all of similar size.

NAME
Collected near Philadelphia, Pennsylvania, by John Bartram, and sent to Europe, where a drawing of it was published by Dillenius in 1741. Linnaeus named it *Lycopodium obscurum* in 1753.

NORTH AMERICA

MIDWEST

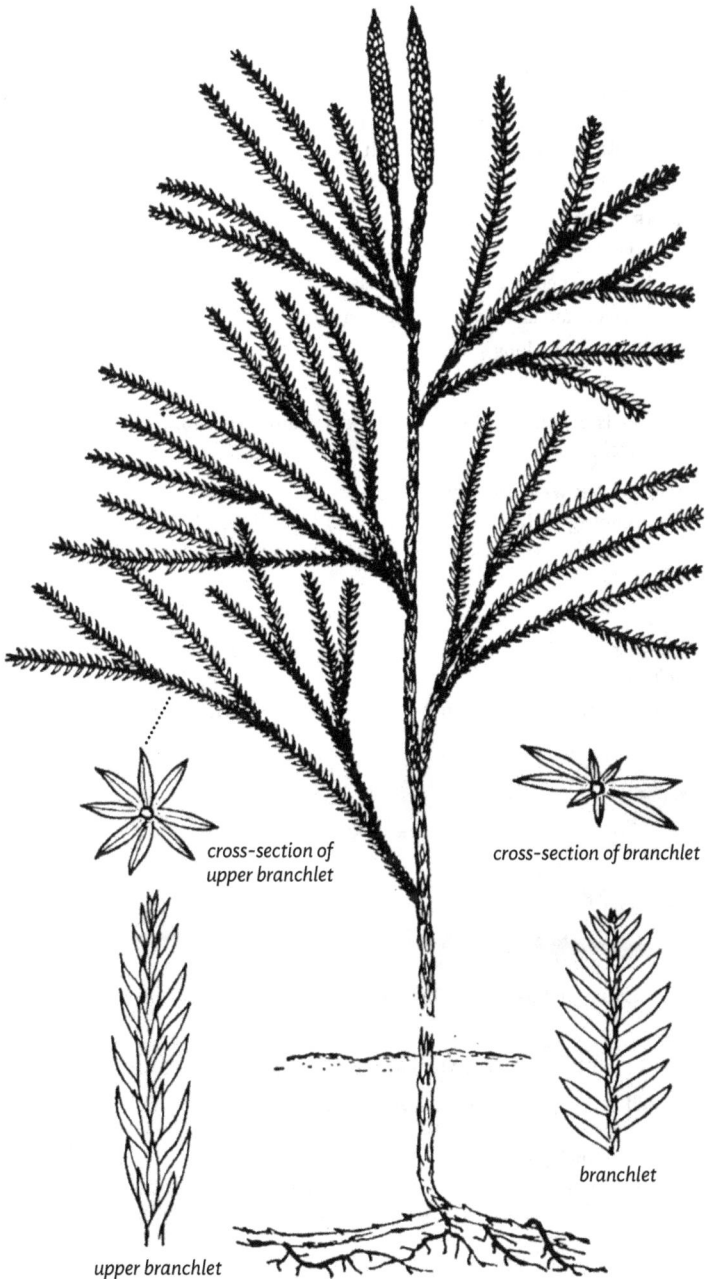

cross-section of
upper branchlet

cross-section of branchlet

upper branchlet

branchlet

Diphasiastrum
CREEPING-CEDAR

Small plants of drier habitats resembling miniature trees; **branches** flattened or 4-angled in cross-section (round in *D. sitchense*); **leaves** 4-ranked (5-ranked in *D. sitchense*), neither spine- nor hair-tipped. **Strobili** (cones) stalked, the stalks branched into segments of equal length. Worldwide, 21 species defined; 6 species in North America, 4 species in our flora.

KEY CHARACTERS
• Plants treelike, the main axis with forked branches.

• **Branches** flattened or 4-angled in cross-section.

• **Leaves** keeled and also unkeeled along midrib.

• **Strobili** (cones) on long stalks.

NAME
Diphasium is another genus in Clubmoss Family, *astrum*, refers to a 'partial resemblance.'

ADDITIONAL MIDWEST SPECIES
Alpine creeping-cedar [*Diphasiastrum alpinum* (L.) Holub], collected long ago in Keewenaw County in Michigan's Upper Peninsula, but now likely no longer present in state.

HYBRIDS
Diphasiastrum × habereri (House) Holub, hybrid between *D. digitatum* and *D. tristachyum.*

Diphasiastrum × zeilleri (Rouy) Holub, hybrid between *D. complanatum* and *D. tristachyum.*

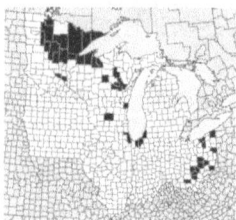

Diphasiastrum xhabereri   *Diphasiastrum xzeilleri*

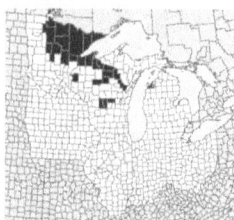

KEY TO DIPHASIASTRUM | CREEPING-CEDAR

1   Strobili 1 (–2) on short to elongated, rarely forked stalks (peduncles); base of strobilus with a few sporophylls scattered along peduncle; stomata on both leaf surfaces; Mich Upper Peninsula only   **1. Diphasiastrum sabinifolium**
SAVIN-LEAF CREEPING-CEDAR, page 392

1   Strobili 2–4 on forked stalks; base of strobilus compact and abruptly distinct from peduncle; stomata on lower leaf surfaces only **2**

2   Horizontal stems almost always buried in soil (4 cm or more deep); ultimate (outermost) branches squarish in cross section, 1–2 mm wide; plants usually blue-green (except in deep shade); lower leaves same size as upper; nw Ind, Mich, Minn, Ohio, Wisc

2. Diphasiastrum tristachyum
DEEP-ROOT CREEPING-CEDAR, page 394

2   Horizontal stems on ground surface, or slightly buried (< 4 cm deep) in surface litter; ultimate (outermost) branches flat in cross section; 2–4 mm wide; plants not blue-green; lower leaves reduced to a leaf base and a short free appendage

. . . . **3**

3   Ultimate branches symmetrically in one plane forming a layered appearance; annual constrictions mostly absent; stalks of strobili staying green until after spore discharge in autumn; regionwide

3. Diphasiastrum digitatum
FAN CREEPING-CEDAR, page 390

3   Ultimate branches not evenly layered, giving an irregular appearance; annual constrictions uniformly present; stalks of strobili losing green color by time of spore discharge in late summer; e Iowa, Mich, Minn, Wisc

4. Diphasiastrum complanatum
TRAILING CREEPING-CEDAR, page 388

**Diphasiastrum
sabinifolium**
SAVIN-LEAF
CREEPING-CEDAR

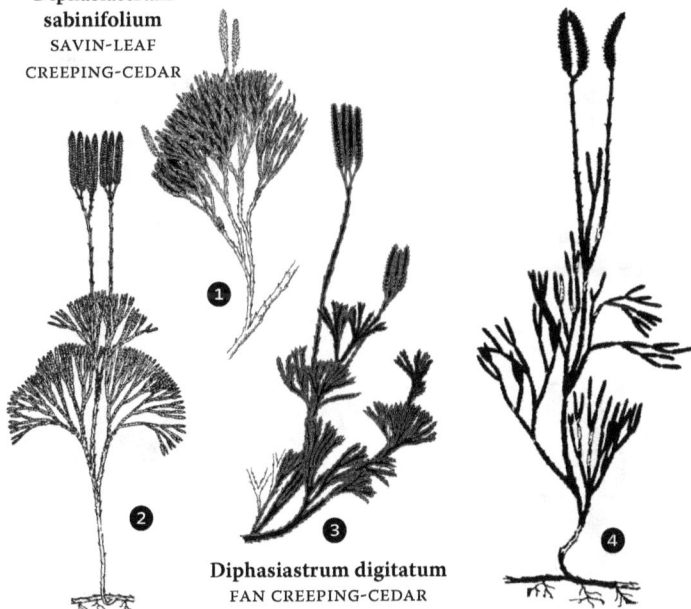

Diphasiastrum digitatum
FAN CREEPING-CEDAR

Diphasiastrum tristachyum
DEEP-ROOT CREEPING-CEDAR

Diphasiastrum complanatum
TRAILING CREEPING-CEDAR

Diphasiastrum complanatum
(L.) Holub
TRAILING CREEPING-CEDAR

FIELD TIPS
- creeping, resembling a small cedar tree
- branches flattened, bright green
- leaves 4-ranked, appressed, of different sizes
- strobili on slender stalks, these sometimes branched

MIDWEST RANGE
e Iowa, Mich, Minn, Wisc.

HABITAT
Woods, thickets, clearings; soils typically dry, sandy and acidic.

DESCRIPTION
Stems Horizontal stems mostly shallowly to deeply underground. Upright stems to about 30-40 cm long, with crowded, irregularly forking branchlets, these flattened, and often strongly constricted between annual growths. Upper surface of branches green and satiny, underside pale and dull.

Leaves 4-ranked, appressed and attached to stem for more than half their length; upper rank of leaves small and very narrow; lateral leaves larger, keeled; underside leaves smallest.

Strobili Mostly 1 or 2 on slender, sometimes forked stalks; the strobili fertile for their entire length.

SYNONYMS
- *Diphasium anceps* (Wallr.) Á. Löve & D. Löve
- *Diphasium complanatum* (L.) Rothm.
- *Diphasium wallrothii* H.P. Fuchs
- *Lycopodium anceps* Wallr.
- *Lycopodium complanatum* L.

SIMILAR SPECIES
Fan creeping-cedar (*D. digitatum*) usually with horizontal stems at or near ground surface; stems and branches lack annual constrictions, and its strobili are often sterile in their upper portion.

NAME
Linnaeus described *Lycopodium complanatum* in 1753.

From Latin, *complanatus*, made level or flat, referring to the flattened leafy branches, similar to those of a cedar tree.

NORTH AMERICA

MIDWEST

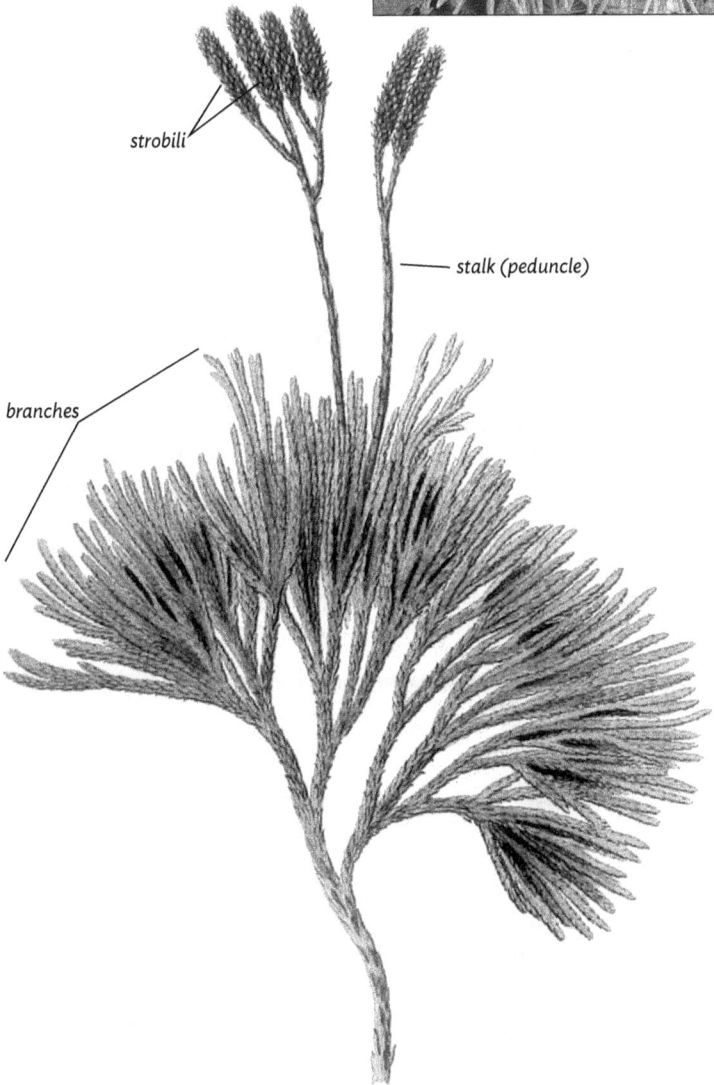

strobili

stalk (peduncle)

branches

Diphasiastrum digitatum
(A. Braun) Holub

FAN CREEPING-CEDAR

FIELD TIPS
- creeping, resembling a small cedar tree, often in large patches
- branches flattened, dark green
- leaves 4-ranked, of different sizes, appressed or spreading at tips
- strobili stalked, the stalks usually twice-forked

MIDWEST RANGE
Ill, Ind, e Iowa, Mich, Minn, Ohio, Wisc.

HABITAT
Sandy woods, clearings, fields, thickets, where often forming extensive patches from the shallow horizontal stems; soils typically sandy and acidic.

DESCRIPTION
Stems Horizontal stems mostly on ground surface or in upper surface litter. Upright stems 30-40 cm long, with fanlike branches, constrictions between annual growth absent or indistinct; upper surface of branches green and shiny, underside pale and dull.

Leaves Tiny, 4-ranked, appressed and attached to stem for more than half their length; upper rank of leaves small and very narrow; lateral leaves larger, keeled, spreading at tip; underside leaves smallest, triangular.

Strobili 2-4, on stalks usually twice-forked from near the same point; tips of the strobili are often sterile (check a number of plants to verify this).

SYNONYMS
- *Diphasium complanatum* (L.) Rothm. subsp. *flabelliforme* (Fernald) Á. Löve & D. Löve
- *Lycopodium complanatum* L. var. *flabelliforme* Fernald
- *Lycopodium digitatum* Dill. ex A. Braun
- *Lycopodium flabelliforme* (Fernald) Blanch.

SIMILAR SPECIES
Trailing creeping-cedar (*D. complanatum*) horizontal stems more deeply buried and not on surface; has distinct annual constrictions on stems and branches, and the strobili are completely fertile.

NOTES
Another of the Philadelphia plants collected by John Bartram and illustrated by Dillenius in 1741. For many years thereafter it was confused with the more northern *Diphasiastrum complanatum*, but Fernald segregated it

NORTH AMERICA

MIDWEST

as a variety in 1909, and Blanchard
raised it to species rank in 1919.

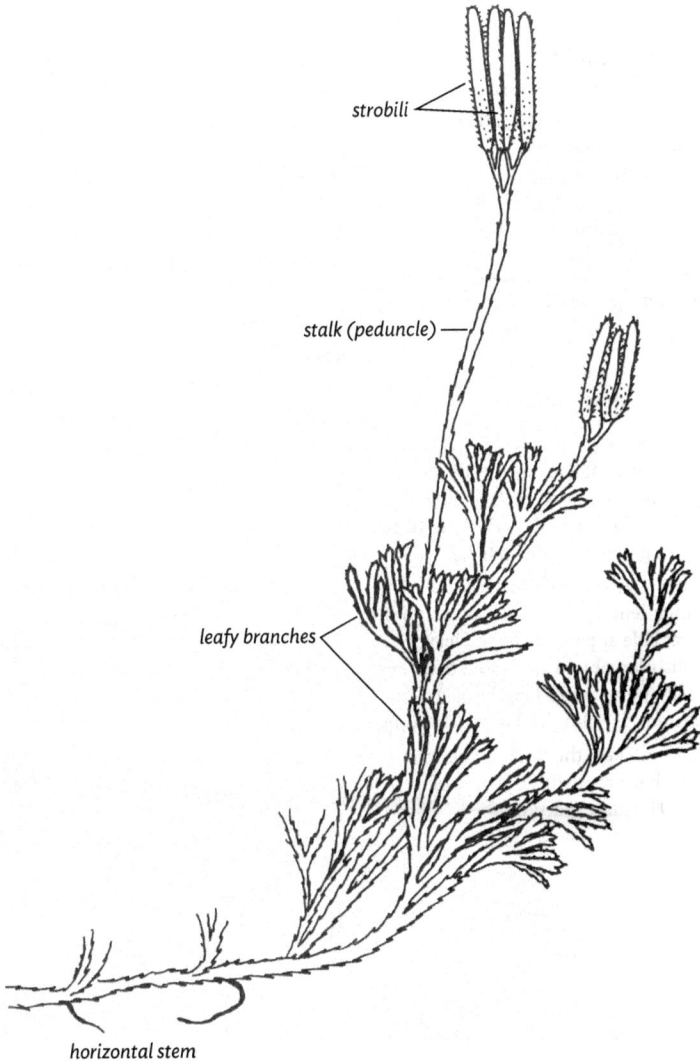

strobili

stalk (peduncle)

leafy branches

horizontal stem

Diphasiastrum sabinifolium
(Willd.) Holub

SAVIN-LEAF CREEPING-CEDAR

FIELD TIPS

- creeping, resembling a small cedar tree
- branches compressed in cross-section
- leaves 4-ranked, of similar size
- strobili on slender stalks, these sometimes branched
- sporophylls on lower portion of strobili widely spaced

MIDWEST RANGE

Mich Upper Peninsula; more common north and east of our region.

HABITAT

Sandy woods and meadows, often where disturbed as in blowout areas.

DESCRIPTION

Stems Horizontal stems mostly creeping on ground surface. Erect stems dichotomously forked, up to 20 cm long; sterile branchlets flattened.

Leaves 4-ranked, awl-like, nearly similar lengths; leaves of upper and lower side appressed; lateral leaves slightly larger and appressed to stem for about half their length, the free portion spreading and incurved at tip.

Strobili Stalked, the stalks sometimes forked; sporophylls near base of strobili widely spaced.

SYNONYMS

- *Lycopodium armatum* Desv.
- *Lycopodium sabinifolium* Willd.

NOTES

Considered to have originated as a hybrid between *Diphasiastrum sitchense* (Rupr.) Holub and *D. tristachyum*. *Diphasiastrum sitchense* differs from *D. sabinifolium* in having sessile rather than stalked strobili, but that species is not known from the Midwest region.

NAME

Sabina is a section of the conifer genus *Juniperus*; the leaves of this species are scalelike and similar to those of juniper.

NORTH AMERICA

MIDWEST

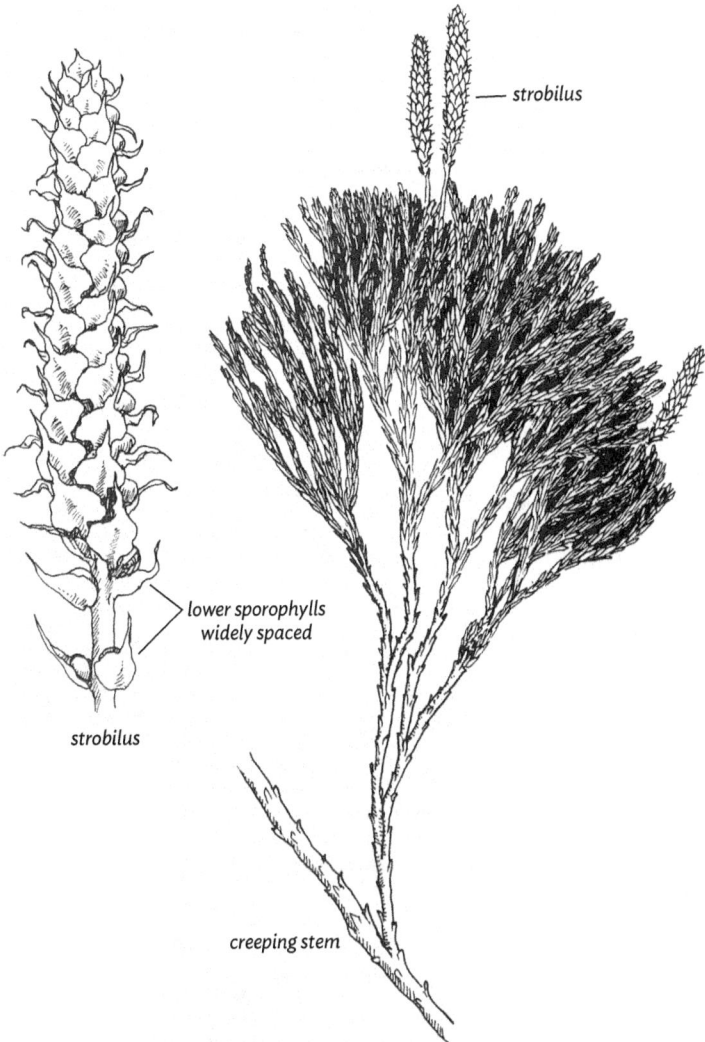

strobilus

lower sporophylls
widely spaced

strobilus

creeping stem

Diphasiastrum tristachyum
(Pursh) Holub

DEEP-ROOT CREEPING-CEDAR

FIELD TIPS
- plants blue-green, in crowded clusters
- branches somewhat 4-angled, fanlike, with annual constrictions
- leaves 4-ranked, appressed, of different sizes
- strobili on slender stalks, these once- or twice-forked, rising above leafy branches

MIDWEST RANGE
nw Ind (threatened), Mich, Minn, Ohio, Wisc.

HABITAT
Dry, sandy fields and open woods; sandy disturbed places.

DESCRIPTION

Stems Horizontal stems usually deeply buried. Upright stems to about 30 cm long. Sterile branches upright to loosely spreading, somewhat 4-angled in cross-section, with small annual constrictions.

Leaves Leaves tiny, narrow, 4-ranked, evergreen, bluish green and usually covered with a fine white powder, attached to stem for more than half their length; upper and lower rank of leaves about half the size of lateral leaves.

Strobili Mostly 3-4 on leafy-bracted stalks, the stalks once- or twice-forked.

SYNONYMS
- *Diphasium chamicyparissus* (A. Braun) Á. Löve & D. Löve
- *Diphasium complanatum* (L.) Rothm. subsp. *chamicyparissus* (A. Braun) Kukkonen
- *Diphasium tristachyum* (Pursh) Rothm.
- *Lycopodium chamicyparissus* A. Braun
- *Lycopodium complanatum* L. subsp. *chamicyparissus* (A. Braun) Nyman
- *Lycopodium tristachyum* Pursh

SIMILAR SPECIES
The blue-green color, squarish stems, and conspicuous constrictions at the juncture between each year's growth help separate **deep-root creeping-cedar** from other members of the genus.

NOTES
This species grows both in Europe and eastern North America, but was first named from plants collected in Virginia, by Pursh in 1814. Later named in Europe *Lycopodium chamaecyparissus*.

Plants in shade tend to be more green

NORTH AMERICA

MIDWEST

in color than blue-green, and
more openly branched.

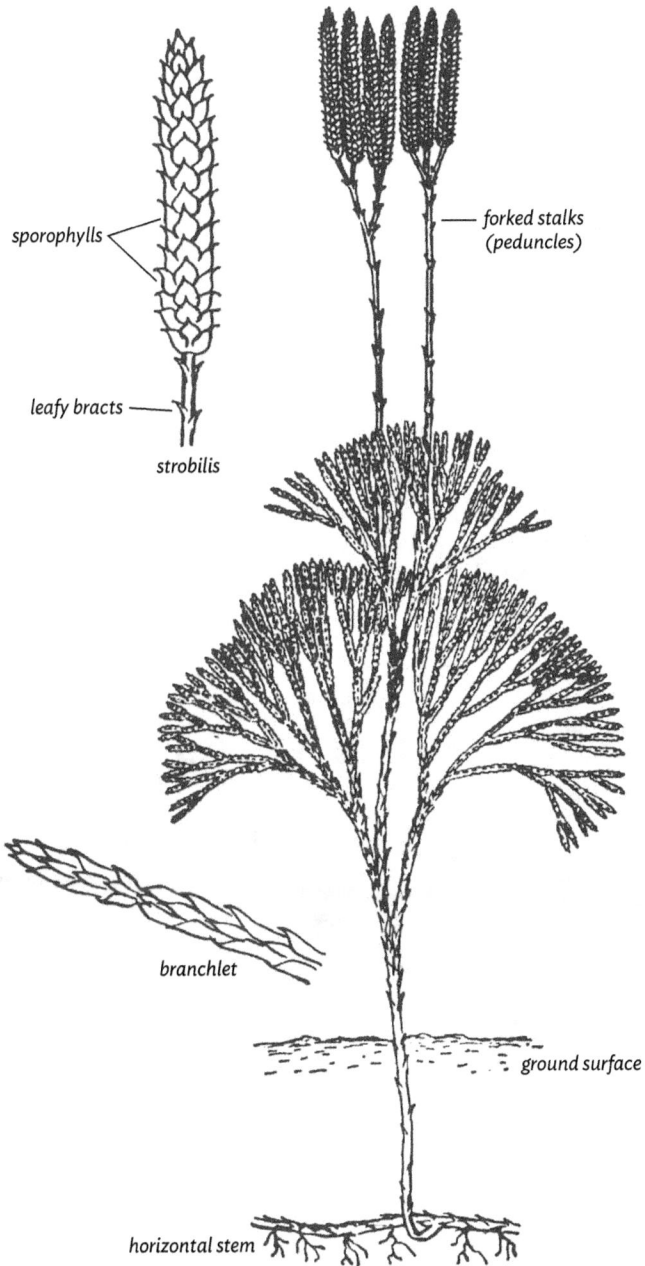

sporophylls

forked stalks
(peduncles)

leafy bracts

strobilis

branchlet

ground surface

horizontal stem

Huperzia
FIR-MOSS

Evergreen plants with small, stalkless, pointed **leaves; horizontal stems** absent. **Sporangia** borne at bases of stem leaves, rather than in terminal strobili (cones) as in the other members of Family. *Huperzia* also produce **gemmae** (small plantlets) in upper portion of stem; if the gemmae land on a favorable site they can generate roots and develop into a new plant, identical to the parent plant. A large genus (as now defined), with over 250 species worldwide (many new species have been defined from Asia); 7 in North America, 4 in our flora.

KEY CHARACTERS

- Plants with bristly upright stems with **sporangia** borne in distinct zones; **strobili** (cones) absent.
- **Horizontal stems** absent (but older stems sometimes lean and appear horizontal).
- **Gemmae** (plantlets) produced in upper parts of mature stems.

HYBRIDS

Huperzia × bartleyi (Cusick) Kartesz & Gandhi, hybrid between *H. lucidula* and *H. porophila*; local in Midwest region.

Huperzia × buttersii (Abbe) Kartesz & Gandhi, hybrid between *H. lucidula* and *H. selago*; ne Minnesota and nw Wisconsin; resembles a slender *H. lucidula*, but distinguished by presence of abortive spores, and scattered stomata on upper leaf surface.

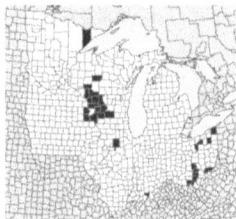

Huperzia xbartleyi

Huperzia xbuttersii

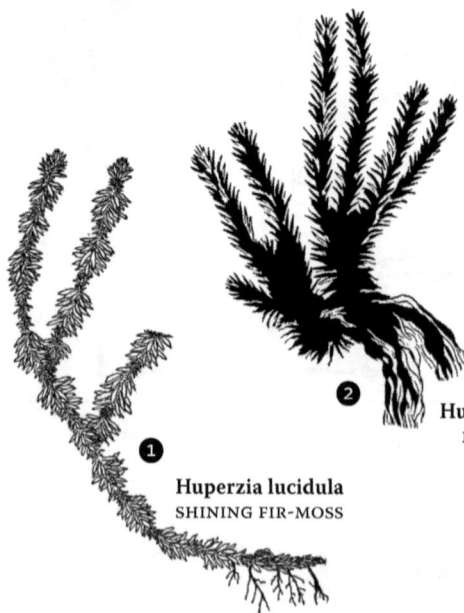

Huperzia porophila
ROCK FIR-MOSS

Huperzia lucidula
SHINING FIR-MOSS

KEY TO HUPERZIA | FIR-MOSS

1   Leaves obovate, widest above the middle, spreading to ± reflexed, the upper portion of at least the larger leaves with distinct teeth; shoots "shaggy" with conspicuous annual constrictions; usually growing on soil; regionwide   **1. Huperzia lucidula**
SHINING FIR-MOSS, page 400

1   Leaves lance-shaped, widest below the middle, leaves (at least those on the upper stem) often ascending, entire or with a few small teeth; annual constrictions absent or faint. **2**

2   Leaves lance-shaped with sides nearly parallel; stomates on upper surface of each leaf number 2-50 (view fresh leaves under 20x lens to see the light-colored, dot-like stomates); Ill, s Ind, ne Iowa, n Mich, Minn, Ohio, sw Wisc   **2. Huperzia porophila**
ROCK FIR-MOSS, page 402

2   Leaves lance-shaped (as above) or ovate or triangular; if leaf shape is inconclusive, then number of stomates on upper leaf surface is greater than 60 . . . . **3**

3   Leaves near base of plant essentially same size as those on upper portion; gemmae formed in a single whorl at end of the annual growth; n Mich, n Minn, n Wisc   **3. Huperzia selago**
FIR-MOSS, page 404

3   Leaves near base of plant conspicuously longer than those on upper portion; gemmae formed throughout upper portions of shoot; Mich, nw Minn, n Wisc   **4. Huperzia appressa**
MOUNTAIN FIR-MOSS, page 398

Huperzia selago
FIR-MOSS

Huperzia appressa
MOUNTAIN FIR-MOSS

Huperzia appressa (Desv.) Á. Löve & D. Löve
MOUNTAIN FIR-MOSS

FIELD TIPS
- short, evergreen plants of rocky places
- leaves yellow-green to green, reduced in size on upper portion of stem
- sporangia and gemmae borne on upper stems

MIDWEST RANGE
Mich, ne Minn, nw Wisc; historical record from central Ohio; in ne North America, typically a mountain species of higher elevations.

HABITAT
Cliffs, talus slopes, where open and exposed, on moss or thin soil.

DESCRIPTION
Stems Short, to only 10 cm long; clustered; annual constrictions absent.

Leaves Ascending, narrowly lance-shaped or with the sides parallel; stomates on both surfaces; margins entire; upper stem leaves smaller than those of lower stem.

Sporangia In distinct zones on upper stems.

SYNONYMS
- *Huperzia appalachiana* Beitel & Mickel

SIMILAR SPECIES
- **Fir-moss** (*Huperzia selago*) produces gemmae in a single whorl near end of each year's growth; gemmae in **mountain fir-moss** are scattered throughout much of stem.

NORTH AMERICA

MIDWEST

Mountain fir-moss
(*Huperzia appressa*)

Huperzia lucidula (Michx.) Trevisan
SHINING FIR-MOSS

FIELD TIPS
- region's most common and widespread fir-moss
- leaves evergreen, shiny, usually widest above middle, margins toothed
- stems with distinct annual constrictions

MIDWEST RANGE
Iowa, Ill, Ind, Mich, Minn, Ohio, Wisc.

HABITAT
Moist deciduous, mixed conifer-hardwood, or conifer forests, often sprawling on mossy boulders, logs, and hummocks; less commonly on shaded, mossy sandstone.

DESCRIPTION
Stems Ascending and sprawling, single or few-forked, to ca. 20 cm long, rooting towards base; annual constrictions evident; lower stem often hidden in leaf litter; sometimes forming a circular "fairy ring" as the stems grow outward.

Leaves Mostly 6-ranked, 7-12 mm long, oblong lance-shaped, widest above middle, spreading or angled downward; stomates only visible on underside; margins toothed, especially above middle of leaf.

Sporangia In upper portion of stem, yellow when mature. **Gemmae** in single whorl from uppermost leaf axils.

SYNONYMS
- *Huperzia selago* (L.) Bernh. ex Schrank & Mart. subsp. *lucidula* (Michx.) Á. Löve & D. Löve
- *Lycopodium lucidulum* Michx.
- *Lycopodium reflexum* Sw., non Lam.
- *Urostachys lucidulus* (Michx.) Herter ex Nessel

NOTES
Shining fir-moss is distinctive and usually readily identified species. Key characters include toothed leaves widest above the middle, stems with conspicuous annual constrictions, and stomata only on lower surface of the leaves (visible under magnification). *Huperzia lucidula* is our most common and widespread fir-moss, usually found in forest soil rather than on rocks.

NAME
In the course of his travels, Michaux observed this plant from "Canada to the Carolina mountains," and named it *Lycopodium lucidulum* in 1803.

NORTH AMERICA

MIDWEST

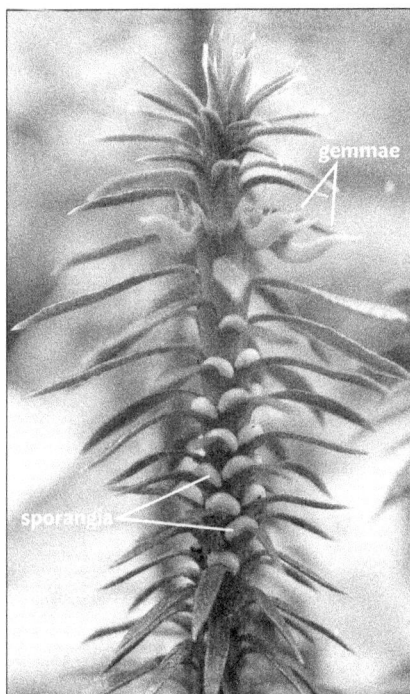

*Upper stem of **shining fir-moss** with gemmae and sporangia.*

Huperzia porophila

(Lloyd & Underwood) Holub

ROCK FIR-MOSS

FIELD TIPS

- shaded sandstone habitats
- stems seldom branched
- leaves narrow, mostly entire, not shiny

MIDWEST RANGE

Ill (threatened), s Ind, ne Iowa (threatened), Mich Upper Peninsula, Minn (threatened), Ohio, Wisc.

HABITAT

Sandstone outcrops and boulders and rocky woods, where cool, moist and shaded.

DESCRIPTION

Stems Clustered, to 15 cm long; annual constrictions faint or absent.

Leaves Narrow, ca. 3-8 mm long; margins nearly entire.

Sporangia Borne from upper leaf axils.

SYNONYMS

- *Huperzia selago* (L.) Bernh. ex Schrank & Mart. var. *patens* (P. Beauv.) Trevis.
- *Huperzia selago* (L.) Bernh. ex Schrank & Mart. var. *porophila* (Lloyd & Underw.) Á. Löve & D. Löve
- *Lycopodium lucidulum* Michx. var. *porophilum* (Lloyd & Underw.) Clute
- *Lycopodium porophilum* Lloyd & Underw.

- *Lycopodium selago* L. subsp. *patens* (P.Beauv.) Calder & Roy L. Taylor
- *Lycopodium selago* L. var. *porophilum* (Lloyd & Underw.) Clute

SIMILAR SPECIES

Rock fir-moss characterized by leaves with nearly parallel sides and stomates on upper leaf surface numbering fewer than 50 (requires microscope).

- Often confused with **shining fir-moss** (*Huperzia lucidula*), but rock fir-moss usually smaller and more compact, the leaves lance-shaped and not shiny, with stomates on upper surface, and the ascending stems seldom branched.

NOTES

This species is of hybrid origin, the parents believed to be *Huperzia lucidula* and *H. selago*.

NORTH AMERICA

MIDWEST

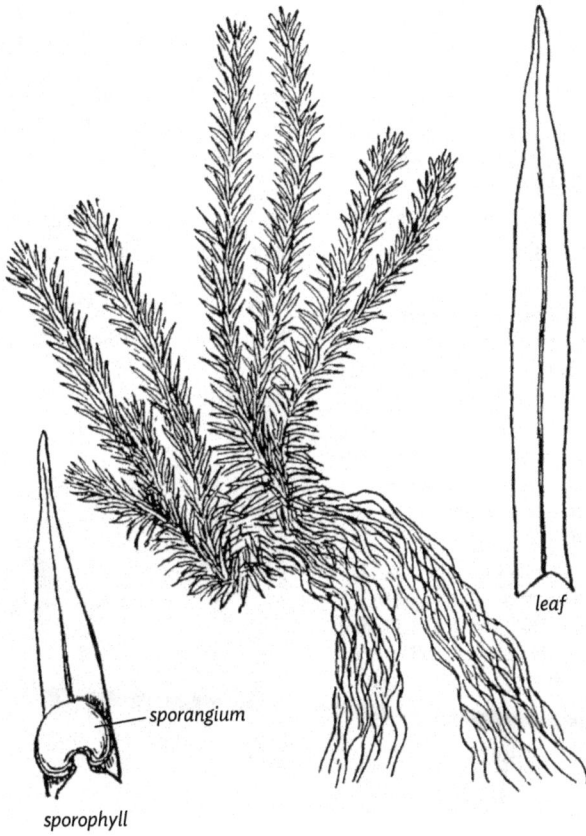

leaf

sporangium

sporophyll

Huperzia selago

(L.) Bernh. ex Mart. & Schrank

FIR-MOSS

FIELD TIPS

- plants in dense, flat-topped clusters
- stems lack annual constrictions
- gemmae at stem tip
- leaves small, entire, sharp-pointed

MIDWEST RANGE

n Mich, n Minn, n Wisc.

HABITAT

Conifer swamps, moist to dry conifer woods; mossy, shaded streambanks, boulders, trails and old roads, moist sandy borrow pits, and moist to wet rock cliffs along Lake Superior.

DESCRIPTION

Stems Horizontal stems short, leafy; erect stems to 15 cm long, branched several times, usually near the base; annual constrictions barely evident.

Leaves Crowded, 8- to 10-ranked, ascending, yellow-green, 3-8 mm long, lance-shaped, with stomates on both surfaces; margins entire or nearly so.

Sporangia In zones in upper stems; **gemmae** (reproductive buds) in a single whorl in upper leaf axils, and tip of stem can appear thicker than rest of stem.

SYNONYMS

- *Lycopodium selago* L.

SIMILAR SPECIES

- Distiguished from **shining fir-moss** (*Huperzia lucidula*) by the entire leaves (rather than leaves toothed at tip).

- Distinguished from **mountain fir-moss** (*Huperzia appressa*) by the longer leaves, and the gemmae restricted to a single whorl near the end of each years growth.

NAME

This plant was well-known to Linnaeus and was assigned genus and species name of *Lycopodium selago* in his Species Plantarum in 1753.

Selago referring to South African plants of genus *Selago*, the plants heather-like.

NORTH AMERICA

MIDWEST

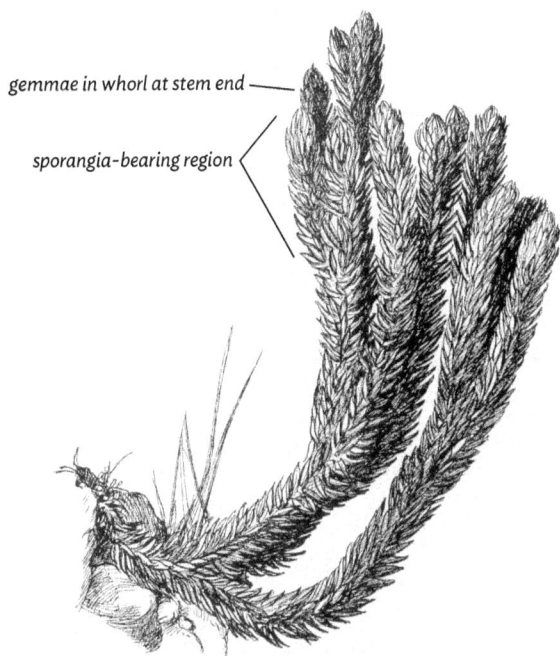

gemmae in whorl at stem end ——

sporangia-bearing region ⟨

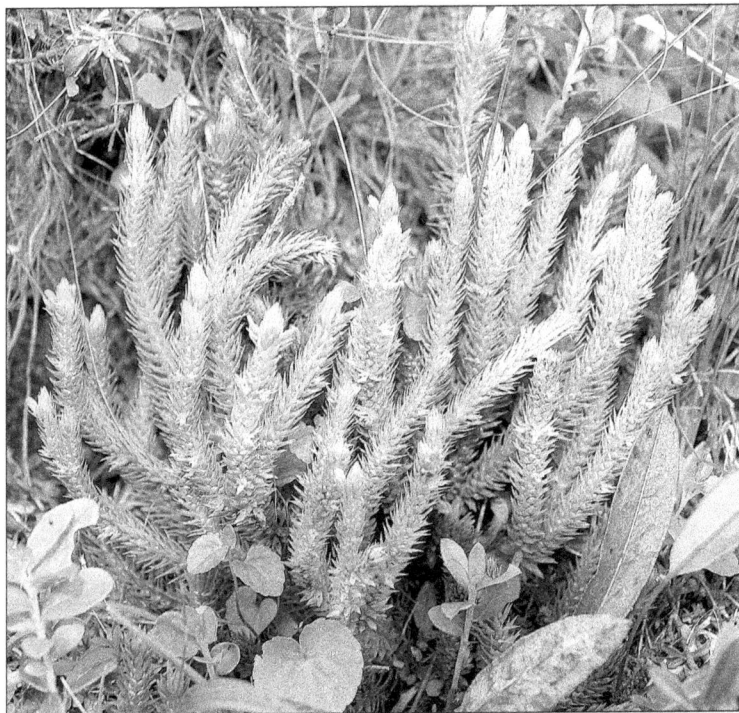

Lycopodiella
BOG CLUBMOSS

Small plants of wet places with **horizontal stems** creeping on ground surface, and fertile, upright stems. **Strobili** (cones) formed on upper part of upright stems, stalkless and covered with leaves. **Gametophytes** grow on surface and are photosynthetic. Worldwide, 25 species in genus; 6 species in North America, 4 species in our flora but only **northern bog clubmoss** (*Lycopodiella inundata*) is common.

KEY CHARACTERS
• Plants small, with slender upright stems tipped by **strobili** (cones).

• **Horizontal stems** not flattened.

• **Leaves** of horizontal stem all alike.

ADDITIONAL MIDWEST SPECIES
• **Southern appressed clubmoss** [*Lycopodiella appressa* (Chapman) Cranfill]; known in Midwest only from wet places in Winnebago County in northern Illinois, more common in eastern and southeastern USA. Previously treated as *Lycopodium inundatum* var. *bigelovii* Tuckerm. Differs from *Lycopodiella inundata* by its taller fertile stems (to about 35 cm long) and its mostly toothed leaves.

• **Prostrate clubmoss** (*Lycopodiella margueritae* J.G. Bruce, W.H. Wagner & Bietel); midwest range nw Ind, Mich Lower Peninsula (threatened), Ohio (endangered), Wisc. Distinguished from *Lycopodiella appressa* and *L. subappressa* by the thicker and large strobili, and the more spreading leaves on the strobili and upright shoots.

• **Northern appressed clubmoss** (*Lycopodiella subappressa* J.G. Bruce, W.H. Wagner & Bietel); midwest range ne Ill, nw Ind (endangered), Mich, nw Ohio (endangered).

Lycopodiella appressa
Southern appressed
clubmoss

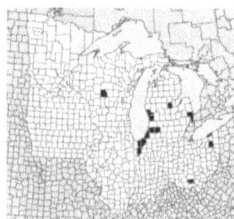

Lycopodiella margueritae
Prostrate clubmoss

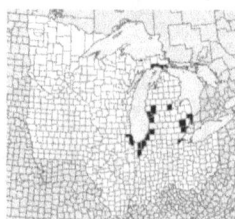

Lycopodiella subappressa
Northern appressed
clubmoss

KEY TO LYCOPODIELLA | CLUBMOSS

1 Horizontal stems less than 1 mm in diameter except at root nodes; horizontal stem leaves usually less than 6 mm long, teeth or bristles absent; erect shoots mostly less than 10 cm tall; n Ill, Ind, e Iowa, Mich, Minn, Ohio, Wisc   **1. Lycopodiella inundata**
NORTHERN BOG CLUBMOSS, page 408

1 Horizontal stems 1–3 mm in diameter; horizontal stem leaves 4–13 mm long, margins toothed or bristle-tipped; tallest erect shoots often more than 10 cm tall. **2**

2 Horizontal stems 1.0–1.7 (–2) mm in diameter; horizontal stem leaves usually 4–6 mm long, angled upward along stem; uncommon in ne Ill, nw Ind, Mich, n Ohio
2. Lycopodiella subappressa
NORTHERN APPRESSED CLUBMOSS, page 406

2 Horizontal stems 1.8–2.2 (–3) mm in diameter; horizontal stem leaves usually 6–13 mm long, spreading (frequently perpendicular to the stem); rare in n Ind, s Mich, Ohio and c Wisc
3. Lycopodiella margueritae
PROSTRATE CLUBMOSS, page 406

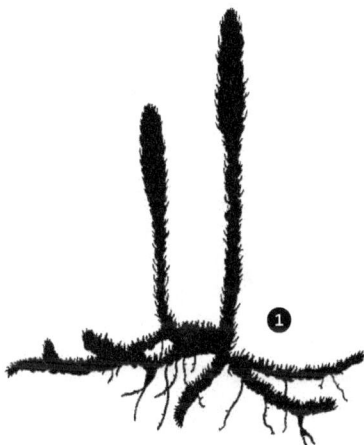

Lycopodiella inundata
NORTHERN BOG CLUBMOSS

Lycopodiella inundata (L.) Holub
NORTHERN BOG CLUBMOSS

FIELD TIPS
- horizontal stems evergreen, creeping on surface
- fertile stems erect, tipped by a 'bushy' strobilus
- wet, often sandy habitats

MIDWEST RANGE
n Ill (endangered), Ind (endangered), e Iowa (endangered), Mich, Minn, Ohio, Wisc.

HABITAT
Moist to wet sandy depressions and borrow pits, bog mats, sandy lake and pond shores.

DESCRIPTION
Stems Horizontal or arching, forking, very slender, leafy, the leaves turned upwards, rooting at intervals; fertile stems upright, to 10 cm long.

Leaves Spiralled in 8 or 10 ranks, narrow and linear, gradually tapered to long tip; margins mostly entire.

Strobili Single, stalkless, wider than stem and appearing like a bushy tail.

SYNONYMS
- *Lepidotis inundata* (L.) C. Borner
- *Lycopodium inundatum* L.

NAME
Being well-known in northern Europe, this plant was named *Lycopodium* *inundatum* by Linnaeus in 1753.

NORTH AMERICA

MIDWEST

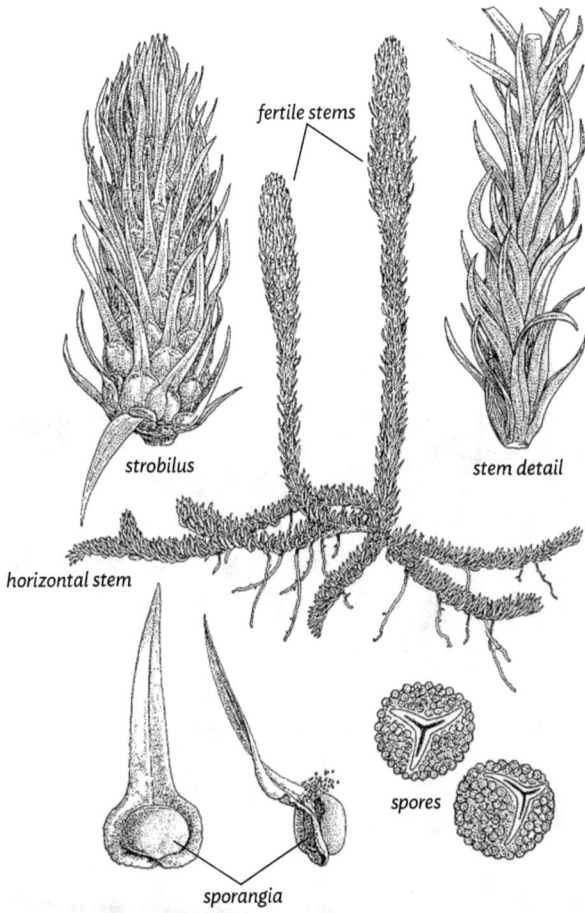

strobilus

fertile stems

stem detail

horizontal stem

sporangia

spores

Lycopodium
GROUND-PINE

Trailing, evergreen clubmosses with clusters of upright leafy stems. **Horizontal stems** on or near ground surface. **Upright stems** branched, but not treelike as in some other members of Clubmoss Family. **Leaves** tipped by a long, hairlike bristle. **Sporangia** borne in 1 or several stalked cones at ends of stems. **Gametophytes** underground, disk-shaped, and nonphotosynthetic. Worldwide, 46 species are recognized, many from Asia; 2 species in North America and in our flora.

KEY CHARACTERS

- **Stems** upright, bristly, with several branches.
- **Branches** round in cross-section.
- **Leaves** tipped with a long, translucent hair.
- **Strobili** (cones) stalked, single or several in a branched cluster.

Running ground-pine (*Lycopodium clavatum*)

NAME

From Greek, *lykos*, a wolf, and *pous*, a foot; perhaps the creeping stems and unequally forked branches suggestive of wolf tracks. Another common name is **staghorn clubmoss** due to resemblance of upright branches to a stag's antlers.

KEY TO LYCOPODIUM | GROUND-PINE

1 Peduncles mostly with 2-3 (-4) clearly stalked strobili; n Ill, n Ind, e Iowa, Mich, Minn, Ohio, Wisc **1. Lycopodium clavatum** RUNNING GROUND-PINE, page 412

1 Peduncles with 1-2 strobili, if paired, then essentially stalkless; ne Ill, Iowa, Mich, Minn, Ohio, n Wisc **2. Lycopodium lagopus** ONE-CONE GROUND-PINE, page 414

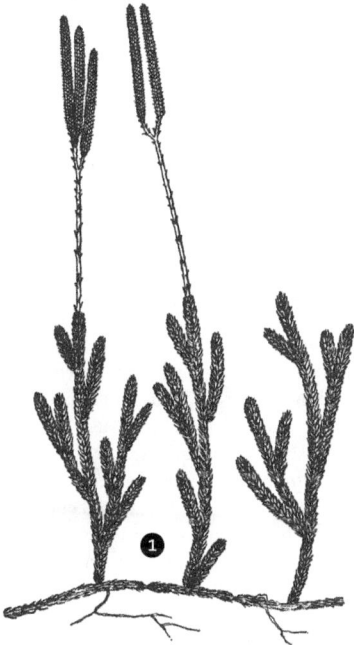

Lycopodium clavatum
RUNNING GROUND-PINE

Lycopodium lagopus
ONE-CONE GROUND-PINE

Lycopodium clavatum L.
RUNNING GROUND-PINE

FIELD TIPS
- trailing, evergreen clubmoss
- stems densely covered with small ascending leaves
- strobili (cones) usually several at ends of long, slender, branched stalk

MIDWEST RANGE
n Ill (endangered), n Ind, e Iowa (endangered), Mich, Minn, Ohio, Wisc.

HABITAT
Dry to moist woods and clearings; soils usually sandy.

DESCRIPTION
Stems Horizontal stems on ground surface, forking, rooting at intervals, with annual constrictions. Erect branches at first simple, then branching; fertile branches with a leafy-bracted stalk bearing 1-5 sessile or short-stalked strobili (cones).

Leaves Linear, tapered to translucent or white hairlike bristle, spreading to ascending and often incurved; margins entire or sometimes toothed.

Strobili On long slender stalks with several branches, each branch tipped by a single strobilus. Bracts of strobili coarsely fringed, at least the lower bracts with white hairlike tips.

SIMILAR SPECIES
Identification of mature plants of running ground-pine is usually easy. The extended soft, hair-like bristles on the leaves help distinguish *Lycopodium* from other members of the family.

The much less common **one-cone ground-pine** (*Lycopodium lagopus*) is distinguished, as the name implies, by having only a single strobilus per vertical stem, versus usually 2 or more in *L. clavatum*. Another useful distinction is the number of lateral branches from each vertical stem: *L. lagopus* generally has 2 or 3 branches, *L. clavatum* has 3-6 branches.

Young or sterile plants may be confused with **interrupted clubmoss** (*Spinulum annotinum*), but the leaves in that species are tapered to a stiff bristle and not to a long hairlike tip (and when mature, the strobili are not stalked).

NAME
This is the plant to which the term clubmoss was originally applied in Europe, and in 1753 Linnaeus accordingly gave it the species name *clavatum* (from Latin, *clava*, club, referring to its clublike stalked cones). It was later found in North America and at high elevations in the tropics.

NORTH AMERICA

MIDWEST

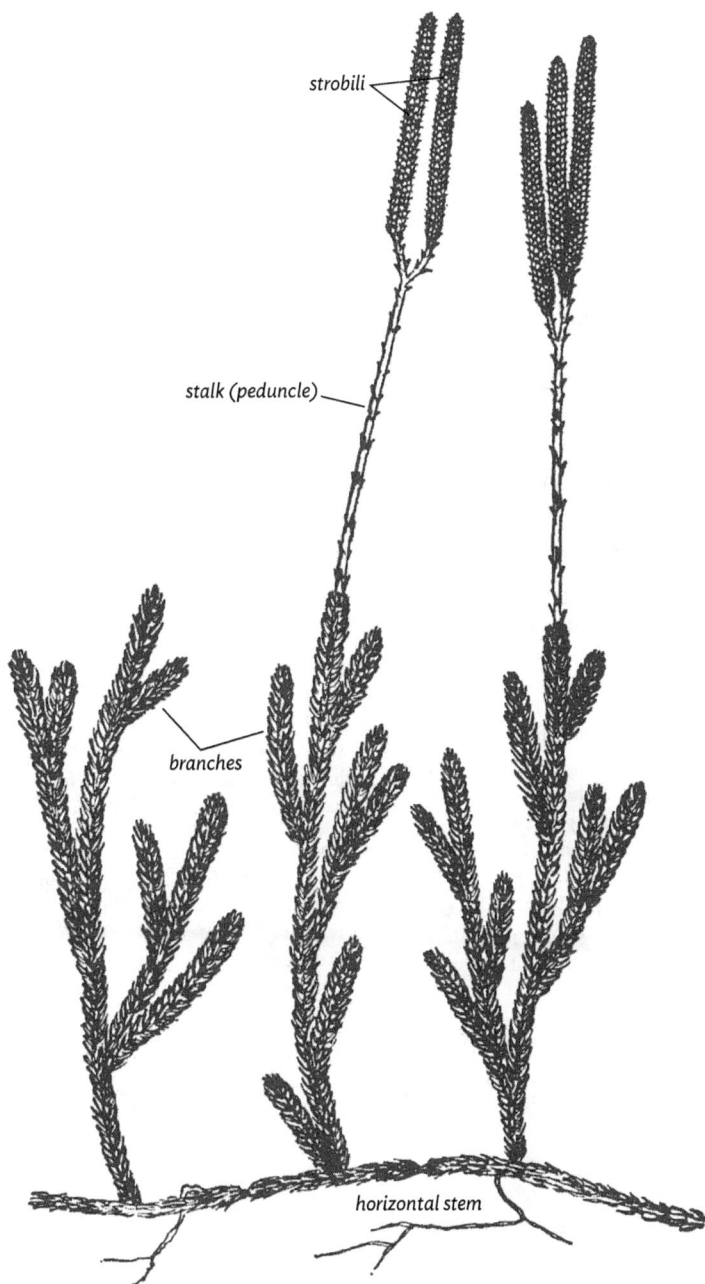

strobili

stalk (peduncle)

branches

horizontal stem

Lycopodium lagopus
(Laestad. ex Hartm.) Zinserl. ex Kuzen

ONE-CONE GROUND-PINE

FIELD TIPS

- trailing, evergreen clubmoss
- stems densely covered with small upright leaves
- strobili (cones) usually single at end of stalk

MIDWEST RANGE

ne Ill, Iowa, Mich, Minn, Ohio, n Wisc.

HABITAT

Sterile, sandy acid soils in open mixed conifer-hardwood or coniferous forests, sandy borrow pits, fields, roadsides.

DESCRIPTION

Stems Horizontal stems on or near ground surface, forking, rooting at intervals, with annual constrictions. Erect branches at first simple, then branching 1-2 times, the branches upright, with noticeable annual constrictions; fertile branches with a leafy-bracted stalk bearing 1-5 sessile or short-stalked strobili (cones).

Leaves Linear, tapered to translucent or white hairlike bristle, ascending to appressed; margins entire or sometimes toothed.

Strobili Single at end of each long, slender mostly unbranched stalk, (rarely with 2 strobili per stalk, these attached at same point). Sporophylls (bracts of strobili) tapered to hairlike tip.

SYNONYMS

- *Lycopodium clavatum* L. subsp. *megastachyon* (Fernald & Bissell) Selin
- *Lycopodium clavatum* L. subsp. *monostachyon* (Hook. & Grev.) Seland.
- *Lycopodium clavatum* L. var. *brevispicatum* Peck
- *Lycopodium clavatum* L. var. *integerrimum* Spring

SIMILAR SPECIES

Similar to **running ground pine** (*Lycopodium clavatum*), but with leaves ordinarily ascending or appressed and strobilus single on a shorter peduncle.

NOTES

In the past, this taxon, with 1 strobilus per stalk and more appressed leaves than in typical running ground-pine, has been considered a subspecies, variety, or form of *L. clavatum*.

NAME

Lagopus from Greek, *lagos*, hare or rabbit, plus *pous*, foot, perhaps referring to the feathery appearance of the branches.

NORTH AMERICA

MIDWEST

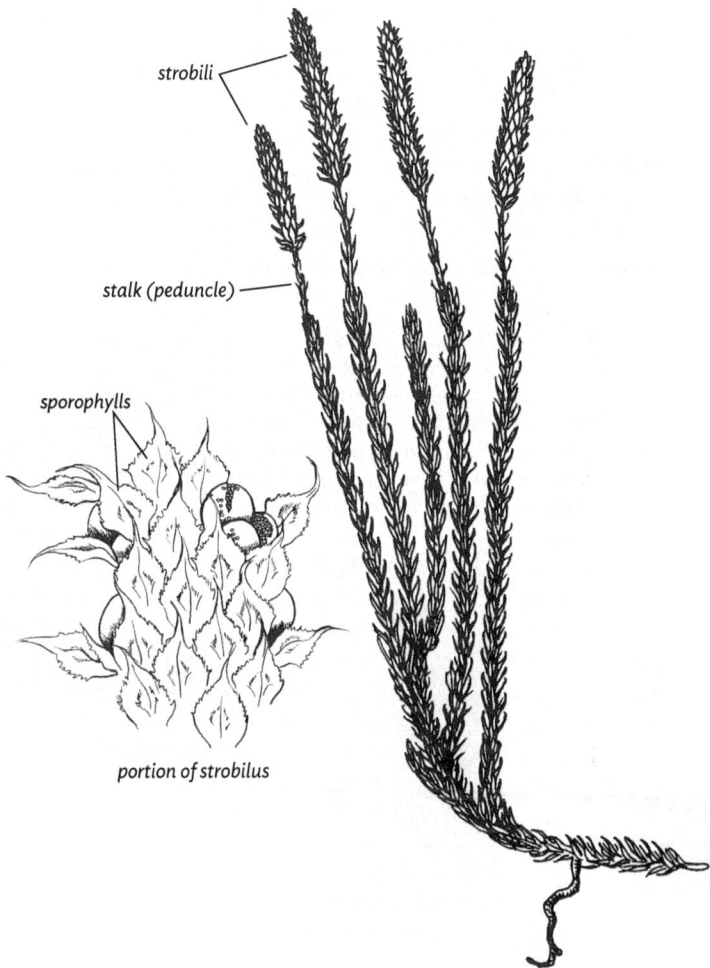

strobili

stalk (peduncle)

sporophylls

portion of strobilus

Spinulum
BRISTLY CLUBMOSS

Evergreen, trailing clubmoss; upright **stems** prickly, with annual constrictions (resulting from shorter leaves being formed at end of one year's growth and beginning of the next year's growth). **Sporangia** borne in stalkless strobili at ends of branches. Worldwide, 3 species, all of which occur in North America; 2 species in our flora.

KEY CHARACTERS
• Plants with upright shoots, these branched near base.
• **Branches** round in cross-section.
• **Leaves** tipped with tiny spine.
• **Strobili** (cones) not stalked.

NAME
From Latin, *spinula*, a small spine, as found at the leaf tips in *Spinulum*.

ADDITIONAL MIDWEST SPECIES
• **Northern clubmoss** [*Spinulum canadense* (Nessel) A. Haines]; synonym *Lycopodium annotinum* var. *pungens* (La Pylaie) Desv.; in our region, known only from Michigan's Upper Peninsula and perhaps best considered a minor variant of *Spinulum annotinum*; **leaves** of erect stems 2.5-6.0 mm long, lance-shaped to lance-oblong, flat, obscurely toothed, strongly ascending to tightly appressed against the stem; this species reported to have usually more than 25 stomata on upper leaf surface, in *S. annotinum,* stomata usually absent.

Spinulum canadense
Northern clubmoss

KEY TO SPINULUM | CLUB-MOSS

1 Longest leaves 6–9 mm long, spreading or reflexed, usually at least obscurely toothed near tip; Mich, Minn, Wisc
1. Spinulum annotinum
INTERRUPTED CLUBMOSS, page 418

1 Longest leaves less than 6 mm long, all but lowest leaves ascending, usually entire; Mich Upper Peninsula only
2. Spinulum canadense
NORTHERN CLUBMOSS, page 416

Spinulum annotinum
INTERRUPTED CLUBMOSS

Spinulum canadense
NORTHERN CLUBMOSS

Spinulum annotinum (L.) Haines
INTERRUPTED CLUBMOSS

FIELD TIPS
• evergreen trailing plants
• stems prickly to touch
• leaves spreading to reflexed
• strobili stalkless, single at stem ends

MIDWEST RANGE
Mich, Minn, Wisc.

HABITAT
Wet to dry forests and clearings; soils often sandy or rocky.

DESCRIPTION
Stems On ground surface or in uppermost humus layer, mostly unbranched, rooting at intervals, with annual constrictions. Erect stems simple or branched near base, 15-20 cm long, with pronounced annual constrictions.

Leaves 8- to 10-ranked, more or less stiff and prickly, spreading to reflexed, widest above middle, tipped with a sharp point; margins finely toothed.

Strobili Stalkless at ends of leafy stems; usually 1 strobilus per branch stem; sporophylls abruptly tapered to a sharp pointed tip.

SYNONYMS
• *Lycopodium annotinum* L.
• *Lycopodium dubium* Zoega

SIMILAR SPECIES
• **Shining fir-moss** (*Huperzia lucidula*) similar, but its leaves not spine tipped and it does not produce strobili.

NAME
Linnaeus named this species *Lycopodium annotinum* from European specimens in 1753, but it was soon found in America as well.

From Latin, *annotinus*, a year-old, perhaps referring to the annual constrictions 'interrupting' the stem.

NORTH AMERICA

MIDWEST

strobilus

sporophylls

portion of strobilus

horizontal stem

SELAGINELLACEAE
Spike-Moss Family
Selaginella
SPIKE-MOSS

Low, creeping moss-like plants with branching stems and few fine roots. **Leaves** simple, overlapping, 4-ranked or spirally arranged, with or without a bristle tip. **Spores** of 2 types, with some sporangia containing megaspores, others with microspores, borne in axils of leaf-like sporophylls of a terminal cone. **Megaspores** germinate into the female, egg-bearing gametophyte; the smaller **microspores** form the male, sperm-bearing gametophyte. Worldwide, the family contains 5 genera and somewhat more than 300 species, mainly in tropical and subtropical regions, and nearly all of which are in genus *Selaginella*. *Selaginella* is the sole genus in North America, with ca. 38 species; 4 species in our flora.

KEY CHARACTERS
- Small, creeping plants.
- **Leaves** small and overlapping on stem.
- **Spores** of 2 types, borne in sporangia in axils of leaf-like sporophylls at ends of stems.

NAME
Diminuitive of *selago*, an ancient name of a clubmoss (Lycopodiaceae), which *Selaginella* resemble.

strobilus

Selaginella rupestris
SAND CLUBMOSS

Selaginella selaginoides
NORTHERN SPIKE-MOSS

KEY TO SELAGINELLA | SPIKE-MOSS

1 Sterile leaves in a spiral around stem, and all of similar length and width . . . **2**

1 Sterile leaves 4-ranked around stem, and of 2 sizes **3**

2 Sterile leaves tipped with a short, rough bristle (this maybe absent in older leaves); strobili appearing 4-angled from the 4-ranks of appressed sporophylls (leavings bearing sporangia); plants of dry, sandy or rocky sites; Ill, Ind, Iowa, Mich, Minn, Wisc

1. Selaginella rupestris
SAND CLUBMOSS, page 424

2 Sterile leaves tapered to a sharp point but not bristle-tipped; strobili cylindric, the sporophylls widely spreading; plants of fens and calcareous shores; n Mich, ne Minn, Wisc (Door County)

3. Selaginella selaginoides
NORTHERN SPIKE-MOSS, page 426

3 Sterile leaves of mid-portion of stem abruptly tapered tip, upper part of leaf underside keeled (like a boat hull); c and s Ill, s Ind, Ohio   **3. Selaginella apoda**

MEADOW SPIKE-MOSS, page 422

3 Sterile leaves of mid-portion of stem long-tapered to tip, not keeled; Ill, Ind, Iowa, Mich, Ohio, Wisc   **2. Selaginella eclipes**

HIDDEN SPIKE-MOSS, page 423

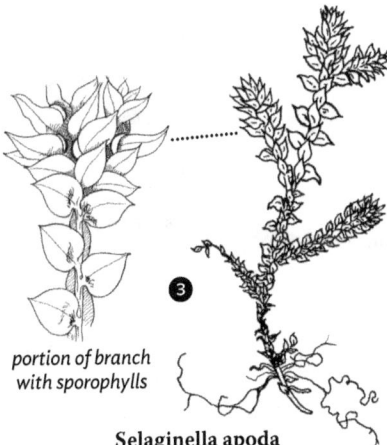

Selaginella eclipes
HIDDEN SPIKE-MOSS

portion of branch with sporophylls

Selaginella apoda
MEADOW SPIKE-MOSS

Selaginella apoda (L.) Spring
MEADOW SPIKE-MOSS

FIELD TIPS
- delicate, creeping, evergreen plant
- leaves thin and translucent, mosslike but with true veins
- stems threadlike, pale green
- southern portions of our region

MIDWEST RANGE
c and s Ill, s Ind, Ohio.

HABITAT
Wet woods, swamps and shores.

DESCRIPTION

Stems Delicate, freely branching, forming small open mats.

Sterile leaves Thin and membranous, yellow-green, 4-ranked; of 2 sizes, the lateral 2 rows larger and spreading, the dorsal and ventral leaves smaller and appressed; with keeled midrib on upper portion of leaf underside, the midrib stopping short of leaf tip; margins very finely toothed.

Strobili Somewhat 4-angled, at ends of branches.

SYNONYMS
- *Selaginella apus* Spring

SIMILAR SPECIES
- **Hidden spike-moss** (*Selaginella eclipes*) similar but median leaves in that species long tapered to a bristly tip. In *S. apoda*, leaves tapered to only a pointed, not bristly, tip.

NAME
This plant was known to Linnaeus from "Carolina, Virginia, and Pennsylvania," and named by him in 1753 *Lycopodium apodum*. Upon its transfer to the genus *Selaginella* by Spring in 1840 the species name was wrongly spelled *apus*, as pointed out by Fernald in 1915.

fertile portion of branch with sporophylls

NORTH AMERICA

MIDWEST

Selaginella eclipes Buck

HIDDEN SPIKE-MOSS

FIELD TIPS
- creeping, forming small mats
- stems with threadlike branches
- leaves 4-ranked, of 2 sizes

MIDWEST RANGE
Ill, Ind, Iowa (endangered), Mich, sw Ohio (threatened), Wisc.

HABITAT
Open areas of fens, wet sandy, rocky, or marly lake shores and river banks, seeps, openings in wet meadows, usually in strongly calcareous sites.

DESCRIPTION
Stems Creeping, forming small mats; fertile branch tips only slightly elevated above ground surface.

Sterile leaves Of 2 types, the lateral leaves elliptic to oval, at right angles to stem; the other leaves smaller, tipped with long transparent bristle; leaves not keeled, the midrib extending to leaf tip; margins with very small teeth (visible under a hand lens).

Strobili Loose, flattened, single or in pairs.

SYNONYMS
- *Selaginella apoda* subsp. *eclipes* (W.R. Buck) Skoda

SIMILAR SPECIES
- **Meadow spike-moss** (*Selaginella apoda*) similar but median leaves tapered to only a pointed, not bristly, tip as in *S. eclipes*. The leaves of meadow spike-moss are also keeled on their underside near the tip.

NOTES
Hidden spike-moss was not described as a distinct species until 1977, and because of many similarities, is perhaps best considered as a subspecies of *Selaginella apoda*.

NORTH AMERICA

MIDWEST

Selaginella rupestris (L.) Spring
SAND CLUBMOSS

FIELD TIPS
- stiff, evergreen, mat-forming plant
- stems covered with appressed leaves
- strobili square-sided
- dry, rock or sandy habitats

MIDWEST RANGE
Ill, Ind, Iowa, Mich, Minn, Wisc;
historical records from Ohio.

HABITAT
Dry rocky balds, sand dunes and open
sandy areas; soils acidic; vegetation
sparse, *Polytrichum* moss may be
present. In Michigan, on rocks, old
dunes and jack pine areas of the
northern Lower Peninsula.

DESCRIPTION
Stems Often forked at ends, covered
with overlapping, appressed leaves;
forming open, evergreen, gray-green
mats.

Sterile leaves Narrowly lance-shaped,
about 3 mm long, green or sometimes
reddish, tipped with white or
transparent bristle, grooved on the
underside midrib; margins distinctly
spiny.

Strobili 4-sided, at ends of branches.

SYNONYMS
- *Lycopodium rupestre* L.

NOTES
Plants bright green when fresh and
moist, turning gray-green when dry.

NAME
In 1753 Linnaeus gave the name
Lycopodium rupestre to a plant said to
grow in "Virginia, Canada, and Siberia."
In 1805 Beauvais proposed the genus
Selaginella for members of the family
having 2 sizes of spores, and in 1838,
Spring placed Linnaeus' plant in the
Spike-Moss Family.

NORTH AMERICA

MIDWEST

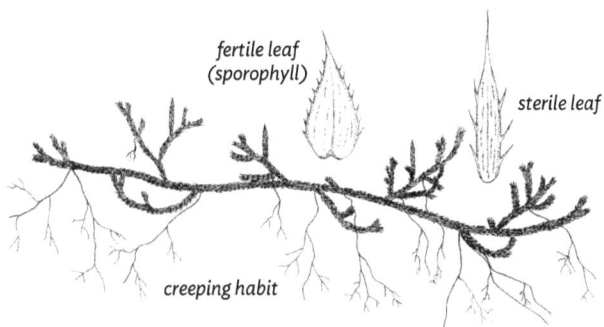

fertile leaf
(sporophyll)

sterile leaf

creeping habit

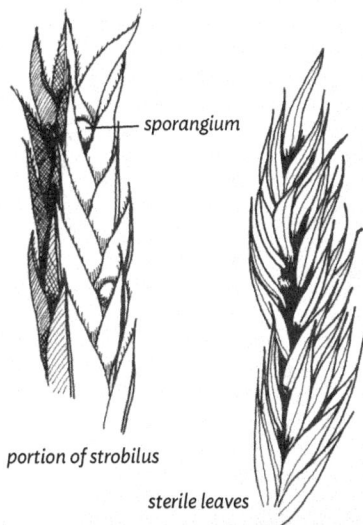

sporangium

portion of strobilus

sterile leaves

Selaginella selaginoides

(L.) Beauv. ex Mart. & Schrank

NORTHERN SPIKE-MOSS

FIELD TIPS

• mat-forming plant
• leaves with spine-like teeth
• strobili cylindric, on upright stems
• northern portions of region

MIDWEST RANGE

n Mich, ne Minn (endangered), Ohio,
Wisc (Door County, endangered); a
circumboreal species more common
north of our region.

HABITAT

Fens and fen-like shores, especially
along the Great Lakes; calcareous
places, in sun or partial shade; often on
mossy hummocks and under trees of
northern white cedar.

DESCRIPTION

Stems Creeping and branching,
forming small mats. Fertile branches
upright, with lower leaves similar to
those of the stem but the leaves
becoming larger upwards to form the
sporophylls of the nearly cylindric
strobilus.

Sterile leaves 2-4 mm long, spreading
or ascending, tapered to pointed tip;
margins coarsely toothed with soft,
spine-like projections.

Strobili Cylindrical at ends of upright
stems.

SIMILAR SPECIES

• Northern spike-moss
similar in habit to
northern bog clubmoss
(*Lycopodiella inundata*) with its
creeping sterile stems and erect fertile
branches tipped by slightly expanded
strobili; however, leaf margins of
northern bog clubmoss smooth and
not spiny-toothed.

NOTES

Plants may be difficult to see as
sometimes partially buried in moss or
hidden by the surrounding taller
vegetation.

NORTH AMERICA

MIDWEST

sporangium

portion of strobilus

habit

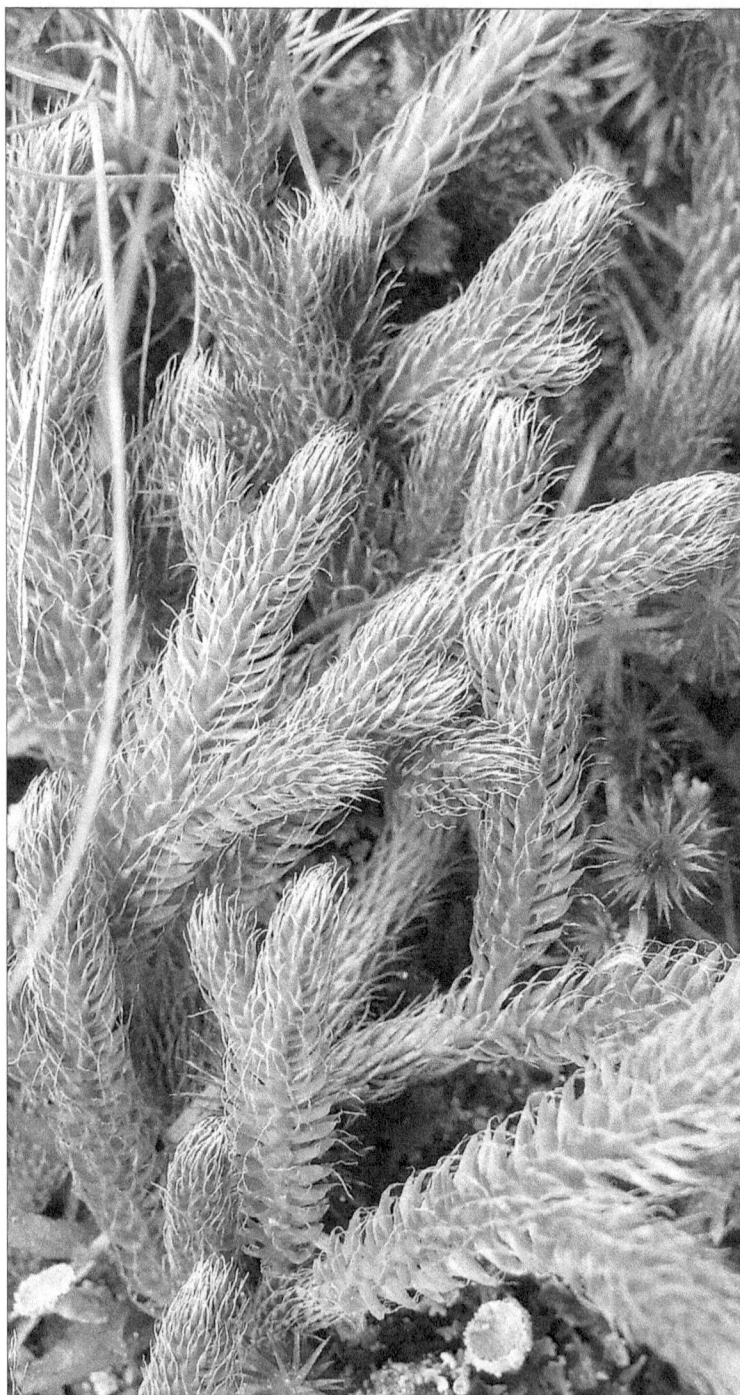

Running ground-pine (*Lycopodium clavatum*), a member of the Clubmoss Family.

References

IDENTIFICATION GUIDEBOOKS

Billington, Cecil. 1952. Ferns of Michigan. Cranbrook Institute of Science.

Cobb, B., C. Lowe and E. Farnsworth. 2005. A field guide to ferns and their related families. New York Houghton Mifflin Company.

Cody, W. J., and D. M. Britton. 1989. Ferns and fern allies of Canada. Canadian Government Publishing Center, Ottawa.

Cranfill, Ray. 1980. Ferns and fern allies of Kentucky. Kentucky Nature Preserves Commission.

Fassett, N. C., D. W. Dunlop, M. E. Diemer, R. M. Tryon. 1953. The ferns and fern allies of Wisconsin. 2nd Edition. University of Wisconsin Press.

Flora of North America Editorial Committee (eds). 1993. Flora of North America, North of Mexico, Vol. 2. Oxford University Press, New York.

Gleason, H. A. and A. Cronquist. 1991. Manual of vascular plants of northeastern United States and adjacent Canada. 2nd edition. New York Botanical Garden, New York.

Hallowell, A. C., and B. G. Hallowell. 2001. Fern finder: A guide to native ferns of central and northeastern United States and eastern Canada. Nature Study Guild Publishers. Rochester, New York.

Harris, J. G. and M. W. Harris. 1994. Plant identification terminology; an illustrated glossary. Spring Lake Publishing. Spring Lake, Utah.

Lellinger, D. B. 1985. A field manual of the ferns and fern-allies of the United States and Canada. Smithsonian Institution Press. Washington, D.C.

Magee, D. W. and H. E. Ahles. 1999. The flora of the northeast: A manual of the vascular flora of New England and adjacent New York. The University Press of Massachusetts.

Mickel, J. T. 1979. How to know the ferns and fern allies. W. C. Brown Company Publishers. Dubuque, Iowa.

Mohlenbrock, R. H. 1999. Ferns. Southern Illinois University Press.

Odum, Eugene C. 1981. Field guide to northeastern ferns. New York State Mueum Bulletin 444. Albany, New York.

Parsons, F. T. 1961. How to know the ferns: A guide to the names, haunts, and habits of our common ferns. Dover Publications, New York.

Peck, James H. 1982. Ferns and fern allies of the driftless area of Illinois, Iowa, Minnesota, and Wisconsin. Milwaukee Public Museum Press. Milwaukee, Wisconsin.

Shaver, J. M. 1970. Ferns of the eastern central states, with special reference to Tennessee. Dover Publications, New York.

Snyder, L. H., Jr., and J. G. Bruce. 1986. Field guide to the ferns and other pteridophytes of Georgia. University of Georgia Press, Athens.

Taylor, W. C. 1984. Arkansas ferns and fern allies. Milwaukee Public Museum, Milwaukee, Wisconsin.

Thomas, G. B. and S. W. Francis. 2004. Wildflowers and ferns of Kentucky. University Press of Kentucky.

Tyron, R. M. and R. C. Moran. 1997. The ferns and allied plants of New England. Massachusetts Audubon Society.

Tyron, Rolla. 1980. Ferns of Minnesota. University Of Minnesota Press.

Wherry, E. T. 1972. The southern fern guide. Doubleday, Garden City, New York.

Wherry, E. T. 1961. The fern guide; northeastern and midland United States and adjacent Canada. Doubleday. New York.

FERN CULTURE

Foster, F. G. 1984. Ferns to know and grow. 3rd ed. Timber Press.

Hoshizaki, B. J. and K. A. Wilson. 1999. The cultivated species of the fern genus *Dryopteris* in the United States. American Fern Journal 89: 1–100.

Hoshizaki, B. J. and R. C. Moran. 2001. Fern grower's manual. Timber Press.

Jones, D. L. 1987. Encyclopaedia of ferns. Timber Press.

Mickel, J. T. 1994. Ferns for American gardens. MacMillan Publishing Company, New York.

Moran, R. C. 2004. A natural history of ferns. Timber Press.

Olsen, Sue. 2007. Encyclopedia of garden ferns. Timber Press.

SELECTED TECHNICAL REPORTS

Beitel, J. M. 1979. The Clubmosses *Lycopodium sitchense* and *L. sabinifolium* in the Upper Great Lakes Region. Michigan Botanist 18: 3–13.

Blasdell, R.F. 1963. A monographic study of the fern genus *Cystopteris*. Memoirs of the Torrey Botanical Club Vol. 21, No. 4.

Chase M. W. and J. L. Reveal. 2009. A phylogenetic classification of the land plants to accompany APG III. Botanical Journal of the Linnean Society 161: 122–127.

Christenhusz, M. J. M., X.-C. Zhang, and H. Schneider. 2011. A linear sequence of extant families and genera of lycophytes and ferns. Phytotaxa 19: 7–54.

Crane, E. H. 1997. A revised circumscription of the genera of the fern family Vittariaceae. Systematic Botany 32: 509–517.

Cranfill, R. 1983. The Distribution of *Woodwardia areolata*. American Fern Journal 73: 46–52.

Drife, D. C., and J. E. Drife. 1990. Oliver A. Farwell's Early Pteridophyte Records from the Keweenaw Peninsula. Michigan Botanist 29: 89–96.

Evrard, C. and C. Van Hove. Taxonomy of the American *Azolla* Species (Azollaceae) A Critical Review. Systematics and Geography of Plants Vol. 74, No. 2 (2004), pp. 301-318.

Hickey, R. J. 1977. The *Lycopodium obscurum* complex in North America. American Fern Journal 67: 45–49.

Johnson, D. M. 1986. Systematics of the New World species of *Marsilea* (Marsileaceae). Systematic Botany Monographs 11: 1–87.

Kramer, K. U., and P. S. Green, eds. 1990. Pteridophytes and Gymnosperms. Vol. 1 of The Families and Genera of Vascular Plants, edited by K. Kubitzki. Springer-Verlag, Berlin.

Lehtonen, S. 2011. Towards resolving the complete fern tree of life. PLoS ONE 6(10): e24851.

Lumpkin, T. A., and D. L. Plucknett. 1980. *Azolla*: Botany, physiology, and use as a green manure. Economic Botany 34: 111–153.

Moore, A. W. 1969. *Azolla*: Biology and agronomic significance. The Botanical Review 35: 17–35.

Moran, R. C. 1982. The *Asplenium trichomanes* complex in the United States and adjacent Canada. American Fern Journal 72: 5–11.

Pippen, R. W. 1966. *Lygodium palmatum*, the climbing fern, in Michigan. Michigan Botanist 5: 64–66.

Rothfels, C. J., M. A. Sundue, Li-Yaung Kuo, A. Larsson, Masahiro Kato, E. Schuettpelz, and K. M. Pryer. A revised family-level classification for eupolypod II ferns (Polypodiidae Polypodiales). Taxon 61 (3), June 2012: 515–533.

Smith, A.R., K.M. Pryer, E. Schuettpelz, P. Korall, H. Schneider, and P. G. Wolf. 2006. A classification for extant ferns. Taxon 55: 705–731.

Stevens, P. F. 2001 onwards. Angiosperm Phylogeny Website. Version 12, July 2012 (and more or less continuously updated since). http://www.mobot.org/MOBOT/research/APweb/.

Tyron, R. M. and A. F. Tyron. Ferns and allied plants with special reference to tropical America. New York Springer-Verlag. 1982.

Wagner, G. M. 1997. *Azolla*: A review of its biology and utilization. Botanical Review 63: 1–26.

Wagner, W. H. Jr. 1971. The southeastern Adder's-Tongue, *Ophioglossum vulgatum* var. *pycnostichum*, found for the first time in Michigan. Michigan Botanist 10: 67–74.

Wagner, W. H. Jr. and D. M. Johnson. 1981. Natural History of the Ebony Spleenwort, *Asplenium platyneuron*, in the Great Lakes Area. Canadian Field-Naturalist 95: 156–166.

Wagner, W. H. Jr., F. S. Wagner, & D. J. Hagenah. 1969. The log fern (*Dryopteris celsa*) and its hybrids in Michigan—A preliminary report. Michigan Botanist 8: 137-145.

Wagner, W.H. Jr., and F.S. Wagner. 1990. Moonworts (*Botrychium* subg. *Botrychium*) of the upper Great Lakes region, U.S and Canada, with descriptions of two new species. Contr. Univ. Mich. Herb. 17: 313-325.

Wagner, W.H. Jr. and F.S. Wagner. 1981. New species of moonworts, *Botrychium* subg. *Botrychium* (Ophioglossaceae), from North America. Amer. Fern J. 71: 20-30.

Wagner, W.H. Jr. and F.S. Wagner. 1983. Two moonworts of the Rocky Mountain; *Botrychium hesperium* and a new species formerly confused with it. 1983. Amer. Fern J. Vol. 73: 53-62.

Frond Types

fertile frond

sterile frond

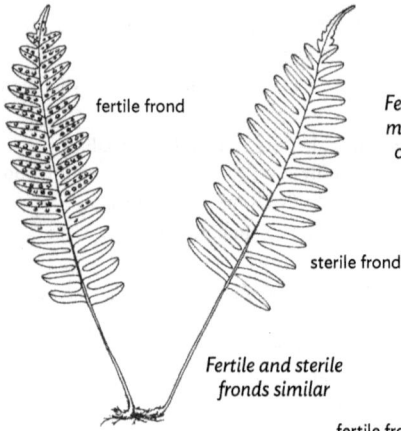

Fertile and sterile blades much different but on a common stipe (stalk)

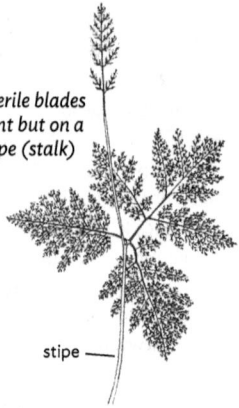

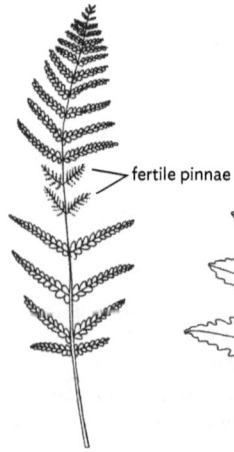

Fertile and sterile fronds similar

stipe

fertile frond

fertile pinnae

fertile pinnae

sterile pinnae

sterile frond

Fertile and sterile pinnae much different but on same frond

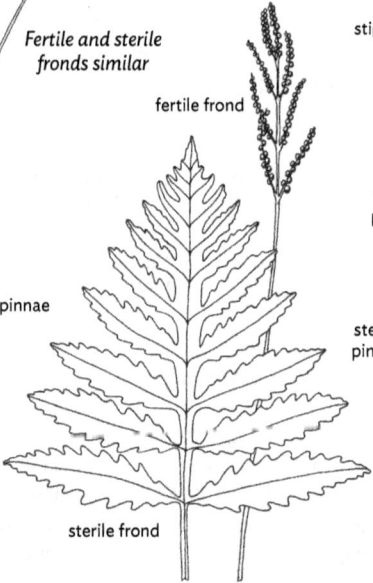

Fertile and sterile fronds much different and borne separately

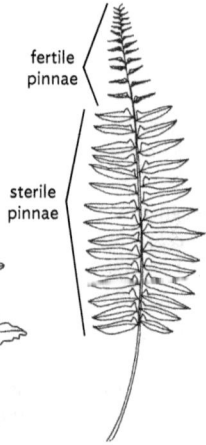

Fertile and sterile pinnae slightly different but on same frond

Glossary

See Introduction (page 4) for an illustrated basic glossary of fern terms.

abaxial of the side or surface of an organ, facing away from the axis, e.g. the lower or dorsal surface of the lamina. Compare to **adaxial**.

acicular stiff and needle-like.

acroscopic pointing towards the apex. Compare to **basiscopic**.

acrostichoid with sporangia densely covering the abaxial surface of the lamina, as in the tropical fern *Acrostichum*.

acuminate gradually tapering to a protracted point.

acute terminating in a sharp or well-defined point.

adaxial of the side or surface of an organ, facing towards the axis, e.g. the upper or ventral surface of the lamina. Compare to **abaxial**.

adnate fused to an organ of a different kind, as a pinna to a rachis.

alate winged.

alete used of a spore which forms alone, i.e. not in diads or tetrads.

allotetraploid having 4 complete sets of homologous chromosomes resulting from interspecies hybridization followed by chromosome doubling; a plant with such a condition.

anadromous a type of venation in which the first set of veins in each segment of the frond arising from the side of the midrib facing toward the frond tip.

anastomosed veins united to form areoles; net-veined.

annular arranged in or forming a ring.

annulus the elastic ring of cells in a sporangium that initiates dehiscence; a complete or partial ring or cluster of thick-walled cells on the spore case functioning to open the spore case.

antheridium (plural **antheridia**) the male sex organ containing the sperm; borne on the underside of the prothallus.

antrorse bent, and pointing towards the apex. Compare to **retrorse**.

apex (plural **apices**) the tip, the point farthest from the point of attachment.

apiculus a small abrupt flexible point at the apex of a pinna or pinnule. adj. **apiculate**.

apogamous applied to ferns in which a sporophyte develops from gametophyte cells other than a fertilized egg.

apogamy a form of asexual reproduction in which new sporophytes are produced directly from the prothallus tissue instead of from the fertilized egg cell (zygote).

apomict a plant that produces viable spores without fertilization.

apospory a form of asexual reproduction in which prothalli are produced directly from young sporophyte tissue instead of from spores.

appressed pressed closely against a surface (or another organ) but not united with it.

arachnoid composed of fine tangled hairs like a cobweb.

archegonium the female sex organ containing the egg; the structure that produces the female gamete or egg. plural archegonia.

areolate having netted veins.

areole a space enclosed by netlike veins (also see **reticulum**).

aristate having a stiff bristle-like tip.

articulate jointed; having joints where separation may occur naturally.

attenuate pinnae gradually tapering to a very narrow, slender point.

auricle an ear-like lobe at the base of a lamina, pinna or pinnule (adj. **auriculate**).

axes collective term referring to the petiole, rachis, costae, costules, and midrib of a frond.

basal pinnae (basal pinna pair) the lowermost pair of pinnae, closest to the common stalk.

basiscopic pointing towards the base; the basiscopic margin of a pinna is that which faces the base of the leaf; the basiscopic margin of a pinnule is that which faces the base of the pinna. Compare to **acroscopic**.

bicolorous having two colors, usually referring to scales in which the central part is darker than the mar-gins. Compare to **concolorous**.

binomial the species name, the two words consisting of the genus name and the specific epithet.

bipinnate twice pinnately branched (**2-pinnate**).

bipinnatifid twice pinnatifid; the divisions of a pinnatifid frond are again pinnatifid (**2-pinnatifid**).

blade the flat, expanded portion of a leaf.

bristle a stiff hair which is more than one cell broad at the base.

bulbiferous bearing bulbils.

bulbil or **bulblet** small bulb-like body borne upon a stem or leaf and serving to vegetatively reproduce the plant.

catadromous a type of venation in which the first basal branch or vein (as on a pinna) arises from the side toward the frond base. Compare to **anadromic**

caudate with a long and thin tip.

caulescent developing an aerial stem or trunk.

cell the basic unit of plant structure consisting, at least when young, of a protoplast surrounded by a wall.

chartaceous thin and papery.

chlorophyll pigment(s) constituting the green coloring matter of plants and absorbing radiant energy in photosynthesis.

ciliate fringed with hairs.

circinate (circinate vernation) coiled in a spiral with the tip innermost, as in the fiddleheads of many ferns.

clathrate latticed or pierced with apertures like a trellis, as in the rhizome scale of *Asplenium* (a hand lens or microscope is needed to see this).

clavate club-shaped.

cleft technically, deeply cut, usually at least half-way to the middle or base.

commissure a juncture or seam; in *Pellaea*, a ±continuous marginal sorus formed when laterally expanded fertile vein endings coalesce.

common stalk the stalk below the junction of the sporophore and trophophore.

compound of a leaf, having the blade divided into 2 or more distinct leaflets.

concolorous colored uniformly; the same color on both sides.

conduplicate folded flat together lengthwise; of developing leaves.

cone tight cluster of highly modified, spore-bearing leaves borne at branch tips.

confluent flowing or running together.

convergent tending to come together, merging.

cordate of a leaf blade, broad and notched at the base; heart-shaped (in 2 dimensions).

coriaceous leathery.

corm in *Isoetes*, the condensed stem, which may be 2-5-lobed.

costa (plural *costae*) the midrib of a pinna.

costule the midrib of a pinnule or segment of lower order, except the central vein of an ultimate segment which is usually termed the midrib.

crenate with small, rounded teeth; scalloped.

crenulate with very small rounded teeth along the margin.

crested having forked tips, usually many; usually refers to the frond, pinnae, or segments.

cristate in ferns, having a tasselled margin to the fronds.

crown the tip of the stem where the leaves arise; usually applied to thick, upright stems.

crozier uncurling frond; a young coiled fern frond; the fiddlehead.

cuneate wedge-shaped, with the narrow part at the point of attachment, usually referring to the base of a pinna or pinnule.

cuspidate tipped with a cusp or a sharp and firm point.

cut (cutting) referring to the presence of dissection or division.

deciduous shed seasonally, not evergreen..

decurrent referring to the bases of leaves or pinnae continuing down the stem or rachis beyond the point of attachment.

decussate borne in pairs alternately at right angles to each other.

deflexed bent abruptly downwards.

deltate or **deltoid** triangular with the sides of about equal length.

dentate of margins, toothed; usually with broad teeth directed outward.

denticulate finely toothed.

dichotomous forked regularly into pairs.

dimorphic having 2 different forms; in ferns the term usually refers to differences in size or shape of the sterile and fertile fronds (or segments). Compare to **monomorphic**.

dioecious having the male and female reproductive structures on separate plants. Compare to **monoecious**.

diploid having 2 of the basic sets of chromosomes in the nucleus; a plant with such a condition. Compare to **haploid, polyploid**.

dissected cut into lobes or lobe-like segments.

distal furthest point; remote from the point of origin or attachment. Compare to **proximal**.

divergent spreading apart.

divided cut to the base or midrib into lobes or segments. As applied to ferns, "divided" is a generic term meaning "dissected" and includes pinnate (cut to midrib) and pinnatifid (cut into lobes without divisions reaching the midrib).

dorsal relating to the back or lower side of a leaf.

echinate spiny, referring to a spore.

eglandular without glands.

elaters in *Equisetum*, appendages of the spore which help in dispersal; elators are formed from the outermost wall layer, coiling and uncoiling as air is dry or moist.

elliptic ellipse shaped, broadest in the center and narrower at the two equal ends.

elongate drawn out, lengthened.

emarginate having a shallow notch in the margin.

emersed standing out of or rising above a surface (as of water).

entire undivided; the margin continuous, not incised, lobed or toothed.

ephemeral lasting only a short time; (of indusium: quickly shed).

epilithic growing on rocks; also epipetric, saxicolous, or rupestral.

epiphyte (epiphytic) a plant that grows upon another plant; an epiphyte uses its host plant only for support; it is not parasitic.

erose of a margin, finely and irregularly eroded or incised.

euphyllophytes the group containing both ferns (monilophytes) and seed-bearing plants (spermatophytes).

eusporangiate having sporangia with walls more than one cell thick. Compare to **leptosporangiate**.

excurrent vein a vein running toward the margin, not the midrib.

exindusiate lacking an indusium.

extirpate to destroy completely; eradicate, in conservation usually refers to a species historically present but now absent from a state.

falcate sickle-shaped, curved and flat.

false indusium an indusium formed by the rolled-over margin of the leaf enclosing the sorus, as in many genera in the Pteridaceae, such as *Adiantum, Cheilanthes* and *Pellaea*.

false veins rows of thickened cells in a leaf/leaflet which are not part of the vascular system, as in some genera of the Hymenophyllaceae.

family taxonomical division comprised of a group of related genera.

farina a meal-like powder, usually white or yellow and found on the lower surfaces of fronds; characteristic of fern genera such as *Argyrochosma, Pentagramma*, and *Pityrogramma* (these not present in our region).

farinose covered with a meal-like powder (**farina**).

fern allies group of vascular plants traditionally thought to be closely related to ferns, based mainly on the spore-bearing characteristic. These comprise the families of horsetails (Equisetaceae), quillworts (Isoetaceae), clubmosses (Lycopodiaceae), whisk ferns (Psilotaceae) and spikemosses (Selaginellaceae). Recent research has shown quillworts, clubmosses and spikemosses to be much less closely related to ferns, while horsetails and whisk ferns are more closely related, in fact essentially part of the main fern grouping (reflected in the organization of this flora).

fertile in ferns usually referring to leaves that bear sori.

fiddlehead a fern leaf, young or in bud, that is coiled in a spiral pattern. Also see **crozier**.

filiform threadlike.

fimbriate of a margin, fringed with fine hairs.

flabellate fan-shaped.

floccose covered with soft tangled woolly hairs.

flora list of all the species growing in a region, or a collective term for all the species growing in a region; also refers to books that identify the plants within a certain geographical area.

frond the whole leaf of a fern, including the blade and stipe (petiole).

fruitdots sori, clusters of sporangia.

fugacious shed or withering away very early.

fused grown to or united with a similar part.

gametophyte a small, usually flat plant bearing the sex organs (archegonia and antheridia) that in turn produce the gametes. Each cell in the body of the gametophyte has one set of chromosomes ($1n$); gametophytes grow from spores.

gene a unit on a chromosome that determines the inheritance of a particular trait.

gemma (plural **gemmae**) a bud or bulbil that detaches from the main plant and develops into a new plant, as in *Huperzia selago*.

gemmiferous bearing gemmae (asexual buds or bulbils).

genus (plural **genera**) term for describing a closely related group of species.

glabrescent becoming glabrous (smooth).

glabrous without hairs or scales; smooth.

gland a structure with a secretory function, embedded or projecting from the surface of the plant.

glandular having glands or functioning as a gland.

glaucous dull green with a bluish white or white, usually waxy, lustre.

globose almost spherical.

haploid with one set of chromosomes in the nucleus. Compare to diploid, polyploid.

hastate spear-shaped; of a leaf blade, narrow and pointed but with two basal lobes spreading approximately at right angles.

herbaceous soft in texture; midway in thickness between membranous and coriaceous, usually applied to the leaf; without a persistent woody stem above ground; dying back to the ground at the end of the growing season; leaf-like in color and texture.

heterophyllous having leaves that are not uniform along a branch, e.g. in *Huperzia*, with long leaves in the lower portions and smaller reduced leaves distally. Compare to **homophyllous**.

heterosporous having 2 kinds of spores (male and female, or microspores and megaspores). Compare to **homosporous**.

hirsute bearing coarse rough relatively long hairs. Compare to **villous**.

homophyllous with all leaves uniform along a branch. Compare to **heterophyllous**.

homosporous producing only one type of spore from which develops a gametophyte producing both male and female gametes. Compare to **heterosporous**.

hyaline translucent, almost like clear glass.

hybrid result of a sexual cross between 2 different taxa.

hydathode an enlarged vein tip on the upper surface of a blade; it often secretes water and minerals; in some ferns, the minerals may accumulate as a white deposit over the hydathode.

imbricate overlapping, like roof tiles.

incised cut deeply, sharply and often irregularly (an intermediate condition between toothed and lobed).

incurved bent or curved inwards or upwards; of leaf margins, curved towards the adaxial surface.

indusiate bearing an indusium.

indusium (plural **indusia**) the covering of a sorus, either a specialised organ or the incurved margin of the blade (also see **false indusium**).

internode the portion of a stem or other structure between two nodes.

introduced a plant that was brought into the country, region, or state (either deliberately or accidently) by humans.

involucre the indusium of members of the Hymenophyllaceae.

jointed able to separate naturally at a certain point, leaving a scar; articulate.

laciniate cut into narrow-pointed lobes.

lacuna (plural **lacunae**) of *Isoetes*, a cavity within the leaves.

lamina (plural **laminae**) the 'blade' of a frond.

lanceolate lance-shaped; long and narrow, but broadest at base and gradually tapering to apex.

lateral situated on or arising from the side of an organ.

leaf frond; used here to include both the "leafy" part and the stipe or stem.

leaf blade the 'leafy' part of the frond excluding the stipe or stem.

leaflet one of the divisions in a compound leaf.

leaf-segment any subdivision of a frond. See **pinna, pinnule, pinnulet**.

leptosporangiate having sporangia with the walls only one cell thick. Compare to **eusporangiate**.

ligulate bearing a ligule; strap-shaped.

ligule a membranous structure towards the base of the upper leaf surface; in *Isoetes* a small triangular or elongate delicate tissue extending slightly above the sporangium.

linear long and narrow, much longer than wide, the sides parallel or nearly so.

lithophytic growing on rock; also see **epilithic**.

lunate shaped like a crescent or half-moon.

lobe a division or segment of an organ; technically, cut not over half-way to the middle or base, the sinuses and apex of segments rounded. Compare to **cleft**.

lobed bearing lobes.

lustrous shiny or glistening, glossy.

lycophytes the group comprising quillworts, clubmosses and spikemosses. Distinct from both euphyllophytes, the group that contains both seed plants (spermatophytes) and ferns (monilophytes).

macrospore the larger kind of spore in Selaginellaceae and *Isoetes*, and in other genera; also called **megaspore**.

marcescent withering without falling off; in contrast to jointed or articulate, which indicate withering and falling off at a joint.

massula (plural **massulae**) group of microspores enclosed in a hardened mucilage.

margin (**marginal**) the edge, as in the edge of a leaf blade.

medial being or occurring in the middle.

megasporangium the larger of the two kinds of sporangia produced in the sexual life cycle of a heterosporous plant; produces megaspores.

megaspore the larger of the two kinds of spores produced in the sexual life cycle of a heterosporous plant, giving rise to the female gametophyte. They may be monomorphic as in *Selaginella*, or polymorphic as in some *Isoetes* species. Compare to **microspore**.

megasporocarp a sporocarp containing megasporangia.

megasporophyll a specialised leaf upon which (or in the axil of which) one or more megasporangia are borne.

meiosis the type of cell division that gives rise to spores; during meiosis, the cell replicates its chromosomes once and divides twice; the result is four cells with only half the chromosome number of the original cell.

micron one thousandth of a millimeter.

microsporangium the smaller of the 2 kinds of sporangia produced in the sexual life cycle of a heterosporous plant; as in *Selaginella* and *Isoetes*.

microspore the smaller of the 2 kinds of spores produced in the sexual life cycle of a heterosporous plant, giving rise to the male gametophyte. Compare to **megaspore**.

microsporocarp a sporocarp containing microsporangia.

midrib the central, and usually the most prominent, vein of a leaf or leaflike organ.

monoecious having the male and female reproductive parts in separate organs but on the same plant. Compare to **dioecious**.

monolete of a spore, bilateral, having a single straight scar.

monomorphic of uniform shape and size. Compare to **dimorphic**.

monilophytes the group comprising all ferns, now known to include horsetails and whisk-ferns; a recently-coined term to distinguish them from spermatophytes (seed-bearing plants).

midrib the central rib or vein of a leaf or other organ.

mucro a sharp abrupt terminal point. adj. **mucronate**.

mycorrhiza an association of a fungus with the root of a higher plant.

native occurring naturally, not introduced by humans.

naturalized an introduced plant that has become established and propogates itself naturally.

node the point on a stem where a leaf emerges.

nodosity in *Adiantum*, a callus or swollen node, often lacking normal coloration, where a pinna or pinnule stalk arises from a rachis.

nothosubspecies (often abbreviated to nothosubsp., nothossp. or n-subsp) hybrid where one or both parents is a subspecies (the term nothospecies is occasionally used to also denote a hybrid between two species).

ob- a prefix signifying the opposite of.

oblanceolate lanceolate with the broadest part near the tip.

obovate ovate or egg-shaped with the broader end at tip.

oblong 2 to 4 times longer than wide and the sides parallel or nearly so.

obtuse blunt or rounded at the tip.

orbicular essentially circular.

ovate egg-shaped in outline and broadest at base.

pallid pale.

palmate having veins, lobes or segments which radiate from a single point, as in maple leaves.

palmatifid of a leaf, deeply (but not completely) divided into several lobes which arise (almost) at the same level.

paraphysis (plural **paraphyses**) a sterile hair occurring among the sporangia of some ferns.

pectinate comblike.

pedate of a palmate or palmately-lobed leaf, having the lateral segments divided again.

pedicel the stalk of a sporangium.

peduncle the stalk of a sporocarp, e.g. in *Marsilea*.

peltate having the stalk attached to the lower surface usually at or near the center; like the handle on an open umbrella.

pendulous hanging or drooping.

perispore the folded membrane of most spores, forming an ornamental external covering.

persistent remaining attached to the plant beyond the expected time of falling.

petiole a leaf stalk, also known as a **stipe**.

phyllopodium (plural **phyllopodia**) a stump-like extension from the rhizome to which the fronds are attached, usually by a distinct abscission layer.

pilose hairy, the hairs soft and clearly separated but not sparse.

pinna (plural **pinnae**) the primary division of a pinnately divided frond; a leaflet; pinnae may be further divided into **pinnules** and again into **pinulules**.

pinna segment division of a pinna, whether cut entirely to the midrib or not.

pinna span the width of a fan-shaped pinna at its distal (outer) margin, measured in degrees of a circle, estimated by imagining a circle with the point of pinna attachment at its center; used most commonly with some species of *Botrychium*.

pinnate having a feather-like arrangement, divided into pinnae, with the pinnae (leaflets) arising at points along the rachis.

pinnate-pinnatifid referring to a blade that is once-pinnate and with the pinnae deeply lobed or cut, but not to their midrib.

pinnatifid cut half to three-fourths to the rachis.

pinnatisect cut almost all the way to the rachis but having the segments confluent with it.

pinnule a division of a pinna (i.e., a pinna lobe); the secondary pinna.

pinnulets divisions of pinnules.

polymorphic having more than 2 distinct morphological variants.

polyploid having more than 2 of the basic sets of chromosomes in the nucleus. Compare to **diploid, haploid**.

procumbent lying flat along the ground.

proliferous having adventitious leaf buds which produce new plants.

prostrate lying flat upon the ground.

prothallus (plural **prothalli**) the gametophyte stage of the fern; this is the independent stage where sexual reproduction takes place; in most ferns, it is a small, flattish, often roughly heart-shaped body.

protostele a simple primitive type of stele having a solid central vascular core.

proximal closest, usually referring to the part of leaf or leaf segment closest to the point of attachment.near to the point or origin of attachment. Compare to **distal**.

pseudo- false; apparent but not genuine.

pteridophyte traditional term encompassing both ferns and "fern allies"; the latter now known to be composed of some groups which are essentially ferns but look unlike them, and others which are not at all closely related. See **Lycophytes, Monilophytes, Euphyllopytes**.

puberulent minutely hairy.

pubescent clothed with short soft erect hairs.

rachis the midrib of a compound frond; i.e., the main axis above the lowermost primary pinna; the upper part of the petiole, bearing the pinnae and continuous with the stipe. Also spelled **rhachis**; plural **rachises** or **rachides**.

radial applied to a rootstock in which the fronds radiate and the roots are borne on all sides of the organ.

receptacle of ferns, the axis bearing the sporangia and sometimes also paraphyses. the tissue upon which the sporangia are borne. The receptacle is bristle-like in *Hymenophyllum* and *Trichomanes*; in most other ferns it is flush with the leaf surface or slightly elevated.

recurved curved or curled downwards or backwards.

reflexed bent sharply downwards or backwards.

reniform kidney-shaped.

reticulate (usually of veins): forming a network; net-veined.

reticulum a network, e.g. of veins. adj. **reticulate**.

retrorse bent, and pointing away from the apex. Compare to **antrorse**.

revolute rolled backward from the margins or apex. In ferns the term usually refers to a leaf margin rolled back to protect the sori (**false indusium**).

rhizoid a thread-like unicellular absorbing structure occurring, in the vascular plants, in gametophytes of ferns and some related plants; also present in many mosses which lack vascular systems.

rhizome the creeping (often underground) or climbing stem of a fern.

rhizophore in *Selaginella*, a leafless stem that produces roots.

rhomboidal diamond-shaped or almost so.

rootstock a short, erect stem.

rupestral growing on rocks; epilithic.

rugose deeply wrinkled.

rugulose covered with minute wrinkles.

saxicolous growing on rocks; epilithic.

scale small, often semi-transparent outgrowth of the outer layer of cells (epidermis), at least 2 cells wide to qualify as a scale, if only one cell wide considered a hair (trichome).

scalloped a series of semi-circles or curves resembling the edge of a scallop; with such a margin.

scandent climbing.

scarious thin, dry-looking, translucent, often whitish.

segment a division or part of a pinna.

septate divided internally by septa.

septum (plural **septa**) a partition.

sericeous clothed with silky hairs.

serrate having sharp, forward-pointing teeth; like a sawblade.

sessile without a stalk.

seta (plural **setae**) a stiff hair or bristle; in ferns these tend to be brown or black.

setose covered with bristles.

siliceous composed of or abounding in silica, as in some *Equisetum*.

simple undivided, not compound; of a frond, not divided into leaflets; of a hair or an inflorescence, not branched.

sinus membrane the membrane of a depression between adjacent lobes in a pinna, as in the Thelypteridaceae; the gap or indentation between teeth or lobes of frond.

soral flap the specialized fertile lobe in *Adiantum*.

sorus (plural **sori**) a cluster of spore cases (**sporangia**).

spathulate spoon-shaped; broad at the tip and narrowed towards the base.

species taxonomic division generally used to describe those plants that will interbreed freely with each other and share a range of visual similarities; the word is used for both the singular and plural.

spermatophytes seed-bearing plants.

spinulose bearing small spines over the surface.

sporangium (plural **sporangia**) a structure within which spores are formed; the globular organ in which the spores are produced.

spore a unicellular or few-celled sexual or asexual reproductive cell that germinates into a prothallus, which in turn gives rise to sexual reproduction.

sporeling tiny fern plant still attached to the gametophyte from which it has developed.

sporocarp a fruiting body containing sporangia; a round structure that contains sporangia within. Characteristic of 2 fern families: Salviniaceae and Marsileaceae; sporocarps are globose and delicate in the Salviniaceae, beanlike and hard in the Marsileaceae.

sporophore the fertile, spore-bearing portion of the frond.

sporophyll a specialised leaflike organ that bears one or more sporangia.

sporophyte the familiar plant that bears roots, stems, and leaves (as opposed to a gametophyte or prothallus); so-called because it is the phase of the life cycle that produces spores; each cell in the body of a sporophyte has two sets of chromosomes (i.e., is $2n$).

sporulation the formation of spores.

stalk an unexpanded, unbranched supporting structure.

stele the vascular system of rhizome or stem, together with leaf traces.

stellate star-shaped.

sterile refers to leaves that do not produce sori and to hybrids in which spores are aborted.

stipe the stalk or petiole of the frond; that portion of the midrib of the frond between the rhizome and the lowermost primary pinna; does not bear pinnae.

stipule a basal appendage of a stipe or petiole, usually paired; in ferns, the term is sometimes applied to the flared leaf bases in *Osmunda*.

stolon runner from the main stem, producing a new plant that roots independently.

stomium the region of a sporangium at which dehiscence occurs and the spores are released.

stramineous straw-colored.

strigose having curved, sharp, forward-pointing hairs.

strobile an inflorescence resembling a spruce or fir cone, partly made up of overlapping bracts or scales.

strobilus a conelike body, as in the Lycopodiaceae and Selaginellaceae, consisting of sporophylls borne close together on the axis. plural strobili.

submersed growing, or adapted to growing, under water.

subspecies (often abbreviated to spp. or subsp.) subdivision of a species.

succulent juicy; fleshy.

sympatric with areas of distribution that coincide or overlap.

synonym an alternative scientific name, but not the one currently accepted by taxonomists.

taxon (plural **taxa**) any members of a specific taxonomic grouping, e.g., species, genus, etc.; one may refer to the *Asplenium trichomanes* and *Asplenium ruta-muraria* taxa (species); *Dryopteris* and *Polystichum* taxa (genus); or the Lycopodiaceae and Selaginellaceae taxa (family).

terete circular or almost so in cross section.

terminal at the tip.

ternate compounded into more or less equal divisions or groups of three.

terrestrial growing on the ground, not in trees or on rocks.

tetraploid having 4 of the basic sets of chromosomes in a nucleus; with $4n$ chromosomes per cell; a plant with such a condition.

tomentum a dense covering of wooly hair.

toothed with small lobes or teeth along the margin.

trichome an epidermal outgrowth, e.g. a hair (branched or unbranched), a papilla.

trigonous three-angled.

tripinnate pinnate with the pinnae and pinnules also pinnate (also **3-pinnate, thrice-pinnate**).

triploid having 3 of the basic sets of chromosomes in the nucleus.

trophophore the sterile, foliaceous portion of the frond.

truncate ending abruptly, as if cut off.

ultimate segment the final, smallest divisions of a frond.

undulate gently wavy; with such a margin.

vallecula applied to the grooves in the intervals between the ridges, as in the stems of *Equisetum*; **vallecular**: pertaining to such grooves.

variety (often abbreviated to **var.**) subdivision of species, but less well-defined than a subspecies.

vascular pertaining to specialized tissue (xylem and phloem) that conducts water, mineral nutrients, and sugars.

vascular bundle the primary fluid-conducting system of a plant.

vein a strand of vascular tissue.

velum a membranous flap-like envelope which partially or wholly covers the sporangium, as in *Isoetes*.

venation the pattern formed by the veins in the leaf.

ventral belonging to the anterior or inner face of an organ, as opposed to dorsal; adaxial.

vernation the arrangement of the unexpanded fronds in a bud.

verticillate arranged in a whorl.

villous covered densely with fine long hairs but not matted. Compare to **hirsute**.

whorl leaves or other plant parts arranged in a circle around the stem.

wing a thin expansion or flat extension of an organ or structure.

xeric characterized by, relating to, or requiring only a small amount of moisture.

xerophyte a plant adapted to dry habitats.

xylem the tissue, in a vascular plant, that conducts water and mineral salts from the roots to the leaves.

zygote a fertilized egg cell; the first cell in the development of a sporophyte.

Index

Synonyms listed in *italics*.